WORKSHOPS IN COMPUTING

Series edited by C. J. van Rijsbergen

AF757261

Springer
London
Berlin
Heidelberg
New York
Barcelona
Budapest
Hong Kong
Milan
Paris
Santa Clara
Singapore
Tokyo

Also in this series

Functional Programming, Glasgow 1993
Proceedings of the 1993 Glasgow Workshop on Functional Programming, Ayr, Scotland, 5–7 July 1993
John T. O'Donnell and Kevin Hammond (Eds)

Z User Workshop, Cambridge 1994
Proceedings of the Eighth Z User Meeting, Cambridge, 29–30 June 1994
J.P. Bowen and J.A. Hall (Eds)

6th Refinement Workshop
Proceedings of the 6th Refinement Workshop, organised by BCS-FACS, London, 5–7 January 1994
David Till (Ed.)

Incompleteness and Uncertainty in Information Systems
Proceedings of the SOFTEKS Workshop on Incompleteness and Uncertainty in Information Systems, Concordia University, Montreal, Canada, 8–9 October 1993
V.S. Alagar, S. Bergler and F.Q. Dong (Eds)

Rough Sets, Fuzzy Sets and Knowledge Discovery
Proceedings of the International Workshop on Rough Sets and Knowledge Discovery (RSKD'93), Banff, Alberta, Canada, 12–15 October 1993
Wojciech P. Ziarko (Ed.)

Algebra of Communicating Processes
Proceeedings of ACP94, the First Workshop on the Algebra of Communicating Processes, Utrecht, The Netherlands, 16–17 May 1994
A. Ponse, C. Verhoef and S.F.M. van Vlijmen (Eds)

Interfaces to Database Systems (IDS94)
Proceedings of the Second International Workshop on Interfaces to Database Systems, Lancaster University, 13–15 July 1994
Pete Sawyer (Ed.)

Persistent Object Systems
Proceedings of the Sixth International Workshop on Persistent Object Systems, Tarascon, Provence, France, 5–9 September 1994
Malcolm Atkinson, David Maier and Véronique Benzaken (Eds)

Functional Programming, Glasgow 1994
Proceedings of the 1994 Glasgow Workshop on Functional Programming, Ayr, Scotland, 12–14 September 1994
Kevin Hammond, David N. Turner and Patrick M. Sansom (Eds)

East/West Database Workshop
Proceedings of the Second International East/West Database Workshop, Klagenfurt, Austria, 25–28 September 1994
J. Eder and L.A. Kalinichenko (Eds)

Asynchronous Digital Circuit Design
G. Birtwistle and A. Davis (Eds)

Neural Computation and Psychology
Proceedings of the 3rd Neural Computation and Psychology Workshop (NCPW3), Stirling, Scotland, 31 August – 2 September 1994
Leslie S. Smith and Peter J.B. Hancock (Eds)

Image Processing for Broadcast and Video Production
Proceedings of the European Workshop on Combined Real and Synthetic Image Processing for Broadcast and Video Production, Hamburg, 23–24 November 1994
Yakup Paker and Sylvia Wilbur (Eds)

Recent Advances in Temporal Databases
Proceedings of the International Workshop on Temporal Databases, Zurich, Switzerland, 17–18 September 1995
James Clifford and Alexander Tuzhilin (Eds)

Structures in Concurrency Theory
Proceedings of the International Workshop on Structures in Concurrency Theory (STRICT), Berlin, 11–13 May 1995
Jörg Desel (Ed.)

Active and Real-Time Database Systems (ARTDB-95)
Proceedings of the First International Workshop on Active and Real-Time Database Systems, Skövde, Sweden, 9–11 June 1995
Mikael Berndtsson and Jörgen Hansson (Eds)

Recent Advances in Temporal Databases
Proceedings of the International Workshop on Temporal Databases, Zurich, Switzerland, 17–18 September 1995
James Clifford and Alexander Tuzhilin (Eds)

continued on back page...

Mark A. Brown and José Carmo (Eds)

Deontic Logic, Agency and Normative Systems

ΔEON '96: Third International Workshop on Deontic Logic in Computer Science, Sesimbra, Portugal, 11–13 January 1996

Published in collaboration with the British Computer Society

BCS

Springer

Mark A. Brown, PhD
Department of Philosophy, 541 Hall of Languages, Syracuse University, Syracuse, NY 13244, USA

José Carmo, PhD
Department of Mathematics, Instituto Superior Técnico, Avenida Rovisco Pais, 1096 Lisboa Codex, Portugal

ISBN-13: 978-3-540-76015-3 e-ISBN-13: 978-1-4471-1488-8
DOI: 10.1007/978-1-4471-1488-8

British Library Cataloguing in Publication Data
A catalogue record for this book is available from the British Library

Apart from any fair dealing for the purposes of research or private study, or criticism or review, as permitted under the Copyright, Designs and Patents Act 1988, this publication may only be reproduced, stored or transmitted, in any form, or by any means, with the prior permission in writing of the publishers, or in the case of reprographic reproduction in accordance with the terms of licences issued by the Copyright Licensing Agency. Enquiries concerning reproduction outside those terms should be sent to the publishers.

© British Computer Society 1996

The use of registered names, trademarks etc. in this publication does not imply, even in the absence of a specific statement, that such names are exempt from the relevant laws and regulations and therefore free for general use.

The publisher makes no representation, express or implied, with regard to the accuracy of the information contained in this book and cannot accept any legal responsibility or liability for any errors or omissions that may be made.

Typesetting: Camera ready by contributors
Printed by the Athenæum Press Ltd., Gateshead
34/3830-543210 Printed on acid-free paper

Preface

This volume presents a variety of papers bearing on the relation between deontic logics, logics of action, and normative systems, i.e. systems of or about interacting agents (computers, human beings, corporations, etc.) whose behaviour is subject to ideal constraints that may not always be fulfilled in practice. The papers range from theoretical studies of the logical and conceptual tools needed, to studies of various applications.

The set of papers collected in this book should be of interest to investigators working in a variety of fields, from philosophy, logic and legal theory to artificial intelligence, computer and management sciences, since it covers topics ranging from theoretical research on foundational issues in deontic and action logics, defeasible reasoning, decision theory, ethical theory, and legal theory, to research on a variety of issues relevant to applications connected with expert systems in the law, document specification, automation of defeasible reasoning, specification of responsibilities and powers in organizations, normative systems specification, confidentiality in database systems, and a host of other applications.

Deontic logic – briefly, the logic of obligation, permission and prohibition, of rights, duties, powers and privileges – may be seen as a branch of applied modal logic that has its origins in the analytic study of law and ethics. Thus it is natural that some of its earliest applications in computer science appeared in connection with the construction of legal expert systems. However, in recent years investigators in other areas of computer science and information technology have become interested in deontic logic as a formal tool. The kinds of applications which have been considered include the specification of fault-tolerant systems and of integrity constraints for databases, the design of electronic contracting systems, the design and specification of system security policies, access control, and other normative aspects of human-computer interaction.

More generally, we may say that deontic logic is useful for the design and specification of any normative system, seen as any set of interacting agents (humans or computers) whose behaviour can be regarded as norm-governed. Norms prescribe how the agents should behave and specify their rights, but also make provision for the possibility that actual behaviour may deviate from the ideal. It is precisely this capacity to describe simultaneously ideal and sub-ideal behaviour that has made deontic logic so attractive and at the same time subject to so many paradoxes.

The development of an information society, in which large numbers of databases – or more generally information systems – are connected in a network and are accessed and modified by very large numbers of users, only reinforces the importance of this research. As we develop policies to ensure the security and integrity of such systems, we must use such concepts as those of rights, permissions, obligations, authority, authorisation, responsibility and delegation, among others. If we are to have any hope of implementing such policies within the system, it is essential that these concepts be precisely understood and defined, and not simply treated informally. On the other hand, taking into account that, at an appropriate level of abstraction, the law, computer systems, businesses, societies, and many other organizational structures may all be seen as instances of, or subject to, normative systems, we may easily see why deontic logic has the wide spectrum of potential application we have mentioned.

Recognition of these connecting interests has led to the decision to organize workshops on this topic at roughly two-year intervals, to promote research and co-operation in this interdisciplinary area. The first workshop (ΔEON'91) was held in Amsterdam (The Netherlands) in December 1991, and the second one (ΔEON'94) was held in Oslo (Norway) in January 1994. This book collects the contributed papers that are to be presented at the third such workshop (ΔEON'96) to be held in Sesimbra (Portugal), from 11 to 13 January 1996.

The previous two workshops have already brought together people working in various aspects of deontic logic. Although we may say that the prime objective of such workshops has been to show the usefulness of deontic logic for computer science and related areas, the reverse direction of influence has also appeared. That is, the interest from other communities in applying deontic logic also has contributed to generating new ideas of how to cope with theoretical and foundational issues in deontic logic. The development of more subtle semantic tools, including the application of dynamic logic to deontic problems, provides just one kind of example. New perspectives on the long-standing "paradoxes" of deontic logic, and a greater concern with the automation of deontic reasoning have become visible as well.

With the third workshop (ΔEON'96) we proposed to broaden the emphasis slightly, to encourage – even more than in the past – contributions in the logic of action and its applications, and also to encourage participation by linguists. Computer scientists are already familiar with logics of action based on dynamic logic, but are less familiar with action/agency operators, developed in philosophical investigations, of the "sees to it that" type. This third workshop will make some of these investigations more accessible to computer scientists. In general, we find that deontic logic and the logics of action, agency and ability, together with epistemic and temporal modalities and with default reasoning, are all important formal tools for the specification of different kinds of normative systems.

The topics on which submissions were invited included a wide range of subjects, ranging from foundational and theoretical issues on deontic logic, normative languages, and the logic of action, to a list of various issues arising from applications in computer science and in public and private administration. The submissions were subject to a process of blind refereeing and twelve were selected based on the high scientific quality and originality of the works. Because of the need to allow ample time for discussions among the participants – one of the main goals of this interdisciplinary workshop – a number of other excellent papers could not be accepted. The workshop is complemented with four invited talks from Nuel Belnap (Pittsburgh University, USA), Andrew Jones (Oslo University, Norway), Krister Segerberg (Uppsala University, Sweden) and Marek Sergot (Imperial College, UK). Their papers are not included in this volume.

The Program Committee consisted of: Mark Brown (Co-Chair, Syracuse University, USA), José Carmo (Co-Chair, IST, Portugal), Frédéric Cuppens (ONERA-CERT, France), José Fiadeiro (FCUL, Portugal), Risto Hilpinen (Turku University, Finland), John Horty (Maryland University, USA), Tom Maibaum (Imperial College, UK), John-Jules Meyer (Utrecht University, The Netherlands), Henry Prakken (Free University, The Netherlands), Martin Sadler (Hewlett Packard, UK), Giovanni Sartor (Bologna University, Italy), Krister Segerberg (Uppsala University, Sweden), Kazimierz Swirydowicz (Poznan University, Poland), Richmond Thomason (Pittsburgh University, USA) and Roel Wieringa (Free University, The Netherlands). José Fiadeiro is also the organizer of the workshop.

The contributions to this volume are organized in alphabetical order by first author. They cover several topics from the call for papers, including foundational issues on deontic operators, deontic conditionals and action logic (in the papers by M. Brown, by J. Horty, by C. Krogh & H. Herrestad and by Y.-H. Tan & L.W.N. van der Torre); actions, decisions and commitments, agency, responsibility and norm-governed interaction, and formal specification of normative systems (in the papers by F. Dignum, J.-J. Ch. Meyer, R.J. Wieringa & R. Kuiper and by F. Santos & J. Carmo); applications to security systems with multiple levels of confidentiality (in the paper by F. Cuppens & R. Demolombe); automation and theoretical aspects of defeasible deontic reasoning (respectively in the papers by A. Artosi, G. Governatori & G. Sartor and by L. Royakkers & F. Dignum); analysis of conceptual relations in legal discourse (L. Allen's paper) and in ethical discourse (P. McNamara's paper); and the use of tools from category theory to provide a theory of norms (A. Johanson's paper). Several papers consider the role of preferences in providing a basis for action in the face of uncertainty or when there is conflict between applicable norms. We believe that this collection of papers will be of great interest to all who are concerned with the various aspects of this interdisciplinary research area.

Mark A. Brown and
José Carmo, editors

Syracuse and Lisbon,
September, 1995

Acknowledgements

This book collects the papers that are to be presented at the Third International Workshop on Deontic Logic in Computer Science (ΔEON'96), to be held at the Hotel do Mar in Sesimbra (Portugal), from 11 to 13 January, 1996.

ΔEON'96 is sponsored by Andersen Consulting and by Caixa Geral de Depósitos (CGD). ΔEON'96 is also supported by Associação Portuguesa para a Inteligência Artificial (APPIA), the British Council, Fundação Luso-Americana para o Desenvolvimento (FLAD) and Junta Nacional de Investigação Científica e Tecnológica (JNICT), as well as by the institutions of the Chairs of the Organizing Committee and Program Committee: the Department of Informatics of the Faculty of Sciences of the University of Lisbon (FCUL), the Departments of Mathematics of the Lisbon Institute of Technology (IST) and of the University of Madeira (UMa), and the Department of Philosophy of Syracuse University. The organizers and Program Committee are grateful for the support received from all these sources.

In addition, we wish to thank all the referees for their excellent work and all the authors that have submitted papers. We are also grateful to Nuel Belnap, Andrew Jones, Krister Segerberg and Marek Sergot for accepting our invitation to give addresses at ΔEON'96. Finally, we want to give special thanks to José Fiadeiro for making possible both the arrangement with Springer-Verlag for the publication of this volume and the Workshop itself, and to Keith van Rijsbergen (editor-in-chief of Springer's Workshops in Computing series) for accepting the publication of this book.

Mark A. Brown and José Carmo

List of Referees

Mark Brown (Syracuse University, USA)
José Carmo (Instituto Superior Técnico, Portugal)
Frédéric Cuppens (Centre D'Etudes et de Recherches de Toulouse, France)
José Fiadeiro (Faculty of Sciences of the University of Lisbon, Portugal)
Risto Hilpinen (Turku University, Finland)
John Horty (University of Maryland, USA)
Andrew Jones (University of Oslo, Norway)
Tom Maibaum (Imperial College, UK)
John-Jules Meyer (Utrecht University, The Netherlands)
Henry Prakken (Free University, The Netherlands)
Martin Sadler (Hewlett Packard, UK)
Giovanni Sartor (Bologna University, Italy)
Jeroen Scheerder (Free University, The Netherlands)
Krister Segerberg (Uppsala University, Sweden)
Marek Sergot (Imperial College, UK)
Kazimierz Swirydowicz (Poznan University, Poland)
Rineke Verbrugge (Free University, The Netherlands)
Richmond Thomason (University of Pittsburgh, USA)
Roel Wieringa (Free University, The Netherlands)

Contents

Authors' Addresses and Positions

Layman E. Allen
Professor of Law and Research Scientist
Law School
University of Michigan
Ann Arbor, MI 48109-1215
USA
e-mail: laymanal@umich.edu

Alberto Artosi
Professor of Philosophy of Science
Department of Philosophy
University of Bologna
via Zamboni 38
40126 Bologna
Italy
Fax: +39-51-258326

Mark A. Brown
Associate Professor
Department of Philosophy
541 Hall of Languages
Syracuse University
Syracuse, NY 13244
USA
e-mail: mabrown@mailbox.syr.edu
and mabrown@cas.syr.edu

José Carmo
Associate Professor
Department of Mathematics
Instituto Superior Técnico
Av. Rovisco Pais
1096 Lisboa Codex
Portugal
e-mail: jcc@inesc.pt

Frédéric Cuppens
Research Engineer
ONERA / CERT
2 Av. Belin
BP no. 4025
31055 Toulouse Cedex
France
e-mail: cuppens@tls-cs.cert.fr

Robert Demolombe
Directeur de Recherche
ONERA / CERT
2 Av. E. Belin
BP no. 4025
31055 Toulouse Cedex
France
e-mail: demolomb@tls-cs.cert.fr

Frank P.M. Dignum
University Lecturer
Department of Mathematics and
Computer Science
Eindhoven University of Technology
P.O. Box 513
5600 MB Eindhoven
The Netherlands
e-mail: dignum@win.tue.nl

Guido Governatori
PhD Student of Computer Science and Law
CIRFID
University of Bologna
via Galliera 3, 40121
Bologna
Italy
e-mail: governat@cirfid.unibo.it

Henning Herrestad
Research Fellow
NRCCL
University of Oslo
P.O. Box 6702 St. Olavs Plass
N-0130 Oslo
Norway
e-mail: henning.herrestad@jus.uio.no

John F. Horty
Associate Professor
Dept. of Philosophy and Institute for
Advanced Computer Studies
University of Maryland
College Park
MD 20742
USA
e-mail: horty@umiacs.umd.edu

Arnold A. Johanson
Professor Emeritus of Mathematics
University of Toledo
1400 Gulf Blvd. #609
Clearwater, FL 34630
USA
Fax: +813-595-0764

Christen Krogh
Research Fellow
NRCCL and Dept. of Philosophy
University of Oslo
P.O. Box 6702 St. Olavs Plass
N-0130 Oslo
Norway
e-mail: christen.krogh@filosofi.uio.no

Ruurd Kuiper
University Lecturer
Department of Mathematics and Computer Science
Eindhoven University of Technology
P.O. Box 513
5600 MB Eindhoven
The Netherlands
e-mail: wsinruur@win.tue.nl

Paul McNamara
Assistant Professor
Department of Philosophy
University of New Hampshire
Durham, NH 03820
USA
e-mail: paulm@christa.unh.edu

John-Jules Ch. Meyer
Professor
Department of Computer Science
Utrecht University
Padualaan 14, De Uithof
P.O. Box 80089
3508 TB Utrecht
The Netherlands
e-mail: jj@cs.ruu.nl

Lambèr Royakkers
PhD Student
Center for Law and Informatization
Tilburg University
P.O. Box 90153
5000 LE Tilburg
The Netherlands
e-mail: L.M.M.Royakkers@kub.nl

Filipe Santos
Assistant and PhD Student
Department of Management Science
ISCTE
Av. das Forças Armadas
1600 Lisboa Codex
Portugal
e-mail: fas@inesc.pt

Giovanni Sartor
IDG - CNR Researcher
IDG
via Panciatichi, 56/16, 50127
Firenze
Italy
e-mail: sartor@cirfid.unibo.it

Yao-Hua Tan
Associate Professor
EURIDIS
Erasmus University Rotterdam
P.O. Box 1738
3000 DR Rotterdam
The Netherlands
e-mail: ytan@euridis.fbk.eur.nl

Leendert W.N. van der Torre
PhD Student
EURIDIS
Tinbergen Institute of the Dept. of Computer Science
Erasmus University Rotterdam
P.O. Box 1738
3000 DR Rotterdam
The Netherlands
e-mail: ltorre@euridis.fbk.eur.nl

Roel J. Wieringa
University Lecturer
Department of Computer Science
Free University
De Boelelaan 1081a
1081 HV Amsterdam
The Netherlands
e-mail: roelw@cs.vu.nl

From the Fundamental Legal Conceptions of Hohfeld to Legal Relations: Refining the Enrichment of Solely Deontic Legal Relations

Layman E. Allen
Law School, University of Michigan
Ann Arbor, United States

Abstract

The Hohfeldian fundamental legal conceptions that deal with solely deontic LEGAL RELATIONS, the duty/privilege and right/no-right pairs, require not only an adequate definition of agency for the person who is obligated or permitted to act, but also the same of patiency for the person to-or-for whom that action is directed. Person-i's DUTY to see-to-it-that that state-of-affairs-s is so for the benefit of person-j is defined here in terms of a deontic OBLIGATION operation and DONE-BY (D2) and DONE-FOR (D4) relations between a state of affairs and persons. This paper is a refinement of the author's earlier efforts to modify, extend, and enrich Hohfeld's fundamental legal conceptions into a more general notion of LEGAL RELATIONS (defined concepts are expressed in all capital letters.) In particular, the agency concept of DONE-BY is being brought into closer conformity with Belnap's emerging stit logic, with the modifications of deontic logic accompanying such changes. The S4-D2 action modal logic considered here is intended to be a part of the A-HOHFELD logic in which LEGAL RELATIONS are defined and from which a representation language called the A-HOHFELD language is derived. The A-HOHFELD language is being used as a representation language for constructing MINT (Multiple INTerpretation) interpretation-assistance systems for helping lawyers to detect alternative structural interpretations of sets of legal rules.

1 Introduction

Logicians and legal scholars have been striving to formalize the fundamental legal conceptions of Wesley N. Hohfeld since the mid-Twentieth Century: Anderson [1] [2]; Fitch [3]; Jones & Sergot [4] [5]; Kanger [6] [7] [8]; Lindahl [9]; Makinson [10]; Thomson [11]; Allen [12] [13]; and Allen & Saxon [14] [15] [16]. This article is part of an effort to modify, extend, and enrich Hohfeld's fundamental legal conceptions into a broader set of LEGAL RELATIONS concepts with the aim of achieving more fully the goal of his pioneering efforts: to provide "lowest common denominators" of legal discourse rich enough to describe all legal relationships between persons and all changes in those relationships [17 at 58-9].

More specifically, addressed here is the enrichment of the solely deontic LEGAL RELATIONS begun in an earlier article [16]. The enrichment is refined by (1) bringing the D2 (DONE-BY) action operator into closer conformity to the emerging "stit" logic (Belnap [18] [19] [20], Belnap & Perloff [21], [22], Perloff [23]) and (2) including DUTIES and PRIVILEGES to be PASSIVE that were omitted in the prior version ([13] and Lindahl [18 at 85-98]. When satisfactorily completed, these enrichments are to be added to the current version of A-HOHFELD logic ([14] [15]) and to the A-HOHFELD representation language based on that logic, which is being used to construct Interpretation-Assistance expert systems for sets of legal rules (Allen & Saxon [15] [16] [24] [25]).

2 Hohfeld's Fundamental Legal Conceptions Transformed to Legal Relations

By way of background to the refinement of the solely deontic LEGAL RELATIONS (SDLR) undertaken here, the transformation from Hohfeld's fundamental legal conceptions to the broader concept of LEGAL RELATIONS is considered briefly. The moves from fundamental legal conceptions to LEGAL RELATIONS include modification, extension and enrichment ([13] [16]).

2.1 Hohfeld's Fundamental Legal Conceptions

In seeking to improve legal thought and expression by clarifying fundamental legal conceptions, Hohfeld regarded the most promising approach to be one of exhibiting the various relations in a scheme of "opposites" and "correlatives" and then providing examples of their scope and application in the appellate court opinions of actual legal cases. He organized into the famed tables of "opposites" and "correlatives", the eight fundamental legal conceptions that he regarded as the "lowest common denominators" of all legal discourse that are capable of describing all legal relationships between persons and changes in such relationships [17 at 30].

	Jural Correlatives		Jural Correlatives	
Jural	right	duty	power	liability
Opposites	no-right	privilege	disability	immunity

The table indicates that "right" is the "opposite" of "no-right" and the "correlative" of "duty" and corresponding relationships for the other terms. All eight are considered to be relationships between two persons -- for example: duty(x,y) and power(x,y).

2.2 Modification of Fundamental Legal Conceptions

The LEGAL RELATIONS that correspond to the first two pairs -- the right/no-right and the duty/ privilege ones -- are defined solely in terms of deontic operators (and some other terms); they are called the SDLR's. They do not involve other conditions; so, they are also referred to as UNCONDITIONAL LEGAL RELATIONS. In the transformations toward LEGAL RELATIONS, the right/no-right and duty/privilege pairs are modified by adding explicitly to the two persons involved in the fundamental legal conception, the statement about the state of affairs that is brought about by one of those persons. The "duty(x,y)" becomes "DUTYsxy", "privilege(x,y)" becomes "PRIVILEGEsxy", etc. This set of four fundamental legal conceptions (and their corresponding LEGAL RELATIONS) are not involved in describing changes in legal states of affairs. They are used only for purposes of describing static legal states at a given point in time.

The other four fundamental legal conceptions, power/disability and liability/immunity, are used not only for describing static legal states but also for describing the changes in legal state from one point in time to another. These four are the only vehicles in Hohfeld's fundamental legal conceptions to account for dynamic change. These dynamic fundamental legal conceptions are modified more extensively in being transformed into LEGAL RELATIONS. Not only is a statement about a state of affairs added, but also one of the persons is replaced by a statement about a LEGAL RELATION in which that person is involved. Thus, "power(x,y)" becomes "POWER_sLR1y", etc.

The LEGAL RELATIONS that correspond to the fundamental legal conceptions are explicitly defined in a system of logic called A-HOHFELD and in a representation language based on that logic that is also called A-HOHFELD ([12] [13] [14] [15] [16]). These defined LEGAL RELATIONS are signalled by expression in all-capital letters.

2.3 Extension of of Fundamental Legal Conceptions

The part dealing with change is the part of Hohfeld's scheme that falls short of the objective of being capable of expressing all legal states and changes. The only means provided to account for change are the power/disability and liability/immunity conceptions. To have the power to create a LEGAL RELATION is to be able to do something that will exercise that power and create that LEGAL RELATION. There is the following sentence in Hohfeld's original article that unmistakeably indicates that he was clearly aware of the fact that changes in legal states arise from circumstances other than the exercise of legal power ([17] at 44-5).

> A change in a given legal relation may result (1) from some superadded fact or group of facts not under the volitional control of a human being (or human beings); or (2) from some superadded fact or group of facts which are under the volitional control of one or more human beings. As

> regards the second class of cases, the person (or persons) whose volitional control is paramount may be said to have the (legal) power to effect the particular change of legal relations that is involved in the problem.

Nevertheless, the Hohfeldian scheme provides only power (and the other dynamic fundamental legal conceptions) to account for legal change, and nowhere else in Hohfeld's writings is there mention of changes in legal state arising from other than exercises of power. This leaves simple examples of legal change -- like that which occurs when an insured barn catches fire as a result of a lightning strike -- unaccounted for by fundamental legal conceptions terminology. There needs to be added some LEGAL RELATION that accounts for legal changes that occur as a result of events that are not actions of a legal person. For each of the eight LEGAL RELATIONS -- RIGHT/NO-RIGHT, DUTY/PRIVILEGE, POWER/DISABILITY, and LIABILITY/IMMUNITY -- there needs to be added a corresponding CONDITIONAL LEGAL RELATION, where the condition that is required to be fulfilled is an event that is not the action of a person. Where the event that fulfills the condition is the action of a person, the LEGAL RELATION involved is a POWER. So, POWER is a special kind of CONDITIONAL LEGAL RELATION, one whose fulfilling condition is the action of a person.

2.4 Summary of Transformed Fundamental Legal Conceptions

The LEGAL RELATIONS that result from this modifying and extending of Hohfeld's fundamental legal conceptions are summarized in Figure 1.

In the A-HOHFELD language, the fundamental legal conceptions are extended to include CONDITIONAL LEGAL RELATIONS for each of the eight LEGAL RELATIONS that correspond to the fundamental legal conceptions. Thus, in A-HOHFELD there is a CONDITIONAL RIGHT, a CONDITIONAL DUTY, a CONDITIONAL NO-RIGHT, a CONDITIONAL PRIVILEGE, a CONDITIONAL POWER, a CONDITIONAL LIABILITY, a CONDITIONAL DISABILITY, and a CONDITIONAL IMMUNITY. The relationship of these added CONDITIONAL LEGAL RELATIONS to the modified fundamental legal conceptions is shown in Figure 1. The modified POWER/ LIABILITY and DISABILITY/IMMUNITY LEGAL RELATIONS are themselves most appropriately regarded as CONDITIONAL LEGAL RELATIONS; they are the capacitive (Makinson's terminology [10 at 407-15]) CONDITIONAL LEGAL RELATIONS. The newly-added ones in this extension of the fundamental legal conceptions are the non-capacitive CONDITIONAL LEGAL RELATIONS. The difference between these two types of CONDITIONAL LEGAL RELATIONS is that for the non-capacitive ones, the event that fulfills the condition of the CONDITIONAL LEGAL RELATION is an event that ***does NOT*** involve the action of a legal person,

but the event that fulfills the condition of the capacitive ones is an event that ***does*** involve the action of a legal person. With capacitive LEGAL RELATIONS the action of a person that results in a change in the legal state of affairs is said to be an exercise of that person's POWER.

Figure 1: LEGAL RELATIONS of the A-HOHFELD Logic and Language

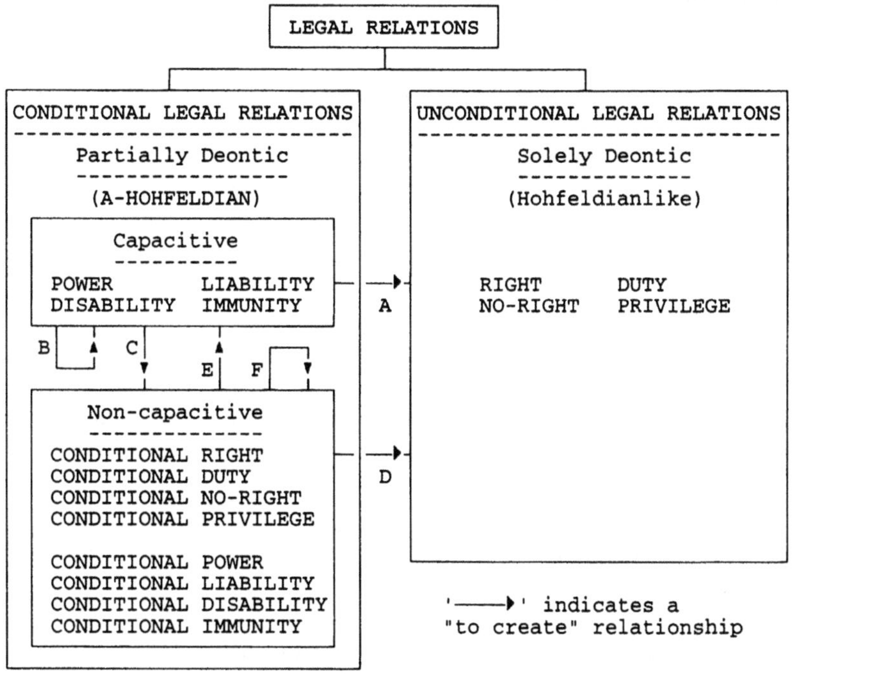

Thus, in this taxonomy of LEGAL RELATIONS there are three general types of such relations: (1) UNCONDITIONAL and CONDITIONAL, where the latter are, in turn, split into (2) capacitive and (3) non-capacitive. Each of the three general types of LEGAL RELATIONS have a "deontic" part -- the UNCONDITIONALS being solely deontic while the CONDITIONALS have a "conditional" part in addition to their deontic part. The concept of OBLIGATION (or its NEGATION) that is incorporated into the definition of the SDLR's of RIGHT/ DUTY and NO-RIGHT/ PRIVILEGE is their deontic part, and one of these four is ultimately involved with each of the CONDITIONAL LEGAL RELATIONS. The conditional parts of the CONDITIONAL LEGAL RELATIONS have to do with the aspects that are fulfilled when POWERS are exercised or conditions are satisfied. Among these three types of LEGAL RELATIONS there are six kinds of "to create" relationships, as shown in Figure 1:

A. Capacitive to create Solely Deontic
B. Capacitive to create Capacitive
C. Capacitive to create Non-capacitive
D. Non-capacitive to create Solely Deontic
E. Non-capacitive to create Capacitive
F. Non-capacitive to create Non-capacitive

Thus, there are capacitive and non-capacitive CONDITIONAL LEGAL RELATIONS that can create themselves (B,F) and both other general types of LEGAL RELATIONS (A,C,D,E), but SDLR's cannot create themselves or any other general type of LEGAL RELATION. Examples of each of these six types of "to create" relationships are the following:

A. POWERS to create RIGHTS
B. POWERS to create POWERS
C. POWERS to create CONDITIONAL POWERS
D. CONDITIONAL RIGHTS to create RIGHTS
E. CONDITIONAL POWERS to create POWERS
F. CONDITIONAL CONDITIONAL RIGHTS to create CONDITIONAL RIGHTS

But there are no

RIGHTS to create RIGHTS or
RIGHTS to create POWERS or
RIGHTS to create CONDITIONAL RIGHTS, etc.

This completes the brief summary of the first part of the broader context in which the main topic of this article occurs -- namely, the refined enrichment of the SDLR's. Next to be considered is the other part of the broader context -- the earlier crude enrichment of the SDLR's.

3 Crude Enrichment of the SDLR's

3.1 Definitions of the Solely Deontic Legal Relations

In the earlier effort to specify the enrichment of the SDLR, the definition of the initial one (namely, DUTY) was in terms of two persons (p1 and p2), a state of affairs brought about by one of them (s), a deontic operator (O), an action operator (D2), and a patient operator (D4), and the other three were in terms of DUTY, NEGATION, and each other. Thus, the following is the definition of DUTY:

DUTYsp1p2 =df OD2D4sp1p2.

What this is intended to say comes out in somewhat stilted prose if the order in which the symbols appear is followed. The definition is intended to say:

The duty that state-of-affairs-s be brought about for the benefit of person-p1 by person-p2

is stipulated to be equal to by definition:

the obligation that something be done by person-p2, namely that state-of-affairs-s be done for the benefit of person-p1.

In somewhat more customary prose:

Person-p2 has a DUTY to person-p1 to do s.

is stipulated to be equal to by definition:

IT IS OBLIGATORY THAT it is brought about that s is so for the benefit of person-p1 by person-p2.

Then, the other three are defined as follows:

action of a person that results in a change in the legal state of affairs is said to be an exercise of that person's POWER.

Figure 1: LEGAL RELATIONS of the A-HOHFELD Logic and Language

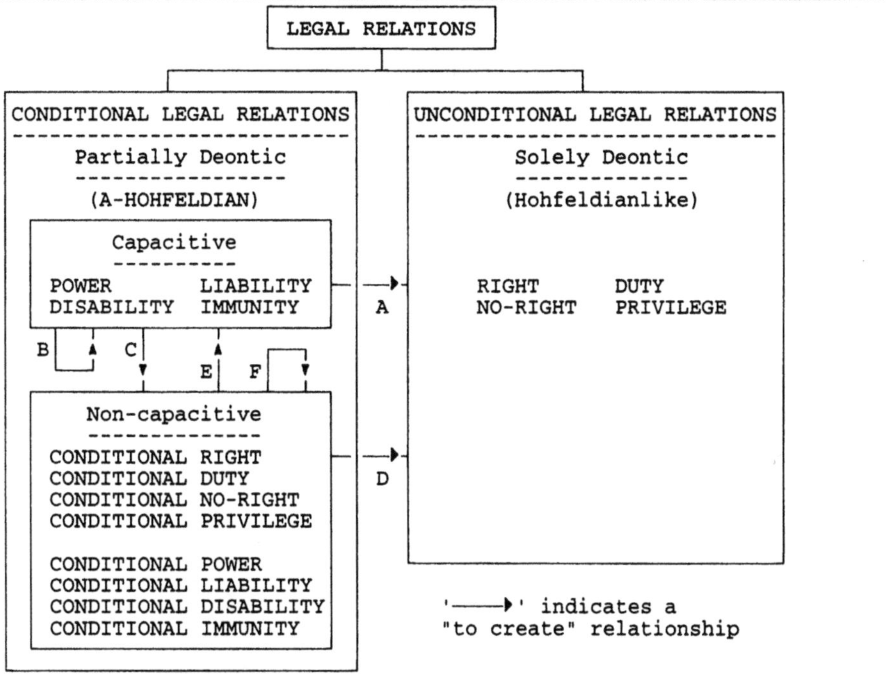

Thus, in this taxonomy of LEGAL RELATIONS there are three general types of such relations: (1) UNCONDITIONAL and CONDITIONAL, where the latter are, in turn, split into (2) capacitive and (3) non-capacitive. Each of the three general types of LEGAL RELATIONS have a "deontic" part -- the UNCONDITIONALS being solely deontic while the CONDITIONALS have a "conditional" part in addition to their deontic part. The concept of OBLIGATION (or its NEGATION) that is incorporated into the definition of the SDLR's of RIGHT/ DUTY and NO-RIGHT/ PRIVILEGE is their deontic part, and one of these four is ultimately involved with each of the CONDITIONAL LEGAL RELATIONS. The conditional parts of the CONDITIONAL LEGAL RELATIONS have to do with the aspects that are fulfilled when POWERS are exercised or conditions are satisfied. Among these three types of LEGAL RELATIONS there are six kinds of "to create" relationships, as shown in Figure 1:

A. Capacitive to create Solely Deontic
B. Capacitive to create Capacitive
C. Capacitive to create Non-capacitive
D. Non-capacitive to create Solely Deontic
E. Non-capacitive to create Capacitive
F. Non-capacitive to create Non-capacitive

S SxOs ————→ SxPs For some x, Ds
weakOBLIGATION strongPERMISSION

4. There is at least one person x who is a member of group-G that has the OBLIGATION to stit s is so.
5. There is at least one person that has the OBLIGATION to stit s is so.

There is a corresponding set of five PERMISSION propositions expressed by the five expressions in the right column of Figure 3. The x variable in the quantifiers in Figure 3 appears as a variable for one of the persons in a DUTY or PRIVILEGE statement; another variable y is present in a corresponding set of quantified deontic expressions for the other person in such a statement. This will provide elements for a two-dimensional 5x5 matrix of quantified O statements (and a like 5x5 matrix of P statements) as summarized in Figure 4.

Figure 4: Strength Diagram of Two-Person OBLIGATION Statements

	UyOs	Uy∈GOs	Sy=jOs	Sy∈GOs	SyOs
UxOs	UU →	UUg →	USj →	USg →	US
Ux∈GOs	Ug →	UgUg →	UgSj →	USg →	US
Sx=iOs	SiU →	SiUg →	SiSj →	SiSg →	SiS
Sx∈GOs	SgU →	SgUg →	SgSj →	SgSg →	SgS
SxOx	SU →	SUg →	SSj →	SSg →	SS

(Each entry also points downward (↓) to the entry below it.)

3.3 Multiple Strengths of DUTY and PRIVILEGE Statements

Quantification in combination with deontic operators determine the inferential strength of various kinds of DUTIES and PRIVILEGES defined in such terms. There are 25 kinds defined for purposes of analyzing LEGAL RELATIONS.

3.4 The Strongest Kind of DUTY Statement

Each of the 25 elements in this 5x5 matrix indicates a different kind of DUTY statement. The UU (Universal- Universal) entry in the northwest corner of Figure 4 is the strongest kind of DUTY statement and the one from which all other 24 kinds of DUTIES represented by the other elements can be inferred. It represents this kind of DUTY statement: UxUyOD2D4sxy. This is equivalent to:

UxUyDUTYsxy. The latter, in turn, is meant to be interpreted as saying: Everyone has a DUTY to everybody else to see to it that state-of-affairs-s is so.

3.5 The Most Important Kind of DUTY Statement

On the other hand, the SiSj (two-individuals) entry squarely in the center of the 25-element matrix represents the following statement about individual persons i and j: OD2D4sij. This is equivalent to: DUTYsij. The latter, in turn, is meant to be interpreted as saying: Person-j has a DUTY to person-i to see to it that state-of-affairs-s is so. This Individual-Individual kind of DUTY statement is squarely in the center in terms of inference strength, as well as in geographical position in Figure 4; it can be inferred from 12 of the other kinds of DUTIES, and 12 of the other kinds can be inferred from it. This is the kind of DUTY that ultimately matters; this is the kind of DUTY that legal systems enforce. For all other kinds of DUTIES there are questions of classification and application about which there is authorized discretion in the decision-making before they affect individuals. This is the kind of DUTY that is used to exemplify the refined enrichment of the SDLR's in this article.

3.6 The Weakest Kind of DUTY Statement

The kind of DUTY at the other extreme in the southeast corner, the SS (Existential-Existential) kind of DUTY that is the weakest kind of all: this kind can be inferred from all 24 other kinds, but none of them can be inferred from it. It represents this kind of DUTY statement: SxSyOD2D4sxy. This is equivalent to: SxSyDUTYsxy. The latter, which, in turn, is meant to be interpreted as saying: Someone (not otherwise identified by membership in any identifiable group) has a DUTY to somebody else (also not identifed by membership in any indentifiable group) to see to it that state-of-affairs-s is so.

3.7 Probable Legal Significance of the Extreme Kinds of DUTIES

This SS kind of DUTY statement makes assertions that are trivially true with respect to legal systems, but they have no effects upon the citizenry of any legal domain. This is a kind of DUTY that is almost certainly too indefinite to be significant in law, as are the other four kinds in the bottom row and the kinds in the right-most column of Figure 4. On the other hand, the nine kinds at the other extremes, the top row and the left-most column are probably too definite to ever be legally significant. They make universal assertions, and in so doing, allow no exceptions and no discretion. But discretion is the essence of legal decision-making systems. There are exceptions to every legal rule, even to the most extreme ones about killing other human beings. This leaves the "central" nine kinds of DUTIES and PRIVILEGES as the ones that are most likely to be significant in the legal domain.

3.8 Corresponding Kinds of PRIVILEGE Statements

There is a 5x5 matrix of 25 kinds of PRIVILEGE statements that correspond to the 25 kinds of DUTY statements of Figure 4. The 25 entries in this matrix of PRIVILEGE entries are related to each other inferentially in exactly the same way that the DUTY entries in the DUTY matrix are related to each other. Also, each of these 25 kinds of PRIVILEGE statements can be inferred from the kind of DUTY statement in the corresponding position in Figure 4.

3.9 Enrichment of Individual-Individual SDLR in Earlier Effort

In the earlier effort at enriching the SDLR's, the context of the rich diversity of 25 different kinds of SDLR was established as indicated above. There, Makinson's [10 at 404-7] neat choice-pairs notational device was used to systematically enlarge the solely deontic possibilities between i and j with respect to s and Ns in the Individual-Individual SDLR. By an equivalent of the Makinson choice-pairs notation, the expression:

```
+   +    +    +  ij      (where '+' indicates the presence of 'N',
↕ O ↕ D2 ↕ D4 ↕ s ↕             '-' indicates its absence, and
-   -    -    -  ji             '↕' indicates alternation)
```

is shorthand for the 32 LEGAL RELATIONS in Columns A and D of Table 1.

Table 1: 32 Sets of Equivalent Solely Deontic LEGAL RELATIONS

A. Deontic Statements	B. DUTY Statements	C. RIGHT Statements	D. Deontic Statements	E. PRIVILEGE Statements	F. NO-RIGHT Statements
1 OD2D4sij	sDUTYsij	sRIGHTsji	17 NOD2D4sij	wPRIVILEGENsij	wNO-RIGHTsji
2 OD2D4sji	sDUTYsji	sRIGHTsij	18 NOD2D4sji	wPRIVILEGENsji	wNO-RIGHTsij
3 OD2D4Nsij	sDUTYNsij	sRIGHTNsji	19 NOD2D4Nsij	wPRIVILEGEsij	wNO-RIGHTNsji
4 OD2D4Nsji	sDUTYNsji	sRIGHTNsij	20 NOD2D4Nsji	wPRIVILEGEsji	wNO-RIGHTNsij
5 OD2ND4sij	lDUTYNsij	lRIGHTNsji	21 NOD2ND4sij	rPRIVILEGEsij	rNO-RIGHTNsji
6 OD2ND4sji	lDUTYNsji	lRIGHTNsij	22 NOD2ND4sji	rPRIVILEGEsji	rNO-RIGHTNsij
7 OD2ND4Nsij	lDUTYsij	lRIGHTsji	23 NOD2ND4Nsij	rPRIVILEGENsij	rNO-RIGHTsji
8 OD2ND4Nsji	lDUTYsji	lRIGHTsij	24 NOD2ND4Nsji	rPRIVILEGENsji	rNO-RIGHTsij
9 OND2D4sij	wDUTYNsij	wRIGHTNsji	25 NOND2D4sij	sPRIVILEGEsij	sNO-RIGHTNsji
10 OND2D4sji	wDUTYNsji	wRIGHTNsij	26 NOND2D4sji	sPRIVILEGEsji	sNO-RIGHTNsij
11 OND2D4Nsij	wDUTYsij	wRIGHTsji	27 NOND2D4Nsij	sPRIVILEGENsij	sNO-RIGHTsji
12 OND2D4Nsji	wDUTYsji	wRIGHTsij	28 NOND2D4Nsji	sPRIVILEGENsji	sNO-RIGHTsij
13 OND2ND4sij	rDUTYsij	rRIGHTsji	29 NOND2ND4sij	lPRIVILEGENsij	lNO-RIGHTsji
14 OND2ND4sji	rDUTYsji	rRIGHTsij	30 NOND2ND4sji	lPRIVILEGENsji	lNO-RIGHTsij
15 OND2ND4Nsij	rDUTYNsij	rRIGHTNsji	31 NOND2ND4Nsij	lPRIVILEGEsij	lNO-RIGHTNsji
16 OND2ND4Nsji	rDUTYNsji	rRIGHTNsij	32 NOND2ND4Nsji	lPRIVILEGEsji	lNO-RIGHTNsij

The 32 generated deontic-action-patienthood statements in Columns A and D of Table 1 are either the definitions of, or are equivalent to, the corresponding 32 DUTY/ PRIVILEGE statements in Columns B and E. Similarly, the 32 RIGHTS/NO-RIGHTS statements in Columns C and F are equivalent to the corresponding statements in both Columns A,D and Columns B,E.

The lower-case letters that precede each of the DUTY/PRIVILEGE and RIGHT/NO-RIGHT expressions above stand for the following abbreviations:

s--strong, l--left(intermediate), r--right(intermediate), w--weak

which have the following inference-strength relationships:

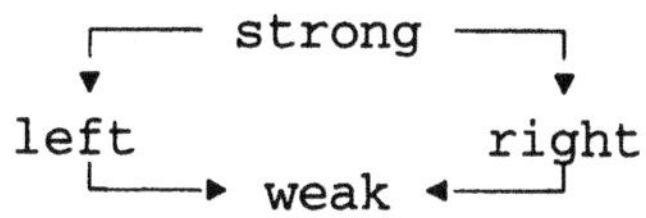

which accounts for all of the inference-strength relationships between the expressions summarized in Figure 5, except for the following four:

LR1→LR25, LR2→LR26, LR3→LR27, and LR4→LR28.

These four result from the deducibility of PERMISSION from OBLIGATION.

The inference relationships among the 32 SDLR's listed in Table 1 are summarized in Figure 5. For example, the inference diagram shows that the following can be inferred from LR1: LR7, LR13, LR11, LR19, LR25, LR31, and LR21, which is to say, among other things, that both LR25 strongPRIVILEGEsij and LR19 weakPRIVILEGEsij can be inferred from: LR1 strongDUTYsij.

This account of the crude enrichment of the earlier effort will facilitate comparison with the refinements being added in this article. It also completes the more extended summary of the final part of the broader context in which the refined enrichment of the SDLR's are undertaken here. Next to be considered are the two aspects of the refinement -- making the DONE-BY action operator more stit-like, and including options for PASSIVITY.

Figure 5: Strength Diagram of 32 SDLR's Between Individual-i and Individual-j

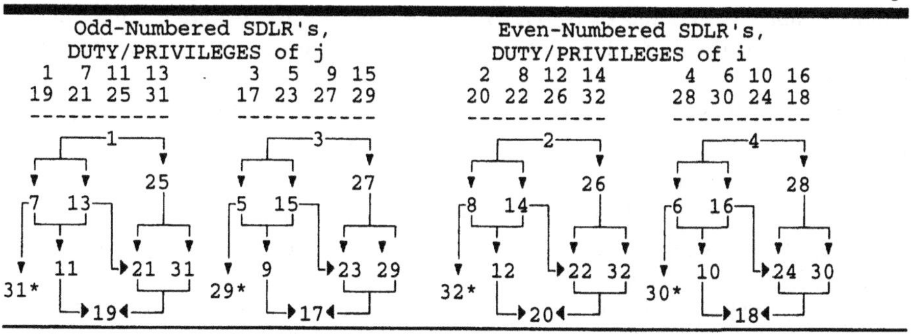

4 Refinement of DONE-BY Operator to be More stit-like

The formulation of an action modal logic involving the DONE-BY operator here is intended to conform closely to the stit logic being developed by Belnap, Perloff, and others, but it would be startling if the two corresponded in every respect. The formulation here clearly differs in the trivial respects that the notation is different, and that 'D2' is the converse of 'stit' (or a close approximation). There is a significant substantive difference due to the presence of the D2oIoD2i rule schema in the formulation presented here. Another source of possible inadvertent differences is that stit logic is formulated semantically; the approach here to D2 (and its associated D4) is syntactic. The formulation below, to be incorporated into the full A-HOHFELD logic, is intended to replace the theory of action, Ac, presented in [14 at 398-403]. The presentation here is simplified by omitting the subscript restrictions, but they, of course, must be included in the final version, if it is to remain a relevance logic (the proofs given here take them into account).

4.1 Syntactic Formulation of Action Modal Logic (S4-D2)

Formation Rule: If x is an individual variable ranging over persons, and w is a wff, then D2wx and D4wx and are wffs.

Transformation Rule Schemas:
In alphabetical order of names as follows: capital letters, small letters, (,), and _.
D2NNoD2i: D2NNwx —:▸ D2wx From 'D2NNwx', it is valid to infer 'D2wx'. I.e., if NNw is DONE-BY x, then it is valid to infer that w is DONE-BY x.
D2NoND2i: D2Nwx —:▸ ND2wx From 'D2Nwx', it is valid to infer 'ND2wx'.
D2o: D2wx —:▸ w From 'D2wx', it is valid to infer 'w'.
D2oD2D2i: D2wx —:▸ D2D2wxx From 'w is DONE-BY x', it is valid to infer '(w is DONE-BY x) is DONE-BY x'.
D2oD2NNi: D2wx —:▸ D2NNwx From 'D2wx', it is valid to infer 'D2NNwx'.
D2oIoD2i: D2vx, Ivw —:▸ D2wx From 'D2vx' and 'Ivw', it is valid to infer 'D2wx' -- i.e., if v is DONE-BY x, AND v NATURALLY IMPLIES w, then it is valid to infer that w is DONE-BY x. This is likely to be the most controversial schema of S4-D2; it is ruled out by the negative conditions of Belnap and Horty's a-stit and d-stit, although condoned by the c-stit of Chellas. See [28].
D2oNND2i: D2wx —:▸ NND2wx From 'D2wx', it is valid to infer 'NND2wx'.
D4NNoD4i: D4NNwx —:▸ D4wx From 'D4NNwx', it is valid to infer 'D4wx'.
D4NoND4i: D4Nwx —:▸ ND4wx From 'D4Nwx', it is valid to infer 'ND4wx'.
D4o: D4wx —:▸ w From 'D4wx', it is valid to infer 'w'.
D4oD4D4i: D4wx —:▸ D4D4wxx From 'w is DONE-BY x', it is valid to infer '(w is DONE-BY x) is DONE-BY x'.
D4oD4NNi: D4wx —:▸ D4NNwx From 'D4wx', it is valid to infer 'D4NNwx'.
D4oIoD4i: D4vx, Ivw —:▸ D4wx From 'D4vx' and 'Ivw', it is valid to infer 'D4wx'.
D4oNND4i: D4wx —:▸ NND4wx From 'D4wx', it is valid to infer 'NND4wx'.
KD2_D2oD2Ki: K.D2vx.K2wx —:▸ D2-Kvw-x From 'K.D2vx.D2wx' it is valid to infer 'D2-Kvw-x'.
OD2oNoND2i: OD2wx, Nw —:▸ ND2wx From 'OD2wx' and 'Nw', it is valid to infer 'ND2wx'.
OD4oNoND4i: OD4wx, Nw —:▸ ND4wx From 'OD4wx' and 'Nw', it is valid to infer 'ND4wx'.

Definitions
PASd: PASwx =df K.ND2wx.ND2Nwx
REFd: REFwx =df D2ND2w
INACTIVEd: INACTIVEwx =df K.REFwx.REF_Nwx
Fd: FD2wx =df O-REF_D2wxx

Pd: PD2wx =df NFD2wx
INDIFFERENTd: INDIFFERENTwx =df K.PD2Nwx.PD2wx
Qd: Qvw =df K.Ivw.Iwv

Since the action modal logic being formulated here is intended as a replacement

Figure 6: 190 Pairs of REF and D2 Rule Schemas (1+2+3+ ... +19 = 190)

1	→ REFsi	20,D2o,D2o,D2o
2	→ D2ND2sii	1,REFd
3	→ D2ND2D2siii	2,F6:P2→3
	See APPENDIX of Proofs	for proof of F6:P2→3
4	→ REF_D2sii	3,REFd
5	→ D2REFsii	2,D2oD2D2i,REFd
6	→ D2D2ND2D2siiii	4,REFd,D2oD2D2i
7	→ D2REF_D2siii	6,REFd
8	→ D2D2D2ND2D2siiiii	6,D2oD2D2i
9	→ D2D2REF_D2siiii	8,REFd
10	→ D2ND2D2D2siiii	3,F6:P2→3
11	→ REF_D2D2siii	10,REFd
12	→ D2ND2D2D2D2siiiii	10,F6:P2→3
13	→ REF_D2D2D2siiii	12,REFd
14	→ D2REF_D2D2siiii	11,F6:1→5
15	→ D2D2REF_D2D2siiiii	14,D2oD2D2i
16	→ D2D2REF_D2D2D2siiiiii	15,F6:9→15
17	→ D2D2D2REF_D2D2siiiiii	15,D2oD2D2i
18	→ D2D2D2REF_D2D2D2siiiiiii	16,D2oD2D2i
19	→ D2D2REFsiii	18,F6:P18→19
20	→ D2D2D2REFsiiii	19,D2oD2D2i

for the earlier action theory, Ac, in A-HOHFELD logic [14], other rule schemas and definitions of that logic are used in the proofs here. Specifically, the following rule schemas and definitions from A-HOHFELD logic are used.

Io:	Ivw, v → w	NNo:	NNw → w
IoIN_Ni:	Ivw → I.Nw.Nv	RI:	Ivw → I(... Ivw)
IoNoNi:	Ivw, Nw → Nv		RI =ab IoI(I)i
I()oIi:	I(v: ... w)→ Ivw	Od:	Ow =df I.Nw.V

4.2 DONE-BY Equivalencies and REFRAIN Equivalencies

There are summarized in Figure 6 190 pairs of derived rules schemas involving 20 D2 and REF statements that are equivalent to a REFRAIN-statement in the S4-D2 action modal logic formulated here. The strategy for proving the 190 pairs of rule schemas of Figure 6 is to show that each expression can be inferred from the previous one and that the first one can be inferred from the last, that is:

```
1→2→3→4→5→6→7→8→9→10→11
↑                     |
|                     ↓
20←19←18←17←16←15←14←13 ←12
```

This is achieved in Figure 6 by proving the following pattern, where the ←→ relationships are (1) either proved or are so when the justifications for the → relationship are solely REFd and/or D2oD2D2i, (2) 19→18 is provable using D2o, 1→4, and D2oD2D2i, and (3) 16→15 is provable by inspection of the joint inferences of 15, 17, 18, and 16 in the pattern.

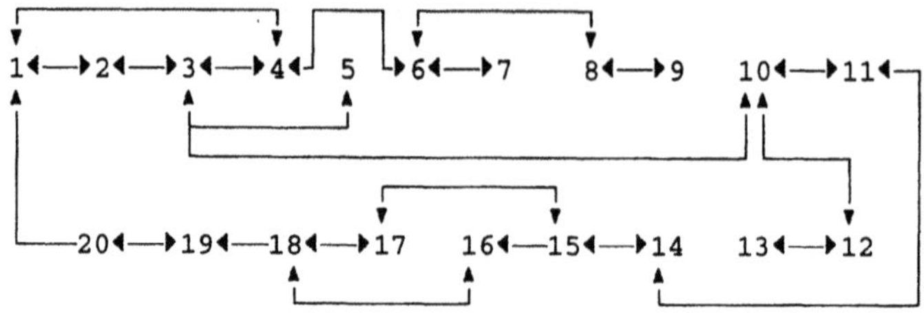

The patterns of relationships between the REF's and the D2's of Figure 6 are somewhat more apparent if they are re-ordered as follows:

```
1—▸4  REFsi—▸REF_D2si                1—▸5  REFsi—▸D2_REFsi
1—▸11 REFsi—▸REF_D2D2si              1—▸18 REFsi—▸2D2D2_REFsi
1—▸13 REFsi—▸REF_D2D2D2si ...        1—▸19 REFsi—▸D2D2D2_REFsi ...
                 1—▸7  REFsi—▸D2_REF_D2si
                 1—▸15 REFsi—▸D2D2_REF_D2D2si
                 1—▸17 REFsi—▸D2D2D2_REF_D2D2D2si ...
      1—▸9  REFsi—▸D2D2_REF_D2si
      1—▸18 REFsi—▸D2D2D2_REF_D2D2si ...
                              1—▸14 REFsi—▸D2_REF_D2D2si
                              1—▸16 REFsi—▸D2D2_REF_D2D2D2 ...
```

What is apparent is that from a REFRAIN-statement one can infer a REFRAIN-statement with any arbitrary number of D2's either in front of it or behind it or in combinations. The names of the 190 pairs of rule schemas can be constructed whenever (if-ever?) needed. For example, the names for the F6:6◂——▸13 pair would be: F6:D2D2ND2D2oREF_D2D2D2i and F6:REF_D2D2D2oD2D2ND2D2i.

A similar set of pairs of rules schemas can be generated for D2 and REF that can be inferred from a DONE-BY statement as shown in Figure 7, and something more interesting that just another collection of rule schemas emerges from such an effort. The proofs of all except 3——▸4 and 7——▸8 are either 1-step or 2-step proofs with justifications of REFd and/or D2OD2D2i, which means that the inferences for these can go both ways (by using D2o). The justifications for these six inferences are made on the diagram of Figure 7. There is only a separate proof given for 3——▸4; The D2oI(s,D2NoND2i)oD2i used in justifying 7——▸8 is proved below. The pattern of proofs from 1 to 9 and around to 1 again is shown in Figure 8.

Figure 7: 36 Pairs of D2 and REF Rule Schemas (1+2+3+ ... +8 = 36)

```
1 ┌──▸D2si                    9,REF-REF-D2oD2i   ??
2 ├──▸D2D2sii                 1,D2oD2D2i
3 ├──▸D2D2D2siii              2,D2oD2D2i
4 ├──▸D2ND2ND2siii            3, (See F7:P3—▸4.)
5 ├──▸REF_ND2sii              4,REFd
6 ├──▸D2D2ND2ND2siiii         4,D2oD2D2i
7 ├──▸D2REF_ND2siii           6,REFd
8 ├──▸D2ND2D2ND2siiii         6,D2oI(s,D2NoND2i)oD2i
  │            See proof of D2oI(s,D2NoND2i)oD2i below.
```

```
9 └──▶REF_REFsii                  8,REFd,REFd
```

The inference of 1 from 9, of course, would be by means of REF_REFoD2i, which in this notation is one-half of the Belnap Refref conjecture. [19 at 139] The completion of the loop of Figure 8 means that all 9 expressions in Figure 7 are mutually provable from each other. But, should it be possible, for example, to infer 1 from 5 -- D2si from REF_ND2sii? Or any of the statements in Figure 8 that are above other statements in Figure 8? Surely it is possible to remain INACTIVEsi, i.e., both REF_ND2sii and REF_D2sii, but 5──▶1 suggests otherwise. The 36 pairs of rule schemas of Figure 7 are provable if REF_REF_D2i is added to the basis of S4-D2, but that includes 26 undesirable rule schemas such as F7:5──▶1, along with the desirable 46 rule schemas (including the other half of the Refref conjecture).

Figure 8: D2 and REF Pattern of Inferences of Fig. 7
With Refref, 72 Rule Schemas & Without One-Half of Refref, 46 Rule Schemas

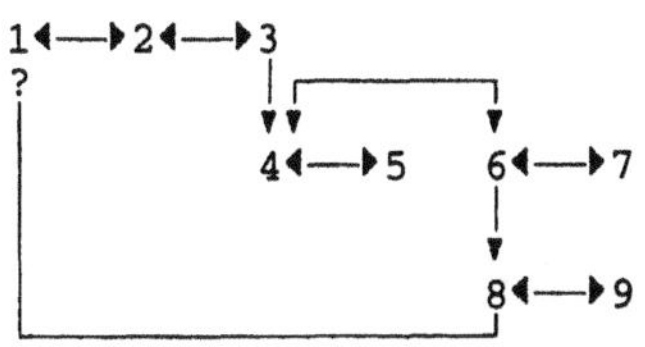

	Pairs	Single Schemas
	3	3 x 6 = 18
	6	4 x 2 = 8
	1	2 x 0 = 0
	--	---
Totals without Refref:	10	26
Grand Total		(2 x 10) + 26 = 46

This indicates that in order to achieve what seems to be the sound intuition of the Belnap Refref conjecture, the S4-D2 action modal logic will need to be otherwise weakened -- possibly by removing or weakening D2oD2D2i, D2oNND2i, or D2oIoD2i.

The proof of D2oI(s,D2NoND2i)oD2i also gives rise to something of interest. Consider the following proof of D2oI()oD2i, which is more general. It is a proof of just one rule -- for sentence variables, s and t. The fact that a proof of every wff v and w would be of the same form, makes it, in effect, a proof of the rule schema: D2vx, I(v: ... w) ──▶ D2wx

```
      Proof ofD2oI()oD2i:
      | D2si, I(s: ... t) ──▶ D2ti
    ──┼──────────────────────────────────────────────
      | The Proof       Justification    Subscripts
      | ---------       -------------    ----------
     1| D2si            s                1
     2|_I(s: ... t)     s                2
     3| Ist             2,I()oIi         2
     4| D2ti            1,3,D2oIoD2i     12
```

Now consider the following proof of D2oI(s,D2NoND2i)oD2i, which is a particular instance of D2oI()oD2i:

```
Proof of D2oI(s,D2NoND2i)oD2i
| D2D2NND2D2siiii ──▶ D2ND2ND2D2siiii
```

```
1 | _D2D2NND2D2siiii   s                  1
2 |  Ia|_D2NND2D2siii  s                  2
  |   b| ND2ND2D2siii  a,D2NoND2i         2
3 | D2ND2ND2D2siiii    1,2,D2oI()oD2i     1
```

Thus, D2oI(s,D2NoND2i)o2i is D2oI()oD2i with the justification sequence of s,D2NoND2i in the subproof. Any sequence of justifications that justifies a valid inference in S4-D2 can be inserted in the subproof of D2oI()oD2i and determine a new rule schema. The same thing is also true of I()oND2oND2i and any other rule schema that contains a subproof; this insertion of sequences of justifications in a subproof provides a powerful method for generating additional derived rule schemas.

With this summary of the relationship between this S4 type of DONE-BY and REFRAIN completed, it is appropriate to turn attention to the comparison between (1) the relationship between PERMISSION and OBLIGATION in not-required-to-be-agentive deontic logic, i.e., when PERMISSION is defined in terms of OBLIGATION and NEGATION, and (2) the relationship between PERMISSION and OBLIGATION in agentive deontic logic, i.e., when PERMISSION is defined in terms of OBLIGATION, NEGATION, and REFRAIN.

4.3 Non-Agentive OBLIGATION-PERMISSION Relationships

The choice-pairs generation of the alternatives for OD2si and PD2si by:

```
+   +    +                   +   +    +
↕ O ↕ D2 ↕ si       and      ↕ P ↕ D2 ↕ si
-   -    -                   -   -    -
```

are shown in Table 2 (where Pw =df NONw, and the deontic operators are NOT required to be agentive), and the resulting relationship between OBLIGATION and PERMISSION is summarized in Figure 9. Note the similarity of strongDUTY/weak-DUTY of Table 1 and Figure 5 and the strongOBLIGATION/weakOBLIGATION of Figure 9, as well as the parallels between their strongPRIVILEGE/ weakPRIVILEGE and strongPERMISSION/ weak-PERMISSION. Just as strongPRIVILEGE and weakPRIVILEGE can be inferred from strongDUTY and weakDUTY, so also can strongPERMISSION and weakPERMISSION be inferred from strongOBLIGATION and weakOBLIGATION.

Table 2: Equivalent (Not-Required-to-be- Agentive) Deontic Statements (O and P)

O1 OD2si	P7 NPND2si	O5 NOD2si	P3 PND2si
O2 OD2Nsi	P8 NPND2Nsi	O6 NOD2Nsi	P4 PND2Nsi
O3 OND2si	P5 NPD2si	O7 NOND2si	P1 PD2si
O4 OND2Nsi	P6 NPD2Nsi	O8 NOND2Nsi	P2 PD2Nsi

There are other similarities; just as weakDUTY and weakPRIVILEGE can be inferred from strongDUTY and strongPRIVILEGE, so also can weakOBLIGATION and weakPERMISSION be inferred from strongOBLIGATION and strongPERMISSION. Also, there are corresponding incompatibles. As strongDUTY to s is incompatible with strongDUTY to Ns, so also is strongOBLIGATION to s incompatible with strongOBLIGATION to Ns. Similarly, strongPERMISSION to s is incompatible with weakOBLIGATION to Ns just as weakPRIVILEGE to s is incompatible with weakDUTY to Ns, and strongOBLIGATION to s is incompatible with weakPERMISSION to Ns just as strongDUTY to s is incompatible with weakPRIVILEGE to Ns.

Figure 9: Relationships Among Not-Required-to-be-Agentive Deontic Statements

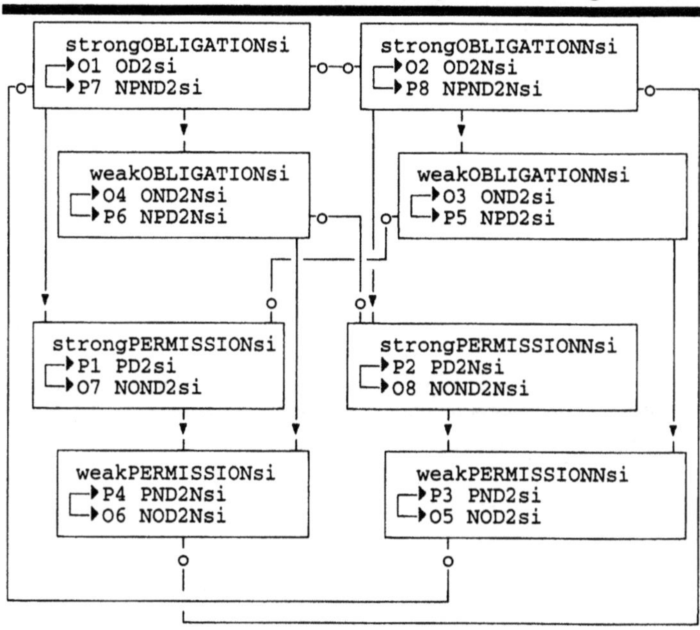

NOTATION
'A←→B' indicates "A always can be inferred from B, AND B always can be inferred from A".
'A——→B' indicates "A cannot always be inferred from B, AND B always can be inferred from A".
'Ao—oB' indicates "A never can be inferred from B, AND B never can be inferred from A".

4.4 Agentive OBLIGATION-PERMISSION Relationships

The choice-pairs generation of the alternatives for OD2si and PD2si are shown in Table 3 (where PD2wx =df NFD2wx, FD2wx =df O_REF_D2wxx, and REFwx =df D2ND2wxx so that PD2wx ←→ NOD2ND2D2wxxx) , and the resulting relationships between OBLIGATION and PERMISSION for agentive deontic operators are summarized in Figure 10.

A comparison of the results summarized in Figure 10 with those in Figure 9 indicates that there are no differences in the logical relationships among the strong-and-weakOBLIGATIONS and the strong-and-weakPERMISSIONS of the agentive deontic statements of Figure 10 and those of Figure 9, whose deontic operators are not required to be agentive. That this turns out to be the case for an

S4 type of DONE- BY action operator is perhaps a little surprising; perhaps it would not be so for a system M or other weaker type of DONE-BY operator, for which it would not be so that D2wx——▸D2D2wxx.

The summary of the 32 SDLR's of Table 1 becomes Table 1' when the non-agentives, 9-16 and 25-32 are replaced by the corresponding agentive representations.

Table 3: Eight Sets of Equivalent (Agentive) Deontic Statements (O and P)

o1	OD2si			NP-REF_D2sii	p7	NPND2si	na p7' NPD2ND2sii	NP-REF_D2sii
o2	OD2Nsi			NP-REF_D2Nsii	p8	NPND2Nsi	na p8' NPD2ND2Nsii	NP-REF_D2Nsii
o3	OND2si	na o3'	OD2ND2sii	O-REF_D2sii	p5	NPD2si		O-REF_D2sii
o4	OND2Nsi	na o4'	OD2ND2Nsii	O-REF_D2Nsii	p6	NPD2Nsi		O-REF_D2Nsii
o5	NOD2si			P-REF_D2sii	p3	PND2si	na p3' PD2ND2sii	P-REF_D2sii
o6	NOD2Nsi			P-REF_D2Nsii	p4	PND2Nsi	na p4' PD2ND2Nsii	P-REF_D2Nsii
o7	NOND2si	na o7'	NOD2ND2sii	NO-REF_D2sii	p1	PD2si		NO-REF_D2sii
o8	NOND2Nsi	na o8'	NOD2ND2Nsii	NO-REF_D2Nsii	p2	PD2Nsi		NO-REF_D2Nsii

NOTATION

na =ab not agentive. D#' =df the agentive version of the non-agentive Deontic D#.

Examples: o3' OD2ND2sii is agentive version (av) of o3 OND2si, o4' is av of o4, etc.

The inference-strength relationship between the two is: D#'——▸D#.

Thus, the refining of the crude enrichment of the SDLR's of Table 1 turns out not to be so refining after all. When the deontic operators are required to be agentive, the relationships among the 32 sets of LEGAL RELATIONS turn out to be the same

Figure 10: Inference-Strength Relationships Among Agentive Deontic Statements

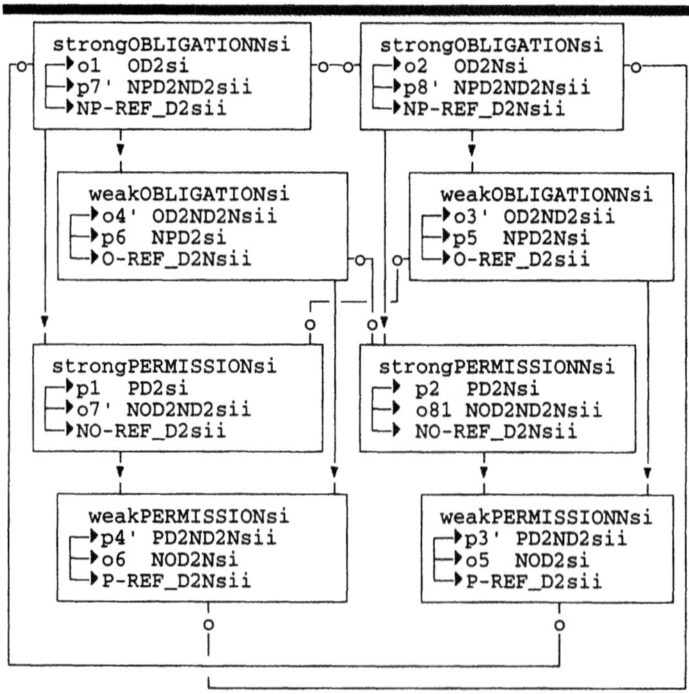

as they were when no such requirements of being agentive were imposed upon the deontic operators. Thus, when requirements of being agentive are added to the deontic operators, the relationships among the LEGAL RELATIONS of Table 1 need only the minor refining of converting 9-16 and 25-32 to agentive alternatives. The relationships among the SDLR's are the same whether the deontic operators are required to be agentive or not. So, the only significant changes that seem to be necessary are to add to Table 1' the SDLR's that indicate that PASSIVITY can be OBLIGATORY and PERMISSIBLE; it is only refinement in this sense of further enrichment that appears to be needed. That is the task next to be considered.

Table 1': 32 Sets of Equivalent SDLR's (Agentive)

	A. OBLIGATION Statements	B. DUTY/PRIVILEGE Statements	C. RIGHT/NO-RIGHT Statements	D. REFRAIN Statements	E. PERMISSION Statements
LR1	OD2D4sij	sDUTYsij	sRIGHTsji		
LR2	OD2D4sji	sDUTYsji	sRIGHTsij		
LR3	OD2D4Nsij	sDUTYNsij	sRIGHTNsji		
LR4	OD2D4Nsji	sDUTYNsji	sRIGHTNsij		
LR5	OD2ND4sij	lDUTYNsij	lRIGHTNsji		
LR6	OD2ND4sji	lDUTYNsji	lRIGHTNsij		
LR7	OD2ND4Nsij	lDUTYsij	lRIGHTsji		
LR8	OD2ND4Nsji	lDUTYsji	lRIGHTsij		
LR9'	OD2ND2D4sijj	wDUTYNsij	wRIGHTNsji	O-REF_D4sij	NPD2D4sij
LR10'	OD2ND2D4sjii	wDUTYNsji	wRIGHTNsij	O-REF_D4sji	NPD2D4sji
LR11'	OD2ND2D4Nsijj	wDUTYsij	wRIGHTsji	O-REF_D4Nsij	NPD2D4Nsij
LR12'	OD2ND2D4Nsjii	wDUTYsji	wRIGHTsij	O-REF_D4Nsji	NPD2D4Nsji
LR13'	OD2ND2ND4sijj	rDUTYsij	rRIGHTsji	O-REF_ND4sij	NPD2ND4sij
LR14'	OD2ND2ND4sjii	rDUTYsji	rRIGHTsij	O-REF_ND4sji	NPD2ND4sji
LR15'	OD2ND2ND4Nsijj	rDUTYNsij	rRIGHTNsji	O-REF_ND4Nsij	NPD2ND4Nsij
LR16'	OD2ND2ND4Nsjii	rDUTYNsji	rRIGHTNsij	O-REF_ND4Nsji	NPD2ND4Nsji
LR17	NOD2D4sij	wPRIVILEGENsij	wNO-RIGHTsji		
LR18	NOD2D4sji	wPRIVILEGENsji	wNO-RIGHTsij		
LR19	NOD2D4Nsij	wPRIVILEGEsij	wNO-RIGHTNsji		
LR20	NOD2D4Nsji	wPRIVILEGEsji	wNO-RIGHTNsij		
LR21	NOD2ND4sij	rPRIVILEGEsij	rNO-RIGHTNsji		
LR22	NOD2ND4sji	rPRIVILEGEsji	rNO-RIGHTNsij		
LR23	NOD2ND4Nsij	rPRIVILEGENsij	rNO-RIGHTsji		
LR24	NOD2ND4Nsji	rPRIVILEGENsji	rNO-RIGHTsij		
LR25'	NOD2ND2D4sijj	sPRIVILEGEsij	sNO-RIGHTNsji	NO-REF_D4sij	PD2D4sij
LR26'	NOD2ND2D4sjii	sPRIVILEGEsji	sNO-RIGHTNsij	NO-REF_D4sji	PD2D4sji
LR27'	NOD2ND2D4Nsijj	sPRIVILEGENsij	sNO-RIGHTsji	NO-REF_D4Nsij	PD2D4Nsij
LR28'	NOD2ND2D4Nsjii	sPRIVILEGENsji	sNO-RIGHTsij	NO-REF_D4Nsji	PD2D4Nsji
LR29'	NOD2ND2ND4sijj	lPRIVILEGENsij	lNO-RIGHTsji	NO-REF_ND4sij	PD2ND4sij
LR30'	NOD2ND2ND4sjii	lPRIVILEGENsji	lNO-RIGHTsij	NO-REF_ND4sji	PD2ND4sji
LR31'	NOD2ND2ND4Nsijj	lPRIVILEGEsij	lNO-RIGHTNsji	NO-REF_ND4Nsij	PD2ND4Nsij
LR32'	NOD2ND2ND4Nsjii	lPRIVILEGEsji	lNO-RIGHTNsij	NO-REF_ND4Nsji	PD2ND4Nsji

5 Refinement of Enrichment: Addition of PASSIVITY

5.1 Overlooked SDLR's: Those that Involve PASSIVITY

Overlooked in the earlier enrichment of the SDLR's were those that involve an agent's being PASSIVE, that is, where both s and Ns are NOT being DONE-BY the agent. The LEGAL RELATIONS involving PASSIVITY are undoubtedly the most pervasive LEGAL RELATIONS. In most legal systems most of the behavior of human agents is unregulated. Those systems are INDIFFERENT, mostly, about whether or not given states of affairs are brought about by such agents. Being INDIFFERENT means that for most states-of- affairs-s, it is neither OBLIGATORY that s be DONE- BY an individual-i nor OBLIGATORY that Ns be DONE-BY such an agent. This, in turn, means that for most states-of-affairs-s, it is PERMITTED both that s NOT be DONE-BY i and that Ns NOT be DONE-BY i. In other words, as far as legal systems are concerned, i's are PERMITTED to be PASSIVE with respect to most s's. So, to understand the freedom afforded in legal systems, it is useful to have precise language to describe and communicate about it. LEGAL RELATIONS involving PASSIVITY should not continue to be overlooked.

5.2 Generation of the SDLR's that Involve PASSIVITY

Since: PASwx =df K.ND2wx.ND2Nwx
the expression to be used for the forced-pairs generation of the SDLR involving PASSIVITY is the following:

```
   +          j +            j
K  ↕ D2 s ... ↕ ↕ D2 Ns  ... ↕
   -          i -            i
```

This forced-pairs generating expression for PASsi results in the four alternatives summarized in Table 4, only one of which needs to be considered for purposes of additions of SDLR's involving PASSIVITY to those of Table 1'.

Table 4: Alternatives Generated When PASSIVITY is Included

	Equivalent to	Comment
1 K.D2sj.D2Nsj	K.D2sj.D2Nsj	contradictory
2 K.D2sj.ND2Nsj	D2sj	Included in Table 1'
3 K.ND2sj.D2Nsj	D2Nsj	Included in Table 1'
4 K.ND2sj.ND2Nsj	K.ND2sj.ND2Nsj	New

Alternative 1 is ruled out because the contradiction KsNs can be inferred from it. Since alternatives 2 and 3 are what is dealt with in Table 1', the only additional alternative that needs to be added to account for PASSIVITY is alternative 4. This leads to the need to explore what additional solely deontic LEGAL RELATIONS are added to those in Table 1' by deontic versions of K.ND2wx.ND2Nwx. These additional solely deontic LEGAL RELATIONS involving O_PAS_D4sij can be generated by using the forced-pairs procedure on the agentive version of OK.ND2D4sij.ND2ND4sij, i.e., the following expression:

```
                            ijj. ........ ijj
+    +     +           + ........... +     ▲
↕ O  ↕ D2  ↕ K.ND2D4   ↕ s    ND2ND4   s   |
-    -     -           - ........... -     ▼
                            jii. ........ jii
```

This generates the 32 sets of equivalent expressions shown in Table 5. The combination of Table 1' and Table 5 completes the specification of the solely deontic LEGAL RELATIONS. The 64 LEGAL RELATIONS listed in the two tables represent all of the different solely deontic LEGAL RELATIONS there are between person-i, person-j, and the state-of-affairs-s and its NEGATION.

5.3 Equivalencies of PASSIVE SDLR's and Other Combinations

From a comparison of their expression as PERMISSION statements or REFRAIN statements, some interesting relationships among various of these SDLR's become apparent. For example,

LR33 ←→ K.LR9'.LR13'	LR37 ←→ K.LR25'.LR29'	LR34 ←→ K.LR10'.LR14'	LR38 ←→ K.LR26'.LR30'
LR35 ←→ K.LR11'.LR15'	LR39 ←→ K.LR27'.LR31'	LR36 ←→ K.LR12'.LR16'	LR40 ←→ K.LR28'.LR32'
LR49 ←→ A.LR25'.LR29'	LR53 ←→ A..LR9'.LR13'	LR50 ←→ A.LR26'.LR30'	LR54 ←→ A.LR10'.LR14'
LR51 ←→ A.LR27'.LR31'	LR55 ←→ A.LR11'.LR15'	LR52 ←→ A.LR28'.LR32'	LR56 ←→ A.LR12'.LR16'

Figure 11 summarizes a start on mapping some of the logical structure among the SDLR's. The odd-numbered SDLR as summarized in Table 5 and Figure 11 are the

Table 5: Sixteen Sets of Equivalent Statements of SDLR's Involving PASSIVITY

	A. Generated Statements	B. PASSIVITY Statements	C. DUTY/PRIVILEGE Statements
LR33	OD2K.ND2D4sij.ND2ND4sijj	OD2_PAS_D4si	spDUTYsij
LR34	OD2K.ND2D4sji.ND2ND4sjii	OD2_PAS_D4sj	spDUTYsji
LR35	OD2K.ND2D4Nsij.ND2ND4Nsijj	OD2_PAS_D4Nsijj	spDUTYNsij
LR36	OD2K.ND2D4Nsji.ND2ND4Nsjii	OD2_PAS_D4Nsjii	spDUTYNsji
LR37	OD2NK.ND2D4sij.ND2ND4sijj	OD2_NPAS_D4sijj	snpDUTYsij
LR38	OD2NK.ND2D4sji.ND2ND4sjii	OD2_NPAS_D4sjii	snpDUTYsji
LR39	OD2NK.ND2D4Nsij.ND2ND4Nsijj	OD2_NPAS_D4Nsijj	snpDUTYNsij
LR40	OD2NK.ND2D4Nsji.ND2ND4Nsjii	OD2_NPAS_D4Nsjii	snpDUTYNsji
LR41	OD2ND2K.ND2D4sij.ND2ND4sijj	O_REF_PAS_D4sijj	wpDUTYNsij
LR42	OD2ND2K.ND2D4sji.ND2ND4sjii	O_REF_PAS_D4sjii	wpDUTYNsji
LR43	OD2ND2K.ND2D4Nsij.ND2ND4Nsijj	O_REF_PAS_D4Nsijj	wpDUTYsij
LR44	OD2ND2K.ND2D4Nsji.ND2ND4Nsjii	O_REF_PAS_D4Nsjii	wpDUTYsji
LR45	OD2ND2NK.ND2D4sij.ND2ND4sijj	O_REF_NPAS_D4sijj	wnpDUTYNsij
LR46	OD2ND2NK.ND2D4sji.ND2ND4sjii	O_REF_NPAS_D4sjii	wnpDUTYNsji
LR47	OD2ND2NK.ND2D4Nsij.ND2ND4Nsijj	O_REF_NPAS_D4Nsijj	wnpDUTYsij
LR48	OD2ND2NK.ND2D4Nsji.ND2ND4Nsjii	O_REF_NPAS_D4Nsjii	wnpDUTYsji
LR49	NOD2K.ND2D4sij.ND2ND4sijj	NOD2_PAS_D4sijj	wpPRIVILEGENsij
LR50	NOD2K.ND2D4sji.ND2ND4sjii	NOD2_PAS_D4sjii	wpPRIVILEGENsji
LR51	NOD2K.ND2D4Nsij.ND2ND4Nsijj	NOD2_PAS_D4Nsijj	wpPRIVILEGEsij
LR52	NOD2K.ND2D4Nsji.ND2ND4Nsjii	NOD2_PAS_D4Nsjii	wpPRIVILEGEsji
LR53	NOD2NK.ND2D4sij.ND2ND4sijj	NOD2_NPAS_D4sijj	wnpPRIVILEGENsij
LR54	NOD2NK.ND2D4sji.ND2ND4sjii	NOD2_NPAS_D4sjii	wnpPRIVILEGENsji
LR55	NOD2NK.ND2D4Nsij.ND2ND4Nsijj	NOD2_NPAS_D4Nsijj	wnpPRIVILEGEsij
LR56	NOD2NK.ND2D4Nsji.ND2ND4Nsjii	NOD2_NPAS_D4Nsjii	wnpPRIVILEGEsji
LR57	NOD2ND2K.ND2D4sij.ND2ND4sijj	NO_REF_PAS_D4sijj	spPRIVILEGEsij
LR58	NOD2ND2K.ND2D4sji.ND2ND4sjii	NO_REF_PAS_D4sjii	spPRIVILEGEsji
LR59	NOD2ND2K.ND2D4Nsij.ND2ND4Nsijj	NO_REF_PAS_D4Nsijj	spPRIVILEGENsij
LR60	NOD2ND2K.ND2D4Nsji.ND2ND4Nsjii	NO_REF_PAS_D4Nsjii	spPRIVILEGENsji
LR61	NOD2ND2NK.ND2D4sij.ND2ND4sijj	NO_REF_NPAS_D4sijj	snpPRIVILEGEsij
LR62	NOD2ND2NK.ND2D4sji.ND2ND4sjii	NO_REF_NPAS_D4sjii	snpPRIVILEGEsji
LR63	NOD2ND2NK.ND2D4Nsij.ND2ND4Nsij	NO_REF_NPAS_D4Nsij	snpPRIVILEGENsij
LR64	NOD2ND2NK.ND2D4Nsji.ND2ND4Nsji	NO_REF_NPAS_D4Nsji	snpPRIVILEGENsji

where spDUTYwxy =df OD2_PAS_D4wxyy (STRONG_PASSIVE_DUTY) --
which is weaker than sDUTY_PAS_wyxy (i.e., OD2D4_PAS_wyxy)
snpDUTYwxy =df OD2_NPAS_D4wxyy (STRONG_NONPASSIVE_DUTY)
wpDUTYwxy =df O_REF_PAS_D4Nwxyy (WEAK_PASSIVE_DUTY)
wnpDUTYwxy =df O_REF_NPAS_D4Nwxyy (WEAK_NONPASSIVE_DUTY)

	D. PERMISSION Statements	E. REFRAIN Statements	F. RIGHT/NO-RIGHT Statements
LR33	NP_REF_PAS_D4sij	NP_REF_PAS_D4sij	spRIGHTsji
LR34	NP_REF_PAS_D4sji	NP_REF_PAS_D4sjii	spRIGHTsij
LR35	NP_REF_PAS_D4Nsi	NP_REF_PAS_D4Nsijj	spRIGHTNsji
LR36	NP_REF_PAS_D4Nsj	NP_REF_PAS_D4Nsjii	spRIGHTNsij
LR37	NP_REF_NPAS_D4sijj	NP_REF_NPAS_D4sijj	snpRIGHTsji
LR38	NP_REF_NPAS_D4sjii	NP_REF_NPAS_D4sjii	snpRIGHTsij
LR39	NP_REF_NPAS_D4Nsij	NP_REF_NPAS_D4Nsijj	snpRIGHTNsji
LR40	NP_REF_NPAS_D4Nsji	NP_REF_NPAS_D4Nsjii	snpRIGHTNsij
LR41	NPD2_PAS_D4sijj	O_REF_PAS_D4sijj	wpRIGHTNsji
LR42	NPD2_PAS_D4sjii	O_REF_PAS_D4sjii	wpRIGHTNsij
LR43	NPD2_PAS_D4Nsijj	O_REF_PAS_D4Nsijj	wpRIGHTsji
LR44	NPD2_PAS_D4Nsjii	O_REF_PAS_D4Nsjii	wpRIGHTsij
LR45	NPD2_NPAS_D4sijj	O_REF_NPAS_D4sijj	wnpRIGHTNsji
LR46	NPD2_NPAS_D4sjii	O_REF_NPAS_D4sjii	wnpRIGHTNsij
LR47	NPD2_NPAS_D4Nsijj	O_REF_A.D2D4Nsij.D2ND4Nsijj	wnpRIGHTsji
LR48	NPD2_NPAS_D4Nsjii	O_REF_A.D2D4Nsji.D2ND4Nsjii	wnpRIGHTsij
LR49	P_REF_PAS_D4sijj	A.NO_REF_D4sij.NO_REF_ND4sij	wpNO-RIGHTsji
LR50	P_REF_PAS_D4sjii	A.NO_REF_D4sji.NO_REF_ND4sji	wpNO-RIGHTsij
LR51	P_REF_PAS_D4Nsijj	A.NO_REF_D4Nsij.NO_REF_ND4Nsij	wpNO-RIGHTNsji
LR52	P_REF_PAS_D4Nsjii	A.NO_REF_D4Nsji.NO_REF_ND4Nsji	wpNO-RIGHTNsij
LR53	P_REF_NPAS_D4sijj	P_REF_NPAS_D4sijj	wnpNO-RIGHTsji
LR54	P_REF_NPAS_D4sjii	P_REF_NPAS_D4sjii	wnpNO-RIGHTsij
LR55	P_REF_NPAS_D4Nsijj	P_REF_NPAS_D4Nsijj	wnpNO-RIGHTNsji

LR56	P_REF_NPAS_D4Nsjii	P_REF_NPAS_D4Nsjii	wnpNO-RIGHTNsij
LR57	PD2_PAS_D4sijj	A.O_REF_D4sij.O_REF_ND4sij	spNO-RIGHTNsji
LR58	PD2_PAS_D4sjii	A.O_REF_D4sji.O_REF_ND4sji	spNO-RIGHTNsij
LR59	PD2_PAS_D4Nsijj	A.O_REF_D4Nsij.O_REF_ND4Nsij	spNO-RIGHTsji
LR60	PD2_PAS_D4Nsjii	A.O_REF_D4Nsji.O_REF_ND4Nsji	spNO-RIGHTsij
LR61	PD2_NPAS_D4sijj	NO_REF_A.D2D4sij.D2ND4sijj	snpNO-RIGHTNsji
LR62	PD2_NPAS_D4sjii	NO_REF_A.D2D4sji.D2ND4sjii	snpNO-RIGHTNsij
LR63	PD2_NPAS_D4Nsijj	NO_REF_A.D2D4Nsij.D2ND4Nsijj	snpNO-RIGHTsji
LR64	PD2_NPAS_D4Nsjii	NO_REF_A.D2D4Nsji.D2ND4Nsjii	snpNO-RIGHTsij

where wpPRIVILEGEwxy =df P_REF_PAS_D4Nwxyy (WEAK_PASSIVE_PRIVILEGE)
wnpPRIVILEGEwxy =df P_REF_NPAS_D4Nwxyy (WEAK_NONPASSIVE_PRIVILEGE)
spPRIVILEGEwxy =df PD2_PAS_D4wxyy (STRONG_PASSIVE_PRIVILEGE)
snpPRIVILEGEwxy =df PD2_NPAS_D4wxyy (STRONG_NONPASSIVE_PRIVILEGE)

DUTIES and PRIVILEGES from person-j's point of view. There are likely to be other interesting inference relationships in addition to those shown in Figure 11. Further analysis is planned.

This brief summary of some of the interesting logical relationships among the SDLR's concludes this refinement of the earlier effort (1) to modify, extend, and enrich the fundamental legal conceptions of Hohfeld, and in doing so, (2) to map out a much larger domain of LEGAL RELATIONS and specify the fundamentally important subset of SDLR's.

Figure 11. Inference Relationships Among the SDLR

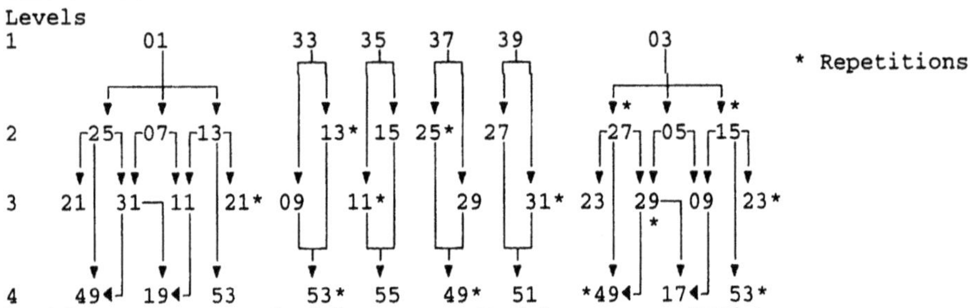

Level 1	Level 2	Level 3	Level 4
LR01 sDUTYsij	LR05 lDUTYNsij	LR11 wDUTYsij	LR19 wPRIVILEGEsij
LR03 sDUTYNsij	LR07 lDUTYsij	LR11 wDUTYsij	LR19 wPRIVILEGEsij
LR33 sDUTY.PASD4sij	LR13 rDUTYsij	LR23 rPRIVILEGENsij	LR49 wPRIVILEGE.PASD4sijj
LR35 sDUTY.PASD4Nsij	LR15 rDUTYNsij	LR21 rPRIVILEGEsij	LR51 wPRIVILEGE.PASD4Nsijj
LR37 wDUTY.PASD4Nsijj	LR25 sPRIVILEGEsij	LR29 lPRIVILEGENsij	LR53 sPRIVILEGE.PASD4Nsijj
LR39 wDUTY.PASD4sijj	LR27 sPRIVILEGENsij	LR31 lPRIVILEGEsij	LR55 sPRIVILEGE.PASD4sijj

6 CONCLUSIONS

The set of lowest common denominators that Hohfeld was seeking with the fundamental legal conceptions seems to be realized in the form of a much richer set of LEGAL RELATIONS. His fundamental insights have been modified, extended, enriched (and now, refined) in the constructing of LEGAL RELATIONS. But now there is an A-HOHFELD language truly rich enough to describe all legal states of affairs and all changes in those states of affairs. With its varying precise expressions for each of the 1600 different SDLR's and equally precise expressions for however many of the infinite number of CONDITIONAL LEGAL RELATIONS that turn out to be useful, the A-HOHFELD language is a tool for precisely describing legal affairs and their changes -- whenever a lawyer, judge, or other concerned person chooses to do so. Its completeness, in the sense of being expressively rich enough to describe all legal states and changes, is not a matter that can be logically proved. One counter-example, however, will be sufficient to disprove the claim. Nearly forty years of asking for counter-examples among legal audiences has not yet yielded one that stands up as a true counter-example. The construction of such a representation language leans heavily upon the underlying system of A-HOHFELD logic, which in turn requires an adequate action logic. This article has been concerned with refinement of the relationship between the deontic components and the action components of LEGAL RELATIONS and completing the story of the SDLR's by including those involving PASSIVITY.

In future explorations of the differences between S4-D2 and the various stit logics with respect to applications in the legal domain, the Horty-Belnap suggestion [28] is appealing (namely, that in light of the interdefinability of cstit and dstit in the presence of historical necessity, the availabilty of both operators for different analytic purposes is advisable). For example, in the case of the exercise of a POWER to offer to sell a unique Rembrandt painting to a specified offeree, the offeror's POWER to make such an offer to that offeree (and thereby create a POWER of acceptance in that offerree) is terminated; after making such an offer, she no longer has the POWER to create a POWER of acceptance in that offeree for such a deal, because the offeree then already has such a POWER of acceptance. This is a clear example of a dstit-type of being DONE-BY in the exercise of such a POWER to offer. On the other hand, were the POWER to offer exercised be one concerning fungible goods (like barrels of oil), the offeror would continue to have the POWER to offer to the same offeree the same amount of the same kind of oil at the same price to be delivered at the same date, despite the fact that the offeree already had a POWER to accept the first offer for such oil. The exercise of such a POWER to offer with respect to fungible goods is a clear example of a cstit type of being DONE-BY where the existent POWER of acceptance by offeree would not preclude the offeror's continuing to have the POWER to create another POWER of acceptance of the offeree by making another offer. Thus, in some instances it might be advisable to have available a dstit type of DONE-BY (with a negative condition semantically and without D2oIoD2i syntactically), while in other instances a cstit type of DONE-BY (without the negative condition semantically and with D2oIoD2i syntactically) is desirable.

And, oh yes, ... another curiousity when deontic operators are agentive and there is closure under consequence. Because of the Q-equivalency of the seven expressions below, there is an interesting relationship between OBLIGATORY PASSIVENESS and OBLIGATORY INACTIVITY -- they can be inferred from each other. Not only can OBLIGATORY PASSIVENESS be inferred from OBLIGATORY INACTIVITY (as expected), but because OBLIGATION is agentive, OBLIGATORY INACTIVITY can be inferred from OBLIGATORY PASSIVENESS.

```
1 ┌─▶ O-PASwx                        7,OoI(s,INACTIVEoPASi)oOi
2 ├─▶ OD2K.ND2wx.ND2Nwx              1,PASd
3 ├─▶ OK.D2ND2wxx.D2ND2Nwxx          2,OoI(D2KoKD2_D2i)oOi
4 ├─▶ K.OD2ND2wxx.OD2ND2Nwxx         3,OKoKO_Oi
5 ├─▶ K.O-REFwx.O-REF_Nwx            4,REFd
6 ├─▶ OK.REFwx.REF_Nwx               5,KOoOKi
```

7 └→ O-INACTIVEwx 6,INACTIVEd

But, of course, INACTIVITY cannot be inferred from PASSIVENESS.

REFERENCES

1. Anderson, Alan Ross. Logic, norms, and roles. Ratio 1962; 4: 36-49.
2. Anderson, Alan Ross. The logic of Hohfeldian propositions. U of Pittsburgh L Rev 1971; 33:28-38.
3. Fitch, Frederic B. A revision of Hohfeld's theory of legal concepts. Logique et Analyse 1967; 10:269-76.
4. Jones, Andrew J.I. & Sergot, Marek. On the role of deontic logic in the characterization of normative systems. Proceedings of the First International Workshop on Deontic Logic in Computer Science (DEON '91) 1991; Amsterdam.
5. Jones, Andrew J.I. & Sergot, Marek. Formal specification of security requirements using the theory of normative positions. Proceedings of Esorics-92 1992; Toulouse.
6. Kanger, Stig. New foundations for ethical theory. 1957; Stockholm. Reprinted in: Hilpinen R (ed) Deontic logic: introductory and systematic readings. D. Reidel, Dordrecht, Holland, 1970, pp. 36-58.
7. Kanger, Stig. Law and logic. Theoria 1972; 38:105-32.
8. Kanger, Stig. On realization of human rights, in action, logic and social theory. Holmstron G & Jones A (ed) Acta Philosophica Fennica 1985; 38:71-78.
9. Lindahl, Lars. Position and change: a study in law and logic. D. Reidel, Dordrecht-Holland/ Boston-U.S.A., 1977.
10. Makinson, David. On the formal representation of rights relations. J of Philosophical Logic 1986; 15:403-25.
11. Thomson, Judith J. The realm of rights. Harvard Univrsity Press, Cambridge, Mass., London, England, 1990
12. Allen, Layman E. Formalizing Hohfeldian analysis to clarify the multiple senses of 'legal right': a powerful lens for the electronic age. S Calif L Rev 1974; 48:428-87.
13. Allen, Layman E. Enriching the deontic fundamental legal conceptions of Hohfeld. To be published in Anniversary anthology in computers and law. Bing J & Torvund O (ed)., TANO-publ., Oslo, 1995.
14. Allen, Layman E. & Saxon, Charles S. Analysis of the logical structure of legal rules by a modernized and formalized version of Hohfeld's fundamental legal conceptions. In: Martino A & Natali F (ed) Automated analysis of legal texts: logic, informatics, law, North-Holland, Amsterdam, 1986, pp 385-450.
15. Allen, Layman E. & Saxon, Charles S. A-Hohfeld: a language for robust structural representation of knowledge in the legal domain to build interpretation-assistance expert systems. In: Meyer JJ & Wieringa R (ed) Deontic logic in computer science: normative system specification. John Wiley & Sons, New York, 1993, 205-24.
16. Allen, Layman E. & Saxon, Charles S. Better language, better thought, better

communication: the A-Hohfeld language for legal analysis. Forthcoming in the Proceedings of the Fifth International Conference on Artificial Intelligence and Law, May 21-24, 1995, University of Maryland, College Park, Md.
17. Hohfeld, Wesley N. Fundamental legal conceptions as applied in judicial reasoning. Yale L J 1913; 23:16-59. Reprinted with a new forward by Arthur L. Corbin, Yale University Press, New Haven, 1964.
18. Belnap, Nuel. Declaratives are not enough. Philosophical Studies 1990; 59: 1-30.
19. Belnap, Nuel. Before refraining: concepts for agency. Erkenntnis 1991a; 34:137-69.
20. Belnap, Nuel. Backwards and forwards in the modal logic of agency. Li Philosophy & Phenomenological Res 1991b; 777-807.
21. Belnap, Nuel & Perloff, Michael. The way of the agent. Studia Logica 1992; 51:463-84.
22. Belnap, Nuel & Perloff, Michael. In the realm of agents. Annals of Mathematics & Artificial Intelligence 1993; 9: 25-48.
23. Perloff, Michael. Stit and the language of agency. Synthese 1991; 86:379-408.

24. Allen, Layman E. & Saxon, Charles S. More IA needed in AI: interpretation assistance for coping with the problem of multiple structural interpretations. Third International Conference on Artificial Intelligence and Law, June 25-28, 1991, St. Catherine's College, Oxford, England, published In the Proceedings of the Conference by the Association for Computing Machinery (ACM).
25. Allen, Layman E. & Saxon, Charles S. Controlling inadvertent ambiguity in the logical structure of legal drafting by means of the prescribed definitions of the A-Hohfeld structural language. Theoria 1994; 9:135-172.
26. Herrestadt, Henning & Krogh, Christen. The right direction. Verso un Sistema Esperto Giuridico Integrale, Firenze, 1993. (To be published in 1994.)
27. Herrestadt, Henning & Krogh, Christen (1995) Deontic logic relativised to bearers and counterparties. To be published in: Bing J & Torvund O (ed) Anniversary anthology in computers and law. TANO-publ., Oslo, 1995.
28. Horty, John F. & Belnap, Nuel. The deliberative stit: a study of action, omission, ability, and obligation. Forthcoming in J. of Philosophical Logic, 1993.

Appendix of Proofs

```
   D2KoKD2_D2i
  | D2Kvwx ──▶ K.D2vx.D2wx
 ─┼──────────────────────────────────
1 | D2Kvwx        s            1
2 |-Ia| Kvw       s            2
  |   | -
  |  b| v         a,Ko         2
  |  c| w         w,Ko         2
3 | I.Kvw.v       2,I()oIi     -
4 | I.Kvw.w       2,I()oIi     -
5 | D2vx          1,3,D2oIoD2i 1
6 | D2wx          1,4,D2oIoD2i 1
7 | K.D2vx.D2wx   5,6,Ki       1
```

Figure 6, Proof of 2──▶3, and 3──▶2

```
    F6:P2◀──▶3
    F6:P2──▶3:  D2ND2D2oD2ND2i
    F6:P3──▶2:  D2ND2oD2ND2D2i
```

```
   F6:P1──▶5
   F6:REFoD2_REFi
   F6:D2_REFoREFi
  | REFsi ──▶ D2REFsii
 ─┼──────────────────────────────
1 |_REFsi        s            1
2 | D2ND2sii     1,REFd       1
3 | D2D2ND2siii  2,D2oD2D2i   1
4 | D2REFsii     3,REFd       1
```

```
   F6:P9──▶15
   F6:D2D2_REF_D2oD2D2_REF_D2D2i
  | D2D2REF_D2siiii ──▶
  |           D2D2REF_D2D2siiiii
 ─┼──────────────────────────────────────
1 |_D2D2REF_D2siiii     s              1
2 | REF_D2sii           1,D2o,D2o      1
3 | REF_D2D2siii        2,F6:1──▶4     1
```

```
 |  D2ND2sii ←→ D2ND2D2siii
-+-------------------------------------------
1|_D2ND2sii              s                1
2| Ia|_ND2si             s                2
 |  b|  I1|_D2D2sii      s                3
 |     2|  D2si          1,D2o            3
 |  c|  ND2D2sii         b,a,I()oNoNi     2
3| D2ND2D2siii           1,2,D2oI()oD2i   1

1|_D2ND2D2siii           s                1
2| Ia|_ND2D2sii          s                2
 |  b|  I1|_D2si         s                3
 |     2|  D2D2sii       1,D2oD2D2i       3
 |  c|  ND2si            b,a,I()oNoNi     2
3| D2ND2sii              1,2,D2oI()oD2i   1
```

```
   F6:P4→1
   F6:REF_D2oREFi
   F6:REFoREF_D2i
 | REF_D2sii → REFsi
-+----------------------------
1|_REF_D2sii    s          1
2| D2ND2D2siii  1,REFd     1
3| D2ND2sii     2,P3→2     1
4| REFsi        3,REFd     1
```

```
   F6:P1→5
   F6:REFoD2_REFi
   F6:D2_REFoREFi
 | REFsi → D2REFsii
-+-------------------------------
1|_REFsi        s             1
2| D2ND2sii     1,REFd        1
3| D2D2ND2siii  2,D2oD2D2i    1
4| D2REFsii     3,REFd        1
```

```
4| D2D2REF_D2D2siiiii 3,F6:P1→5,
 |                      F6:P1→5   1
```

```
   F6:P18→19
   F6:D2D2D2_REF_D2D2D2oD2D2_REFi
 | D2D2D2REF_D2D2D2siiiiiii → D2D2REFsiii
-+------------------------------------------------
1|_D2D2D2REF_D2D2D2siiiiiii s                    1
2| REF_D2D2D2siiii             1,D2o,D2o,D2o     1
3| REFsi                       2,F6:P4→1,
 |                             F6: P4→1,
 |                             F6:P4→1           1
4| D2D2REFsiiii                3,D2oD2D2i,
 |                             D2oD2D2i          1
```

```
   F7:P3→4
   F7:D2D2D2oD2ND2ND2i
   F7:D2ND2ND2oD2D2D2i
 | D2D2D2siii → D2ND2ND2siii
-+------------------------------------------------
1|_D2D2D2siii    s                               1
2| D2D2sii       1,D2o                           1
3| D2NND2sii     2,D2OD2NNi                      1
4| D2D2NND2siii  3,D2oD2D2i                      1
5| D2ND2ND2siii  4,D2oI(s,D2NoND2i)oD2i          1
```

```
   F7:P7→8
   F7:D2_REF_ND2oD2ND2D2ND2i
   F7:D2ND2D2ND2oD2_REF_ND2i
 | D2REF_ND2siii → D2ND2D2ND2siiii
-+------------------------------------------------------
1|_D2REF_ND2siii      s                                1
2| D2D2ND2ND2siiii    1,REFd                           1
3| D2ND2D2ND2siiii    2,D2oI(s,D2NoND2i)oD2i           1
```

Towards a Computational Treatment of Deontic Defeasibility

Alberto Artosi[1], Guido Governatori[2], and Giovanni Sartor[3]
[1] Dipartimento di Filosofia, Università di Bologna,
via Zamboni 38, 40126 Bologna, Italy

[2] CIRFID, Università di Bologna,
via Galliera 3, 40121 Bologna, Italy

[3] IDG
via Panciatichi, 56/16, 50127 Firenze, Italy
e-mail: governat,sartor@cirfid.unibo.it

Abstract

In this paper we describe an algorithmic framework for a multi-modal logic arising from the combination of the system of modal (epistemic) logic devised by Meyer and van der Hoek for dealing with nonmonotonic reasoning with a deontic logic of the Jones and Pörn-type. The idea behind this (somewhat eclectic) formal set-up is to have a modal framework expressive enough to model certain kinds of deontic defeasibility, in particular by taking into account preferences on norms. The appropriate inference mechanism is provided by a tableau-like modal theorem proving system which supports a proof method closely related to the semantics of modal operators. We argue that this system is particularly well-suited for mechanizing nonmonotonic forms of inference in a monotonic multi-modal setting.

1 Introduction

In the last few years application of defeasible (nonmonotonic) reasoning methods to normative rules has become a major topic in the study of norms and normative reasoning. Defeasible reasoning methods have been proven to be able to cope with a variety of important issues. A number of different approaches to formalizing defeasible deontic reasoning have been proposed and several frameworks have been exploited (see e.g. [1, 2, 3, 4, 5, 6, 7]). It turns out, however, that not much effort to account for deontic defeasibility in a computationally oriented manner has been made until now. This can partly be traced to the fact that usual nonmonotonic reasoning methods are not well-suited to computational treatment. On the other hand, it has been argued (see [6]) that computational approaches to deontic defeasibility, e.g. in a logic programming setting, do not fit in with existing inference techniques from the field of non-classical logic theorem proving. In this paper we shall present a computational framework for deontic defeasible reasoning based neither on logic programming nor on current nonmonotonic formalisms but on a system of modal (epistemic) logic devised by Meyer and van der Hoek [8] for dealing with nonmonotonic reasoning by monotonic means. The interesting point is that we extend Meyer and van der Hoek's system to a multi-modal system which subsumes a deontic logic of the Jones and Pörn-type [9, 10]. The idea behind this (somewhat eclectic) formal set-up is to have a modal framework expressive enough to model certain kinds of deontic defeasibility, in particular by taking into account preferences on norms, and that can be easily housed in the computational setting provided by the KEM theorem prover for (normal) modal logics presented in [11]. The paper is organized as follows. In the next two sections we provide a concise overview of Jones-Pörn's and Meyer-van der Hoek's logics. In section 4 we provide some hints about how represent defaults in Meyer van der Hoek's modal setting. In section 5 we shall outline the multi-modal system resulting from combining the above logics. In section 6 we shall provide a brief description of the computational framework KEM. In sections 7, 8 and 9 we shall show how all the above mentioned systems can be accommodated in KEM's computational setting. Finally in section 10 we shall discuss reasoning with preferences in such a setting.

2 Deontic Logic

The deontic basis of our system is provided by Jones and Pörn's [9, 10] deontic logic DL, an extension of standard deontic logic (SDL) which incorporates, besides the normal deontic operators O^i and P^i, the deontic operators O^s and P^s. $\mathsf{O}^i A$ and $\mathsf{P}^i A$ (at a world w) mean (as in SDL): A holds in all, respectively some, of w's deontically ideal versions. $\mathsf{O}^s A$ and $\mathsf{P}^s A$ (at a world w) mean: A holds in all, respectively some, of w's "sub-ideal versions" (intuitively, a sub-ideal version of w is a world where not everything that is ideal in w holds). DL allows us to define the following notions:

- $\mathsf{N}_D A =_{df} (\mathsf{O}^i A \wedge \mathsf{O}^s A)$ (*Deontic Necessity*)
- $\mathsf{O}_\mathsf{T} A =_{df} (\mathsf{O}^i A \wedge \mathsf{P}^s \neg A)$ (*Ought*).

Since DL is a straightforward extension of SDL both $\mathsf{O}^i, \mathsf{P}^i$ and $\mathsf{O}^s, \mathsf{P}^s$ behave as normal KD-modalities. A model for DL is thus a structure:

$$\langle W, R_i, R_s, v \rangle$$

where R_i, $R_s \subseteq W \times W$ are serial (non reflexive) relations on W (intuitive reading: wR_iv = v is an ideal version of w, wR_sv = v is a sub-ideal version of w), subject to the following conditions:

C1: $R_i \cap R_s = \emptyset$;

C2: $\{\langle w, w \rangle : w \in W\} \subseteq R_i \cup R_s$.

The meaning of **C1** and **C2** is that there cannot exist ideal worlds that are also sub-ideal, and every world is either ideal or sub-ideal relative to itself (notice that this amounts to introducing some form of reflexivity in the model). v is as usual with the following clauses for O^i and O^s respectively:

$$\models_w \mathsf{O}^i A \Leftrightarrow \forall v \in W : wR_i v, \models_v A$$

$$\models_w \mathsf{O}^s A \Leftrightarrow \forall v \in W : wR_s v, \models_v A.$$

3 A Modal Logic for Defeasible Reasoning (MDL)

Nonmonotonic reasoning is concerned with reasoning about and with *defeasible* information. In [8] this is modelled by a system of modal logic incorporating, besides the standard modal operators $\Box$ and $\Diamond$, n operators $P_1 \dots P_n$ which refer to "preferred" subsets of the set W of (epistemically) possible worlds. $P_i A$ $(1 \leq i \leq n)$ means: A holds in all the i-preferred worlds (i.e. all the worlds in the set Σ_i). Axiomatically, such a system is obtained by adding to the standard $S5$ system the following axioms:

1. $\Box P_i A \equiv P_i A$
2. $\neg P_i \bot \rightarrow (P_i P_j A \equiv P_j A)$
3. $\neg P_i \bot \rightarrow (P_i \Box A \equiv \Box A)$
4. $\Box A \rightarrow P_i A \, (1 \leq i \leq n)$.

It turns out that this system (Meyer and van der Hoek's $S5P_{(n)}$, henceforth referred to as MDL = *modal default logic*) is that of a multi-modal $K45/S5$ system with $\Box$ and $\Diamond$ behaving as $S5$ modalities and the P_i behaving as $K45$ modalities. A model for MDL is thus a structure

$$< W, \Sigma_1, \dots, \Sigma_n, R, R_1, \dots, R_n, v >$$

where $\Sigma_i \subseteq W$ are subsets (possibly empty) of preferred worlds; $R \subseteq W \times W$ is the standard $S5$ accessibility relation on W; $R_i = \Sigma \times \Sigma_i \subseteq R$ is a $K45$ (i.e. transitive and euclidean) accessibility relation on Σ_i; and v is as usual with the following additional clause:

$$\models_w P_i A \Leftrightarrow \forall v \in W : wR_i v, \models_v A, (1 \leq i \leq n).$$

4 Representing defaults in MDL

According to [8] default reasoning is treated by translating the usual default rules in the MDL language as follows. Reiter's classical rule $\frac{A:B}{C}$ is translated into $A \wedge \Diamond B \rightarrow P_i C$. This is read as "if A is true and B is considered possible, then C is preferred (practically believed)". Similarly, normal defaults become $A \wedge \Diamond B \rightarrow P_i B$, and multiple defaults $A_1 \wedge \Diamond B_1 \rightarrow P_1 C_1$,

$A_2 \wedge \Diamond B_2 \rightarrow P_2C_2 \ldots$ where P_1 and P_2 are preference operators which can be associated either with the same cluster or with distinct preferred sets. In [8] this representation is extended with a mechanism for belief revision to obtain defeasibility. We propose a different solution, where defeasibility is obtained by analyzing defaults and assigning indices to them, according to the following definition. Let the input of our system be a knowledge base consituted of two sets of formulas: a set F of facts, and a set D of normal defaults:

$$\frac{A : B}{B}.$$

Let each default $\frac{A:B}{B}$ in D be translated as

$$\Box(A \rightarrow B)$$

that we abridge as

$$A \Rightarrow B.$$

Let D' be the set of all such formulas and let $S_1, \ldots, S_n$ be all maximal subsets of D' which are consistent with the facts (i.e. $S_i \cup F$ is consistent and for no $S_j, S_i \subseteq S_j$). The key idea is that of introducing an operator P_i in the consequent of every formula in S_i, thus obtaining $A \Rightarrow P_iB$ for each formula $A \Rightarrow B$ in S_i. Let our translated knowledge base be $F \cup D_M$, where D_M denotes the set of modal defaults resulting from D' by assigning preferences.

As a result of this translation, we obtain that (a) the defaults included in each set S_i will be assigned the same preference index while defaults occurring in different sets are assigned different preference indices (obviously defaults occurring in every set will receive all indices); (b) every inconsistency is avoided, since inconsistent defaults are assigned to different preferred sets. This approach recalls the treatment of prioritized defaults of Brewka [12, 13]. Note that if $F \cup D'$ is consistent, then $D_M = D'$. Note also that in our translation we no longer need the justification part of the default, which was modalized in [8] as $\Diamond B$, since our mechanism for assigning preferences takes care of consistency checks.

The above translation enables us to perform a kind of skeptical default inference, as the following example shows.

Example 4.1

$$(\{p, q\}, \{\frac{p : r}{r}, \frac{r : w}{w}, \frac{q : \neg w}{\neg w}\}).$$

Given this knowledge base, our procedure gives rise to the following translation:

$$p, q, p \Rightarrow P_{\{1,2\}}r, r \Rightarrow P_1 w, q \Rightarrow P_2 \neg w.$$

Clearly, the above premises imply both $P_1 w$, and $P_2(\neg w)$. Defeasibility is obtained by revising indices after modifying the knowledge base. Let us assume that a new fact w is added to our knowledge base, so that we obtain the set

$$(\{p, q, w\}, \{\frac{p : r}{r}, \frac{r : w}{w}, \frac{q : \neg w}{\neg w}\}).$$

Clearly, we have now just one set of defaults whose conclusion is consistent with the facts, that is $\{\frac{p:r}{r}, \frac{r:w}{w}\}$. Therefore our translation simply gives:

$$p, q, w, p \Rightarrow P_1 r, r \Rightarrow P_1 w.$$

Note that the model just proposed can be extended with mechanisms for ordering single defaults, and/or for ordering sets of defaults, possibly on the basis of the ordering over single defaults (on the combination of consistency check and ordering, see [12, 13]. We will not consider such mechanisms, which have been much discussed in the literature, but if such methods were adopted it would be possible to restrict admitted set of defaults (and the corresponding preference operators) just to the best (maximal) sets of defaults.

Let us consider a further example to illustrate our method.

Example 4.2 *Acting in self-defence*

The following knowledge base contains two conflicting rules, one saying that committing a tort implies a responsibility, the and the other saying that acting in self-defence implies no responsibility.

1a. John committed a tort.

2a. John acted in self-defence.

3a. People commiting a tort is responsible.

4a. People acting in self-defence is not responsible.

This situation is translated into the following knowledge base

1b. t

2b. d

3b. $\dfrac{t : r}{r}$

4b. $\dfrac{d : \neg r}{\neg r}$.

The corresponding modal translation is

1c. t

2c. d

3c. $t \Rightarrow P_1 r$

4c. $d \Rightarrow P_2 \neg r$.

Clearly, this set safely implies the two consequences $P_1 r$ and $P_2 \neg r$.

5 A multi-modal MDL/DL system for defeasible normative reasoning

The idea of treating defeasible deontic reasoning by combining an existing deontic logic with an already existing nonmonotonic formalism has been suggested by [4, 5]. Following this suggestion, in this section we shall combine MDL with DL to obtain a formalism suited for dealing with defeasible rules in normative reasoning. It is immediate to see that the combination of MDL with DL results in a multi-modal $KD/K45/S5$ system having $\Box, \Diamond, P_1 \ldots P_n$, O^i, P^i, O^s and P^s as its (independent but interacting) modalities. For ease of reference we shall call this system DDL (*defeasible deontic logic*). A models for DDL is a mixed MDL/DL structure

$$\langle W, \Sigma_1, \ldots \Sigma_n, R, R_i, R_s, R_1, \cdots, R_n, v\rangle$$

where $\Sigma_1, \ldots \Sigma_n, R, R_i, R_s$ and $R_j, (1 \leq j \leq n)$ are as before. This means, intuitively, that we have a semantic setting working with several kinds of worlds, i.e.

- possible worlds
- preferred worlds
- deontic worlds $\begin{cases} \text{deontically ideal worlds} \\ \text{deontically sub-ideal worlds} \end{cases}$

for $\Box, P_j, \mathsf{N}_D, \mathsf{O}^i, \mathsf{O}^s$, respectively.

The resources available in DDL allow us to treat some cases involving defeasibility, and others appealing to the DL features. For example, the famous Chisholm paradox can be solved without using defeasibility, but by simply making use of DL.

Example 5.1 *Chisholm Paradox*

1. It ought to be that John does not impregnate Suzy Mae.
2. Not-impregnating Suzy Mae commits John to not-marrying her.
3. Impregnating Suzy Mae commits John to marrying her.
4. John impregnates Suzy Mae.

As it is well-known there have been proposed several formalizations of this celebrated paradox (see [14]). Let us take the following default-like version:

1a. $\mathsf{O}_\top \neg A$

2a. $\dfrac{\neg A : \mathsf{O}_\top \neg B}{\mathsf{O}_\top \neg B}$

3a. $\dfrac{A : \mathsf{O}_\mathsf{T} B}{\mathsf{O}_\mathsf{T} B}$

4a. A

Our $F \cup D'$ translation being consistent we simply obtain

1b. $\mathsf{O}_\mathsf{T} \neg A$

2b. $\neg A \Rightarrow P_1(\mathsf{O}_\mathsf{T} \neg B)$

3b. $A \Rightarrow P_1 \mathsf{O}_\mathsf{T} B$

4b. A

which corresponds substantially to the formalization proposed by Jones and Pörn [9]. In such a case we derive both $\mathsf{O}_\mathsf{T} B$ an $\mathsf{O}^i \mathsf{O}_\mathsf{T} \neg B$ (meaning that in all ideal situations there is the obligation $\mathsf{O}_\mathsf{T} \neg B$, although in the actual, sub-ideal situation, it holds $\mathsf{O}_\mathsf{T} B$).

Let us now consider the following example.

Example 5.2 *Acting in different normative systems*

1a. If Mustafa is muslim he can be poligamous.

2a. If Mustafa is Italian, he cannot be poligamous.

3a. Mustafa is both muslim and Italian.

This set has the following default-like representation:

1b. $\dfrac{m : \mathsf{P}^i p}{\mathsf{P}^i p}$

2b. $\dfrac{i : \neg \mathsf{P}^i p}{\neg \mathsf{P}^i p}$

3b. $m \wedge i$.

The corresponding modal translation is

1c. $m \Rightarrow P_1 \mathsf{P}^i p$

2c. $i \Rightarrow P_2 \neg \mathsf{P}^i p$

3c. $m \wedge i$.

From this we can derive that according to preference P_1 (which gives priority to religious prescriptions) Mustafa can be poligamous ($P_1 \mathsf{P}^i p$), and according to preference P_2 (which gives preference to Italian law) he cannot be poligamous ($P_2 \neg \mathsf{P}^i p$).

Let us finally consider an example which clearly involves not only a conflict, but also the need to choose between different preferences.

Example 5.3 *A question about table manners [2]*

1a. A person must use knife and fork when eating.

2a. A person must not use knife and fork when eating asparagus.

3a. A person is eating asparagus.

This set is translated into

1b. $e \Rightarrow P_1 O_T f$

2b. $e \wedge a \Rightarrow P_2 O_T \neg f$

3b. $e \wedge a$.

In such a case we need to add to our system a mechanism for comparing preferences. Here the preference P_2 should be chosen because it is more specific than P_1.

6 The computational framework *KEM*

In [11, 15] we presented a tableau-like proof system, called *KEM* which has been proved to be able to cope with a wide variety of (normal propositional) modal and multi-modal logics [16]. *KEM* is based on D'Agostino and Mondadori's [17] classical system *KE*, a combination of tableau and natural deduction inference rules which allows for a restricted ("analytic") use of the cut rule. The key feature of *KEM*, besides its being based neither on resolution nor on standard sequent/tableau inference techniques, is that it automatically generates models and checks them for putative contradictions using a label scheme to bookkeep "world" paths. In this section we shall present *KEM* in barest outline. We first recall some basic notions.

Our base language will be a modal propositional language L defined in the usual way. We shall use the letters $A, B, C, \ldots$ to denote arbitrary formulas of L. By a *signed formula* (*S-formula*) we shall mean an expression of the form SA where $S \in \{T, F\}$ and A is a formula of L (intended meaning: $TA = A$ holds, $FA = A$ does not hold at a given word in some Kripke model). We shall denote by X, Y, Z arbitrary signed formulas. As usual X^C will be used to denote the *conjugate* of a S-formula X, i.e. the result of changing S to its opposite (with the exception of the following S-formulas: $T\Box A, F\Diamond A, F\Box A$ and $F\Diamond A$ which also have $T\Diamond\neg A, F\Box\neg A, F\Diamond\neg A, T\Box\neg A$ respectively as their conjugates). Two S-formulas X, Z such that $Z = X^C$, will be called *complementary*.

As we have just said *KEM* approach wants we work with "world" labels. A "world" label is either a constant or a variable "world" symbol or a "structured" sequence of world-symbols we call a "world-path". Intuitively, constant and variable world-symbols stand for worlds and sets of worlds respectively, while a world-path conveys information about access between the worlds in it. We attach labels to S-formulas to yield *labelled signed formulas* (*LS-formulas*), i.e.

pairs of the form X, i where X is a S-formula and i is a label. A LS-formula SA, i means, intuitively, that A is true (false) at the (last) world (on the path represented by) i. In the course of proof search, labels are manipulated in a way closely related to the semantics of modal operators and "matched" using a (specialized, logic-dependent) unification algorithm. That two world-paths i and k are unifiable means, intuitively, that they virtually represent the same path, i.e. any world which you could get by the path i could be reached by the path k and vice versa. LS-formulas whose labels are unifiable turn out to be true (false) at the same world(s) relative to the accessibility relation that holds in the appropriate class of models. In particular two LS-formulas X, X^C whose labels are unifiable stand for formulas which are contradictory "in the same world". These ideas are formalized as follows.

6.1 Label formalism

Let L^n be a multi-modal language with n distinct operators. For the sake of the following discussion let us assume that the semantics of L^n is given by a structure $M = \langle W, \Sigma_1, \ldots, \Sigma_m, R_1, \ldots, R_n, v\rangle$ where W is a (non-empty) set of possible worlds, $\Sigma_i \subseteq W$ $(1 \leq i \leq m)$ and $R_i \subseteq W \times W$ $(1 \leq i \leq n)$. For $1 \leq i \leq n$ let $\Phi^i_C = \{w^i_1, w^i_2, \cdots\}$ and $\Phi^i_V = \{W^i_1, W^i_2, \cdots\}$ be (non empty) sets respectively of constant and variable world-symbols. Let us define

$$\Phi_C = \bigcup_{1\leq i\leq n} \Phi^i_C \text{ and}$$

$$\Phi_V = \bigcup_{1\leq i\leq n} \Phi^i_V .$$

The set $\Im$ of world-labels is now defined as

$$\Im = \bigcup_{1\leq i} \Im_i \text{ where } \Im_i \text{ is :}$$

$$\Im_1 = \Phi_C \cup \Phi_V;$$

$$\Im_2 = \Im_1 \times \Phi_C;$$

$$\Im_{n+1} = \Im_1 \times \Im_n .$$

That is a world-label is either (i) an element of the set Φ_C, or (ii) an element of the set Φ_V, or (iii) a path term (k', k) where (iiia) $k' \in \Phi_C \cup \Phi_V$ and (iiib) $k \in \Phi_C$ or $k = (m', m)$ where (m', m) is a label. According to the above informal explanation, we may think of a label $i \in \Phi_C$ as denoting a (given) world, and a label $i \in \Phi_V$ as denoting a set or worlds (any world) in M. A label $i = (k', k)$ may be viewed as representing a path from k to a (set of) world(s) k' accessible from k. For instance, $(w^i_2, (W^i_1, w^i_1))$ represents a path which takes us to a world w^i_2 accessible via any world accessible from w^i_1 (i.e. accessible from the subpath (W^i_1, w^i_1)) according to R_i (notice that the labels are read from right to left). In what follows we shall use $w_j (W_j)$ to generically refer to a given (any) world quite apart from its i. To facilitate reference we shall sometimes use different names for different kinds of labels.

A bit of terminology. For any label $i = (k', k)$ we call k' the *head* of i, k the *body* of i, and denote them by $h(i)$ and $b(i)$ respectively. Notice that these notions are recursive: if $b(i)$ denotes the body of i, then $b(b(i))$ will denote the body of $b(i)$, $b(b(b(i)))$ will denote the body of $b(b(i))$; and so on. For example, if i is $(w_4, (W_3, (w_3, (W_2, w_1))))$, then $b(i) = (W_3, (w_3, (W_2, w_1)))$, $b(b(i)) = (w_3, (W_2, w_1))$, $b(b(b(i))) = (W_2, w_1)$, $b(b(b(b(i)))) = w_1$. We call each of $b(i)$,$b(b(i))$, etc., a *segment* of i. Let $s(i)$ denote any segment of i (obviously, by definition every segment $s(i)$ of a label i is a label); then $h(s(i))$ will denote the head of $s(i)$. For any label i, we define the length of i, $l(i)$, as the number of world-symbols in i, i.e. $l(i) = n \Leftrightarrow i \in \Im_n$. $s^n(i)$ will denote the segment of i of lenght n, i.e. $s^n(i) = s(i)$ such that $l(s(i)) = n$. For any label $i, l(i) > n$, we define the *countersegment-n* of i, $c^n(i)$ (i.e. what remains of i after deleting $s^n(i)$), as:

$$c^n(i) = h(i) \times (\cdots \times (h(s^k(i)) \times (\cdots \times (h(s^{n+1}, w_0) \cdots) \cdots)(n < k < l(i))$$

We shall call a label i *restricted* if $h(i) \in \Phi_C$, otherwise we call it *unrestricted.* We shall say that a label k is *i-preferred* iff $k \in \Im^i$ where

$$\Im^i = \{k \in \Im : h(k) \text{ is either } w^i_m \text{ or } W^i_m, 1 \leq i \leq n\},$$

and that a label k is *i-ground* $(1 \leq i \leq n)$ iff

1. $\forall s(k) : h(s(k)) \notin \Phi^i_V$, and
2. if $\exists s^m(k) : h(s^m(k)) \in \Phi^i_V$, then $\exists s^j(k), j < m : h(s^j(k)) \in \Phi^i_C$.

6.2 Unification scheme

KEM's label unification scheme involves two kinds of unifications, respectively "high" and "low" unifications. "High" unifications are meant to mirror specific accessibility constraints; they are used to build "low" unifications, which account for the full range of conditions governing the appropriate accessibility relation. We then begin by defining the basic notion of "high" unification. First we define a substitution in the usual way as a function

$$\begin{aligned} \sigma \ &: \ \Phi^0_V \longrightarrow \Im^- \\ &: \ \Phi^i_V \longrightarrow \Im^i, (1 \leq i \leq n) \end{aligned}$$

where $\Im^- = \Im - \Phi_V$. For two labels i, k and a substitution σ, if σ is a unifier of i and k then we shall say that i and k are σ-unifiable. We shall (somewhat unconventionally) use $(i, k)\sigma$ to denote both that i and k are σ-unifiable and the result of their unification. On this basis we define several specialised, logic-dependent notions of both σ "high" (or σ^L) and σ "low" (or σ_L) unification (see [11]). For example, the notion of two labels i, k being σ^K-, σ^D-, and σ^{S5}-unifiable is defined in the following way:

$$
\begin{aligned}
(i,k)\sigma^K &= (i,k)\sigma \iff \\
&\quad \text{at least one of } i \text{ and } k \text{ is restricted, and} \\
&\quad \text{for every } s(i),\ s(k),\ l(s(i)) = l(s(k)),\ (s(i),s(k))\sigma^K \\
(i,k)\sigma^D &= (i,k)\sigma \\
(i,k)\sigma^{S5} &= (h(i),h(k))\sigma.
\end{aligned}
$$

For example, let $i = (w_3,(W_1,w_1))$ and $k = (W_3,(w_2,w_1))$ be two labels to be unified. They σ^K-unify to $(w_3,(w_2,w_1))$ according to the above definition of σ^K-unification. Notice that they are also σ^D-unifiable (this preserves the obvious relation between the K and D logics). Let us consider now two labels $i = (w_3,(W_1,w_1))$ and $k = (W_3,(W_2,w_1))$. They σ^D-unify to $(w_3,(w_2,w_1))$ according to the above definition of σ^D-unification. They are not, however, σ^K-unifiable (according to the second condition of the above definition) since the segments $(W_1,w_1))$, (W_2,w_1) are not σ^K-unifiable (by the first condition of the above definition). The reason is that in the "non idealisable" logic K the "denotations" of W_1 and W_2 may be empty (i.e. there can be no worlds accessible from w_1), which obviously makes their unification impossible, while in the "idealisable" logic D they are not empty, which makes them to be unifiable "on" any constant. As to the notion of σ^{S5}-unification, take $i = (w_3,(W_1,w_1))$ as before, and $k = (W_2,(W_1,(w_2,w_1)))$. These labels σ^{S5}-unify to w_3 according to the above definition of σ^{S5}-unification, which for a label of the form (m',m) amounts to deleting m from the path to a world m' (since, if access is "ubiquitous", then the "way to" m' is irrelevant).

For $L = K, D, S5$ the notion of two labels i,k being σ_L-unifiable is defined quite simply as:

$$
\begin{aligned}
(i,k)\sigma_K &= (i,k)\sigma^K \\
(i,k)\sigma_D &= (i,k)\sigma^D \\
(i,k)\sigma_{S5} &= (i,k)\sigma^{S5}.
\end{aligned}
$$

Notice that in the simple cases above $(i,k)\sigma_L = (i,k)\sigma^L$ (i.e. high and low unifications are alike). For a more complex case see section 9 below.

6.3 Inference rules

The following formulation uses a generalized "α,β,ν_i,π_i" form of Smullyan-Fitting's "α,β,ν,π" unifying notation.

$$
\frac{\alpha, i}{\alpha_1, i} \qquad\qquad \frac{\alpha, i}{\alpha_2, i} \qquad\qquad (\alpha)
$$

$$\frac{\begin{array}{c}\beta, i\\ \beta_1^C, k\end{array}}{\beta_2, (i,k)\sigma_L}[(i,k)\sigma_L] \qquad\qquad \frac{\begin{array}{c}\beta, i\\ \beta_1^C, k\end{array}}{\beta_2, (i,k)\sigma_L}[(i,k)\sigma_L] \qquad (\beta)$$

$$\frac{\nu_i, i}{\nu_0, (m, i)}[m \in \Phi_V^i \text{ and new}] \qquad (\nu_i)$$

$$\frac{\pi_i, i}{\pi_0, (m, i)}[m \in \Phi_C^i \text{ and new}] \qquad (\pi_i)$$

$$\frac{}{X, i \qquad X^C, i}[i \text{ restricted}] \qquad (PB)$$

$$\frac{\begin{array}{c}X, i\\ X^C, k\end{array}}{\times (i,k)\sigma_L}[(i,k)\sigma_L] \qquad (PNC)$$

Here the α-rules are just the familiar linear branch-expansion rules of the tableau method, while the β-rules correspond to such common natural inference patterns as *modus ponens*, *modus tollens*, etc. (i, k, m stand for arbitrary labels). The rules for the modal operators are as usual. "m new" in the proviso for the ν_i- and π_i-rule means: m must not have occurred in any label yet used. Notice that in all inferences via an α-rule the label of the premise carries over unchanged to the conclusion, and in all inferences via a β-rule the labels of the premises must be σ_L-unifiable, so that the conclusion inherits their unification. PB (the "Principle of Bivalence") represents the (LS-version of the) semantic counterpart of the cut rule of the sequent calculus (intuitive meaning: a formula A is either true or false in any *given* world, whence the requirement that j should be restricted). PNC (the "Principle of Non-Contradiction") corresponds to the familiar branch-closure rule of the tableau method, saying that from the occurrence of a pair of LS-formulas X, i, X^C, k such that $(i,k)\sigma_L$ (let us call them *σ_L-complementary*) on a branch we may infer the closure ("$\times$") of the branch. The $(i,k)\sigma_L$ in the "conclusion" of PNC means that the contradiction holds "in the same world".

6.4 Proof search

Let $\Gamma = \{X_1, \ldots, X_m\}$ be a set of S-formulas. Then $\mathcal{T}$ is a *KEM-tree for* Γ if there exists a finite sequence $(\mathcal{T}_\infty, \mathcal{T}_\in, \ldots, \mathcal{T}_\backslash)$ such that (i) $\mathcal{T}_\infty$ is a 1-branch tree consisting of $\{X_1, i \ldots, X_m, i\}$, where i is an arbitrary constant label; (ii) $\mathcal{T}_\backslash = \mathcal{T}$, and (iii) for each $i < n, \mathcal{T}_{)+\infty}$ results from $\mathcal{T}_)$ by an application of a rule of KEM. A branch τ of a KEM-tree $\mathcal{T}$ of LS-formulas is said to be *σ_L-closed* if it ends with an application of PNC, open otherwise. A KEM-tree $\mathcal{T}$ is said to be $\sigma_L -$ *closed* if all its branches are σ_L-closed. As usual with tableau methods, a set Γ of formulas is checked for consistency by constructing a KEM-tree for Γ. It is worth noting that each KEM-tree is a (class of) Hintikka's model(s) where the labels denote worlds (i.e. Hintikka's modal sets), and the unifications behave according to the conditions placed on the appropriate accessibility relations. By a $KEM-proof$ of a formula A we mean a σ_L-closed KEM-tree for FA, i. Moreover we say that a formula A is a *KEM-consequence of a set of formulas* Γ if A occurs in all the open branches of a KEM-tree for Γ. We now describe a sistematic procedure for KEM. First we define the following notions.

Given a branch τ of a KEM-tree, we shall call an LS-formula X, i *E-analysed in* τ if either (i) X is of type α and both α_1, i and α_2, i occur in τ; or (ii) X is of type β and one of the following conditions is satisfied: (a) if β_1^C, k occurs in τ and $(i,k)\sigma_L$, then also $\beta_2, (i,k)\sigma_L$ occurs in τ, (b) if β_2^C, k occurs in τ and $(i,k)\sigma_L$, then also $\beta_1, (i,k)\sigma_L$ occurs in τ; or (iii) X is of type ν_i and $\nu_0, (m,i)$ occurs in τ for some $m \in \Phi_V$ not previously occurring in τ, or (iv) X is of type π_i and $\pi_0, (m,i)$ occurs in τ for some $m \in \Phi_C$ not previously occurring in τ.

We shall call a branch τ of a KEM-tree *E-completed* if every LS-formula in it is E-analysed and it contains no complementary formulas which are not σ_L-complementary. We shall say a branch τ of a KEM-tree *completed* if it is E-completed and all the LS-formulas of type β in it either are analysed or cannot be analysed. We shall call a KEM-tree *completed* if every branch is completed.

At each stage of proof search (i) we choose an open non completed branch τ. If τ is not E-completed, then (ii) we apply the 1-premise rules until τ becomes E-completed. If the resulting branch τ' is neither closed nor completed, then (iii) we apply the 2-premise rules until τ becomes E-completed. If the resulting branch τ' is neither closed nor completed, then (iv) we choose an LS-formula of type β which is not yet analysed in the branch and apply PB so that the resulting LS-formulas are β_1, i' and β_1^C, i' (or, equivalently β_2, i' and β_2^C, i'), where $i = i'$ if i is restricted (and already occurring when $h(i) \in \Phi_C^h$), otherwise i' is obtained from i by instantiating $h(i)$ to a constant not occurring in i; (v) ("Modal PB") if the branch is not E-completed nor closed, because of complementary formulas which are not σ_L-complementary, then we have to see whether a restricted label unifying with both the labels of the complementary formulas occurs previously in the branch; if such a label exists, or can be built using already existing labels and the unification rules, then the branch is closed, (vi) we repeat the procedure in each branch generated by PB.

The above procedure is based on on a (deterministic) procedure working for *canonical* KEM-tree. A KEM-tree is said to be canonical if it is generated by applying the rules of KEM in the following fixed order: first the α-, ν_i- and π_i-rule, then the β-rule and PNC, and finally PB. Two interesting properties of canonical KEM-trees are (i) that a canonical KEM-tree always terminates, since for each formula there are a finite number of subformulas and the number of labels which can occur in the KEM-tree for a formula A (of L) is limited by the number of modal operators belonging to A, and (ii) that for each closed KEM-tree a closed canonical KEM-tree exists. Proofs of termination and completeness for canonical KEM-trees are given in [16].

7 *KEM* for *MDL*

To extend *KEM* to *MDL* we introduce n sets Φ^i_C $(1 \leq i \leq n)$ of preferred constant world–symbols and as many sets Φ^i_V of preferred variable world–symbols. Moreover we define the following specialised notions of σ^L- and σ_L-unification for dealing with these sets.

$$
\begin{aligned}
(i,k)\sigma^{MDL} \quad = \quad & (h(i),h(k))\sigma \text{ iff} \\
& i, \text{ or } k \text{ is } i\text{-ground}, 1 \leq i \leq n, \text{ or} \\
& \exists s(i), s(k) : h(s(i)), h(s(k)) \in \Phi^i, \text{ and} \\
& (h(s(i)), h(s(k))\sigma^{MDL}.
\end{aligned}
$$

On this basis we define the corresponding low unification as:

$$(i,k)\sigma_{MDL} = (i,k)\sigma^{MDL}.$$

Here we give an example proof of a characteristic axiom of *MDL*.

$$
\begin{array}{ll}
1.\ F\neg P_i\bot \rightarrow (P_iP_jA \equiv P_jA) & w_1 \\
2.\ T\neg P_i\bot & w_1 \\
3.\ FP_iP_jA \equiv P_jA & w_1 \\
4.\ FP_i\bot & w_1 \\
5.\ F\bot & (w^i_2, w_1)
\end{array}
$$

$$
\begin{array}{ll|ll}
6.\ TP_iP_jA & w_1 & 7.\ FP_iP_jA & w_1 \\
8.\ FP_jA & w_1 & 13.\ TP_jA & w_1 \\
9.\ TP_jA & (W^i_1, w_1) & 14.\ FP_jA & (w^i_4, w_1) \\
10.\ TA & (W^j_2, (W^i_1, w_1)) & 15.\ FA & (w^j_5, (w^i_4, w_1)) \\
11.\ FA & (w^j_3, w_1) & 16.\ TA & (W^j_3, w_1) \\
12.\ \times & w^j_3 & 17.\ \times & w^j_5
\end{array}
$$

The core of the proof is in the application of *PB* to (3) to obtain (6) and (7). Is should also be noted that the label in (10) is i-ground due to the label in (5), which allows us to close the branch ((10) and (11) are σ_{MDL}-complementary). In the right branch the label of (15) is obviously i-ground, which makes the branch to be closed ((15) and (16) are σ_{MDL}-complementary).

8 *KEM* for *DL*

In this section we show how to employ *KEM* to deal with *MDL*. We first introduce three kinds of world symbols respectively for:

1. Universal deontic worlds: $\Phi_N = \{N_1, N_2, \cdots\}$ and $\Phi_n = \{n_1, n_2, \cdots\}$
2. Ideal worlds: $\Phi_D = \{D_1, D_2, \cdots\}$ and $\Phi_d = \{d_1, d_2, \cdots\}$
3. Sub–ideal worlds: $\Phi_S = \{S_1, S_2, \cdots\}$ and $\Phi_s = \{s_1, s_2, \cdots\}$.

We define

$$\sigma : \Phi_N \rightarrow \Im^d \cup \Im^s$$

In order to get the appropriate unifications we need the following substitution:

$$\sigma^{\#}\Phi_W = \sigma\Phi_W.$$

$$\begin{aligned} \sigma^{\#} \quad &: \quad \Phi_N \rightarrow \Phi^d \cup \Phi^s \\ &: \quad \Phi_S \rightarrow \Phi^s \\ &: \quad \Phi_D \rightarrow \Phi^d \end{aligned}$$

where

$$\Phi^d = \{i^r : r = d\} \qquad\qquad \Phi^s = \{i^r : r = s\}$$

$$\begin{aligned} (i,k)\sigma^W &= (t \times (b(s(i)), b(k))\sigma^J \iff \\ & \quad l(k) > 1, \exists s(i) : \forall s'(i), l(s'(i)) > l(s(i)), \\ & \quad (h(s'(i)), h(k))\sigma^{\#} = (h(s(i)), h(k))\sigma = t \text{ and} \\ & \quad (b(s(i)), b(k))\sigma^J \text{ otherwise}, l(k) = 1 \\ (i,k)\sigma^W &= t \end{aligned}$$

or

$$\begin{aligned} (i,k)\sigma^W &= (t \times (b(i), b(s(k)))\sigma^J \iff \\ & \quad l(i) > 1, \exists s(k) : \forall s'(k), l(s'(k)) > l(s(k)), \\ & \quad (h(i), h(s'(k)))\sigma^{\#} = (h(i), h(s(k)))\sigma = t \text{ and} \\ & \quad (b(i), b(s(k)))\sigma^J \text{ otherwise}, l(i) = 1 \\ (i,k)\sigma^W &= t \end{aligned}$$

where

$$(i,k)\sigma^J = (i,k)\sigma^D \text{ or } (i,k)\sigma^W.$$

We are now able to characterize DL by the following notions of σ^{DL}-, σ_{DL}-unification:

$$(i,k)\sigma^{DL} = \begin{cases} (i,k)\sigma^D \\ (i,k)\sigma^W \end{cases}$$

and

$$(i,k)\sigma_{DL} = (i,k)\sigma^{DL}.$$

To complete the KEM characterization of DL we need the following additional inference rules which are meant to capture the "meaning" of N_D, **C1** and **C2** respectively:

$$\frac{F\mathsf{N}_D A, i}{F(\mathsf{O}^i A \wedge \mathsf{O}^s A), i} \qquad \frac{\begin{array}{c} X,(S,i) \\ X,(D,k) \end{array}}{X,(W,(i,k)\sigma_{DL})}[(i,k)\sigma_{DL}] \qquad \frac{\begin{array}{c} \nu_i, i \\ \nu_0^C, k \end{array}}{\begin{array}{c} \nu_i, j^r \\ \nu_0^C, j^r \end{array}}[(i,k)\sigma_{DL} = j]$$

where

$$\begin{array}{ll} i^r = i^s & \text{if } \nu_i = T\mathsf{O}^i A\,(F\mathsf{P}^i A) \\ i^r = i^d & \text{if } \nu_i = T\mathsf{O}^s A\,(F\mathsf{P}^s A) \end{array}$$

and

$$i^x = i : h(i) \in \Phi^x, (x \in \{d, s\}).$$

Obviously each $\Phi^r_X \subseteq \Phi_X$. Besides the usual closure rule we introduce the following

$$\frac{i \in \Phi^i, i \in \Phi^s}{\times}.$$

stating that no world can be at the same time an ideal and a sub-ideal version of itself.

What follows is an example proof of a characteristic theorem of DL.

1. $F(\mathsf{O}^i A \wedge \neg A) \rightarrow \mathsf{P}^s \neg A$	w_1
2. $T\mathsf{O}^i A \wedge \neg A$	w_1
3. $F\mathsf{P}^s \neg A$	w_1
4. $T\mathsf{O}^i A$	w_1
5. FA	w_1
6. $T\mathsf{O}^i A$	w_1^s
7. FA	w_1^s
8. TA	S_1, w_1^s
9. $\times$	

The steps leading to the nodes (1)–(5) are straightforward. The nodes (6)–(7) come from the application of the rule for **C2** since the world denoted by w_1 is a sub–ideal version of itself. Closure follows immediately from (7) and (8), which are σ_{DL}-complementary (their labels σ_{DL}-unify because of $(S_1, w_1^s)\sigma^{\#}$).

9 *KEM* for *DDL*

In order to extend KEM to deal with DDL we have only to define the following special unifications:

$$\begin{aligned} (i,k)\sigma_{DDL} &= (i,k)\sigma^{S5} \iff \\ & \quad h(i) \text{ or } h(k) \in \Phi^0_V \\ & \quad i, k \text{ are } i\text{-ground, and} \\ & \quad (i,k)\sigma^{S5} \notin \Phi_d \cup \Phi_s;\ \text{otherwise} \\ (i,k)\sigma_{DDL} &= (h(i), h(k))\sigma \times (s^1(i), s^1(k))\sigma \end{aligned}$$

or

$$\begin{aligned} (i,k)\sigma_{DDL} &= (c^n(i), c^m(k))\sigma^X \text{ where} \\ & \quad w_0 = (s^n(i), s^m(k))\sigma_{DDL} \end{aligned}$$

where

$$(i,k)\sigma^X = \begin{cases} (i,k)\sigma^{MDL} \\ (i,k)\sigma^{DL} \end{cases}.$$

The rules of inference for DDL result now from the combination of the rules for MDL and DL.

10 Dealing with preferences

In this section we show how the computational framework developed in the preceeding sections can be used to solve contradictions in the way described in section 4. As it well known, tableau methods, and in particular KEM, can be used as model building systems [11]. In what follows we exploit the obvious fact that the set of formulas in each branch τ of a KEM-tree is consistent (i.e. it has a model) if τ is open, otherwise it is inconsistent. Since we check sets of premises for consistency, we first run KEM on the given set and, finally, we solve contradictions, on each closed branch, by assigning a different preference to the consequents of the translated defaults implying contradictions. To use practically this procedure we have to keep trace of dependencies according to the following definition.

Definition 10.1

- *Each formula depends on itself;*
- *a formula B depends on A either if it is obtained through an application of the α- ν_i- and π_i-rules or it is obtained through an application of KEM's rules on formulas depending on A;*
- *a formula C depends on A, B if it is obtained through an application of a β-rule where A, B are its premises;*
- *if C depends on A, B then C depends on A and C depends on B*

In the course of model construction we shall only keep trace of the dependencies from the premises, i.e. we avoid all the intermediate formulas. Obviously in order to solve pairs of complementary formulas and to assign preferences accordingly, we have to take into account only the premises which essentially imply one of them.

Let A and B be two complementary formulas in a closed branch τ; let $\mathcal{C}_{\mathcal{A}}$ be the set of premises from which A depends; let $\mathcal{C}_{\mathcal{B}}$ be the set of premises from which B depends; let $\mathcal{D}(\mathcal{C}_{\mathcal{A}} \cup \mathcal{C}_{\mathcal{B}})$ be the defaults in $\mathcal{C}_{\mathcal{A}} \cup \mathcal{C}_{\mathcal{B}}$. Note that, since facts are assumed to be consistent, the set $\mathcal{D}(\mathcal{C}_{\mathcal{A}} \cup \mathcal{C}_{\mathcal{B}})$, which we call a *culprit* set, is responsible for the inconsistency,

We say that a set $S \subseteq D$ is *conflict free* if it does not contain any culprit sets. We use KEM inference procedure to find all the $\mathcal{D}(\mathcal{C}_{\mathcal{A}} \cup \mathcal{C}_{\mathcal{B}}) \subseteq \mathcal{D}$, and we use such information to build all maximal conflict free sets S of defaults. Finally we assign preferences in such a way that all formulas in each S have the same preference.

We give an example of how we establish consistent subsets and assign preferences.

Example 10.2

Let us consider the following knowledge base (F, D), where $F = \{p\}$ and $D = \{p \rightarrow q, p \rightarrow r, p \rightarrow s, p \rightarrow (\neg q \vee \neg r) \wedge \neg s\}$.

1. Tq	1
2. Tr	2
3. Ts	3
4. $T(\neg q \vee \neg r) \wedge \neg s$	4
5. $T \neg q \vee \neg r$	4
6. Fs	4
7. Fq	2, 4
8. Fr	1, 4

On the right column we wrote the dependencies (for the sake of economy we have deleted the irrelevant steps). We obtain the following inconsistent sets of defaults: $\{1, 2, 4\}$ and $\{3, 4\}$. Therefore the maximal consistent subsets and their associate preferences are:

$$P_1 \leadsto \{1, 2, 3\}$$

$$P_2 \leadsto \{1, 4\}$$

$$P_3 \leadsto \{2, 4\}.$$

It is worth noting that the ability of KEM to determine the inconsistent subsets of a contradictory one is due to its inference rules, in particular PB and the β-rules.

After assigning preferences, we can query the system whether a formula X is a consequence of our knowledge base simply by running a refutation KEM-tree for the set consisting of the premises and X^C.

11 Final Remarks

It was not the objective of this paper to develop a theory of defeasible deontic reasoning. Our motivation was rather practical. We sought for computationally tractable and easily implementable theorem proving techniques suitable for dealing with some pieces of normative reasoning involving defeasible rules. The preceding discussion was thus mainly aimed at showing the potential scope of application of the method. In effect, we believe that the method for solving contradictions by assigning preferences outlined in section 10 nicely exploits the computational and proof-theoretical advantages offerred by the modal theorem proving system KEM. As we have argued elsewere this system enjoys most of the features a suitable proof search system for modal (and in general non-classical) logics should have. In contrast with (both clausal and non-clausal) resolution methods, and in general "translation-based" methods [18, 19], it works for the full modal language (thus avoiding any preprocessing of the input formulas), and it is flexible enough to be extended to cover any setting having a Kripke-model based semantics (this is clearly shown by our treatment of Jones and Pörn logic DL where the rules specific for such a logic should take care not only of the propositional and modal part but also of the structure of the labels and the relationship between labels and formulas; for example we added another closure rule $\frac{i \in \Phi^{i}, i \in \Phi^{s}}{\times}$ which states that no world can be at the same time an ideal and a sub-ideal version of itself; this result is achieved by determining when a deontic word is ideally (sub-ideally) reflexive (i^{r}) by means of another peculiar inference rule). From this perspective our methods is similar to sequent or tableau proof methods ([20, 22]). Nevertheless, it has several advantages over most tableau/sequent based theorem proving methods: being based on D'Agostino an Mondadori's classical proof system KE, it eliminates the typical redundancy of the standard cut-free methods and, thanks to its label unification scheme, it offers a simple and efficient solution to the permutation problem which notoriously arises at the level of the usual tableau-sequent rules for the modal operators. However, unlike e.g. Wallen's [21] connection method, it uses a natural and easily implementable style of proof construction, and so it appears to provide an adequate basis for combining both efficiency and naturalness. (As to the implementation the reader is referred to [15] where a Prolog implementation is provided, and to [23] where some related issues are discussed).

References

[1] Asher N., Bonevac D. *Prima Facie* Obligation. In: Jones A.J.I., Sergot M. (eds.) *∆eon '94*. Tano, 1993, pp 19–36.

[2] Horty J.F. Moral Dilemmas and Nonmonotonic Logic. *Journal of Philosophical Logic* 1994; 23: 35–65.

[3] Jones A.J.I. On the Logic of Deontic Conditionals. *Ratio Juris* 1991; 3: 355–66.

[4] Mc Carty L.T. Defeasible Deontic Reasoning. *Fundamenta Informaticae* 1994; 21: 125–148.

[5] Prakken H. Two Approaches to Defeasible Reasoning. In: Jones A.J.I., Sergot M. (eds.) *∆eon '94*. Tano, 1993, pp 281–295.

[6] Ryu Y.U., Lee R.M. Defeasible Deontic Reasoning: A Logic Programming Model. In: *Proceedings of the First International Workshop on Deontic Logic in Computer Science.* Amsterdam, 1991, pp 347–363.

[7] Sartor G. The Structure of Norm Conditions and Nonmonotonic Reasoning in Law. In: *Proceedings of the Third International Conference on Artificial Intelligence and Law.* ACM Press, 1991, pp 155–164.

[8] Meyer J.-J.Ch., van der Hoek V. A Modal Logic for Nonmonotonic Reasoning. In: van der Hoek W., Meyer J.-J.Ch., Tan Y. H., Witteveen C.

(eds.) *Non-Monotonic Reasoning and Partial Semantics.* Ellis Horwood, N.Y., 1992, pp 37–77.

[9] Jones A.J.I., Pörn I. Ideality, Sub-ideality and Deontic Logic. *Synthese,* 1985; 65: 275–290.

[10] Jones A.J.I., Pörn I. "Ought" and "Must". *Synthese,* 66, 1986: 89–93.

[11] Artosi A., Governatori G. Labelled Model Modal Logic. In *Proceedings of the CADE-12 Workshop on Automated Model Building.* Nancy, 1994, pp 11–17.

[12] Brewka G. Belief Revision in a Framework for Default Reasoning. In: Fuhrmann A., Morreau M. (eds.) *The Logic of Theory Change. Workshop.* Springer-Verlag, Berlin, 1989, pp 206–222.

[13] Brewka G. *Nonmonotonic Reasoning. Logical Foundations of Commonsense.* Cambridge University Press, Cambridge, 1991.

[14] Åqvist L.. *Introduction to Deontic Logic and the Theory of Normative Systems.* Bibliopolis, Napoli, 1987.

[15] Artosi A., Cattabriga P., Governatori G. An Automated Approach to Deontic Reasoning. In: Breuker J. (ed.), *Proceeding of the Workshop ECAI Artificial Normative Reasoning.* Amsterdam, 1994, pp 132–145.

[16] Governatori G. Labelled Tableau for Multi-Modal Logics. In: Baumgartner P., Hähnle R., Posegga J. (eds.) *Theorem Proving with Analytic Tableaux and Related Methods.* Springer-Verlag, Berlin, 1995, pp 79-95 (Lecture Notes in Artificial Intelligence no.918).

[17] D'Agostino M., Mondadori M. The Taming of the Cut. *Journal of Logic and Computation* 1994; 4: 285–319.

[18] Auffray Y., Enjalbert P. Modal Theorem Proving: An Equational Viewpoint. *Journal of Logic and Computation* 1992; 2: 247-259.

[19] Ohlbach H.J. Semantics-Based Translation Methods for Modal Logics. *Journal of Logic and Computation* 1991,1: 691–746.

[20] Fitting M. First Order Modal Tableaux. *Journal of Automated Reasoning* 1988; 4: 191–213.

[21] Wallen L.A. *Automated Deduction in Non-Classical Logics.* The MIT Press, Cambridge (Mass.), 1990.

[22] Catach L. Tableaux: A General Theorem Prover for Modal Logic. *Journal of Automated Reasoning* 1991; 7: 489–510.

[23] Artosi A., Cattabriga P., Governatori G. A Prolog Implementation of *KEM*. In: Alpuente, Sessa M.I. (eds.) *Proceedings of GULP-PRODE'95.* Universitá di Salerno, Salerno, 1995, pp 395–400.

Doing As We Ought: Towards A Logic of Simply Dischargeable Obligations

Mark A. Brown
Philosophy Department, Syracuse University
Syracuse, NY, USA

Abstract

This paper investigates **diachronic deontic logic**, based on models with **forward-branching time.** Using such models, supplemented by a **choice function** to model human agency, and an **obligation function** to introduce normative features, it is possible to develop a rich language with tense logic, **since** and **until** operators, and operators expressing **action**, **ability**, and **obligation**. It then becomes possible to model, not only the interrelationships among our obligations, but also their interrelationships with the actions which give rise to them, the actions which fulfill them, and the abilities these actions require.

1 Diachronic Deontic Logic

Most systems of deontic logic consider a very static, synchronic, picture: they consider only what logical connections may exist at a given time among various coexisting obligations and prohibitions, and between these commitments and various relevant non-normative facts. (Notable exceptions are Åqvist and Hoepelman [1], and Thomason [2], [3].) But while it may be a reasonable approximation to the truth to consider prohibitions to be stable and static (on the view that most prohibitions arise from general ethical principles or standing laws) the actual role of obligations in our lives is certainly more dynamic. This shows up particularly when we begin to discuss *discharging* our obligations. Some obligations, to be sure, are standing obligations which can be honored, but cannot ever be fully discharged—cannot be brought to an end and put behind us. But many are not like that. We make promises, enter into contracts, incur debts, receive orders and commands. Through such events, from time to time we undertake various new obligations, or have obligations imposed on us by circumstance or external authority; these obligations may grow in number, and newly acquired obligations may conflict with older ones not yet discharged; circumstances may change in such a way that obligations which were initially compatible unexpectedly come into conflict; from time to time we act in such a way as to discharge one or more of our obligations, and when we do so that obligation ceases to exist—our normative burden is, at least for the moment, lessened.

The amount of time that elapses between incurring an obligation and discharging it can vary considerably. Some obligations, it may happen, get discharged almost as soon as they are incurred (I ask to see the photo of your grandchildren; I hand it back after taking a brief look); many eventually get discharged, but only after some significant lapse of time (I take out a mortgage on my new home and pay it back in monthly installments over a period of twenty years); others, it may happen, do not ever get discharged, perhaps through negligence or refusal (I borrow your book and forget to return it, or, having become angry at you for some reason, decide to "punish" you by refusing to return it); in some cases we may take actions which, whether intentionally or otherwise, preclude all possibility of fulfilling one or more of our obligations (I borrow your book, but lose it, or destroy it). In some such cases we may feel it natural to view the original obligation as having been transformed into a new one (having lost your book, I should compensate you; note that it would be inappropriate to say the original obligation was to return the book or compensate you, which would in effect have given me an option to purchase it).

To do justice even to the mere expression of basic facts within this complex and dynamic picture, we need a more complex language than is normally provided for deontic logic. First, if we are to take account of how actions discharge obligations then we will need to combine a logic of obligation with a logic of action, i.e. we will need to combine a logic of what we *ought* to do with a logic of what we *do* do. We could, of course, combine a more or less standard deontic logic (better yet: a deontic logic which allows for conflicting obligations) with a more or less standard logic of action. But merely grafting the one onto the other will still leave us with a static system, and will not enable us to speak, for example, of discharging an obligation. For when an obligation is discharged, a formula becomes false which once was true. (A moment ago I owed you a thousand *escudos*. Now I don't.) To deal with this, we need to bring time more explicitly into the picture.

2 Some Subtleties

As soon as we begin to pay close attention to the place of time in our account of the coming and going of our obligations, we find that many of these matters are more subtle than we normally notice. For example, we might contemplate either of two ways of viewing the discharging of an obligation, ones that differ slightly in the temporal relations involved. One possibility is to hold that an obligation is in force up to *and including* the first moment at which an action which fulfills that obligation is taken, and is thereafter discharged. Another is to hold that the obligation is in force up to *but not including* that moment, and is at that moment discharged and nullified. I'm not at all sure, as yet, that the difference between these is significant, but clearly it might be, and investigation will be required in order to be sure. Pending further insight, I choose the latter for exploration here.

There are further subtleties to be considered. Ordinarily, we imagine that I cannot be under an obligation to perform actions I am not able to perform: we accept the slogan

No responsibility without ability.

But when we consider the matter diachronically, it is more complex than this simple slogan would suggest. I am not relieved of my obligation to repay your loan by the fact that I cannot repay it right now. For example, at the restaurant I discover I have left my wallet at home, and I borrow money from you in order to pay my bill. Immediately after borrowing the money, I pay the bill, thus rendering myself again moneyless, and making myself incapable—for now—of repaying you the money I've borrowed. It was, of course, part of the *point* of my borrowing money from you that I would use it in a way that would leave me unable to repay it immediately. That inability does not, however, relieve me of the obligation to repay it eventually, in the way in which legal bankruptcy might. In short: in some circumstances, my present inability to fulfill my obligation may simply result in sustaining the obligation, not in rendering it null and void.

So in our diachronic setting we will need to reconsider the relation between ability and responsibility. To do so, of course, we will need to be using a language in which it is possible to make claims about ability. So now it appears we need a combined logic of tense, action, ability, and obligation.

There's more: suppose I borrowed a thousand *escudos* at 17.00 yesterday, and hand you a thousand *escudos* now at 16.00 today. It will then be correct to say that I (once) had an obligation to give you a thousand *escudos* and that I now do give you a thousand *escudos*, but more will have to be involved if I am to be correct in saying I now discharge my debt. If it happens that I already paid you a thousand *escudos* at 15.30 today, then what I am doing now does *not* discharge my debt—that had already been accomplished by 15.30 today, and that debt no longer existed at 16.00. Whatever I'm doing now has some other significance. Perhaps, in fact, *I'm* now loaning *you* a thousand *escudos*. So in order to be able to describe what I now do as discharging that obligation, it will be important to be able to say that I was continuously under that obligation from 17.00 yesterday until now, and thus that I have not already discharged it at some time *between* the time of incurring the obligation and now. But that calls for us to be able to say that (at a certain time) I *incur* an obligation, not merely that (at various times) I *have* one, and it also involves being able to identify certain time *periods* (such as the period beginning when an obligation was incurred and continuing to the present), not just the past and future relative to a given moment. The notion of *incurring* is important in other ways: after discharging an obligation to bring it about that A, I am no longer under an obligation to bring it about that A *unless* I have incurred a new obligation to bring it about that A. This last is an important qualification, and one which we cannot express without being able to talk about incurring obligations. Some way will have to be found of expressing this notion.

If particular actions are identified, in part, by the spatiotemporal bounds within which they occur, and in part by (some of) the details of the events that occur within those bounds, then it seems unreasonable to think that my obligations are obligations to perform specific actions. I owe you a thousand *escudos*. If my repaying you today would constitute a different action from my repaying you tomorrow, yet my obligation is equally satisfied by repayment on either occasion, then it is not either of the particular actions (or any of an infinity of other similar ones) which my obligation requires of me. It is, rather some *kind* of action that is required of me: I need to perform some action—any action—of a kind which will get a thousand *escudos* transferred legitimately from me to you, but *which* such action I perform is perhaps immaterial. The situation is similar to that of the person who goes to the store to buy a loaf of bread. There is no particular loaf of bread in question here; *any* decent loaf will do. It appears that it is the outcome which is critical here, not the identity of the particular action which produces that outcome.

Following through on this thought a bit further it appears that, paradoxical as it may sound, even the action expressions in our formalism, i.e. the ones specific to the logic of action embedded in our system, should not be construed as directing attention to particular actions. Rather they identify a *kind* of action, and identify it by its outcome. We have just noted that when I am obligated to pay you a thousand *escudos* there is no one action which is specifically what is required of me. Equally, when I do repay you there may well be a specific action which I do perform, and whose performance is sufficient to justify saying that I have repaid you, but performance of that particular action is by no means a necessary condition of my repaying you. Any of a variety of similar acts would have done as well. Consequently, when I report that I have repaid you, it would be more appropriate to construe that as a report that I have performed *some* (unspecified) action with a certain kind of outcome, than to construe it as a report of the performance of a particular action—a report which would then have been false had I performed any other instead. Indeed it is difficult to imagine that ordinary language is even capable of establishing a reference to any truly particular action. These considerations suggest that our formal language need not (and perhaps even should not) have names for actions. "*Wovon man nicht sprechen kann, darüber muß man schweigen.*."

The consequence, for us, is that rather than adopt a logic with action terms, we should adopt a logic with action assertions, forming such assertions by attaching action operators to formulas describing the relevant outcomes.

However, one of the reasons people consider deontic systems with action terms is that deontic systems based instead on action assertions commonly suffer a deficiency which is associated with Ross's paradox. When we use a formula $\mathcal{O}P$ to express a claim about our obligations, the operator $\mathcal{O}$ can be of either of (at least) two types, bearing natural, but different, interpretations:

Type 1: $\mathcal{O}P$ bears the interpretation that I have an obligation whose fulfillment will have the truth of P as a necessary consequence, *but perhaps not as a sufficient condition*; or

Type 2: $\mathcal{O}P$ bears the interpretation that I have an obligation whose fulfillment will have the truth of P as a necessary consequence *and as a sufficient condition.*

Typical deontic systems based on action assertions warrant the inference from $\mathcal{O}P$ to $\mathcal{O}Q$ whenever Q is a logical consequence of P. Such systems only sustain the first interpretation, and so their operators are of type 1. But the literature provides abundant evidence that systems with operators of type 1 are easily misconstrued as bearing the second interpretation. If the formula A is interpreted to mean that I mail your letter, and B that I burn it, then since $A \vee B$ follows from A, such systems will license the inference from OA to $O(A \vee B)$. It is very natural, and seemingly harmless, to read OA into English as the claim that I have an obligation to mail your letter. For this particular sentence, the difference between a type 1 operator and a type 2 operator is not striking, since in fact A seems to express necessary and sufficient conditions for the fulfillment of the obligation in question. But since this reading seems to treat the operator as of type 2, by parity we must read the formula $O(A \vee B)$ as meaning that I have an obligation to mail or burn your letter, construed as an obligation which can be fulfilled by mailing or burning your letter, and therefore can be fulfilled by burning it. That (as Ross [4] pointed out) is absurd. In particular, we don't want someone to be able to justify burning a letter by saying that they are fulfilling an obligation. Holding consistently to the first interpretation makes Ross's "paradox" harmless, of course, but leaves us somewhat dissatisfied, since on this interpretation deontic formulas do not specify the precise conditions for the fulfillment of any obligation, even when we know and can express such conditions and wish to do so.

In any system in which we wish to discuss discharging obligations, then, we will need an operator of type 2, in order to link the precise characterization of the conditions for discharging the obligation with the form of the statement expressing that obligation. Such an operator cannot be permitted to license inferences of the sort involved in Ross's paradox. Yet we may well also want to be able to speak less specifically of our obligations, to say that we have an obligation which cannot be fulfilled without its becoming true that P, without having to specify precisely what kind of obligation that is. In short, we may well want an operator of type 1, alongside one of type 2. A little later I will discuss the formal connections between the two operators thus conceived.

3 Other Complications

Certain kinds of obligations present special challenges. Among these are our standing obligations, by which I mean those continuing obligations which by their nature can never be nullified by our actions, such as the obligation to honor our parents. No matter what I do to honor my parents, there will never come a time (during my lifetime, anyway, and perhaps not even after my death) when I do not still stand under this obligation, and therefore I will never have fully discharged it. Fitting such obligations smoothly into the same picture as obligations which can be discharged presents a considerable challenge.

A different problem arises when we try to deal with cumulative obligations, by which I mean obligations which can be incurred again, and with additional effect, before their first occurrence has been discharged. Suppose, for example, I borrow 1000 *escudos* from you on Monday, thereby incurring an obligation to pay you 1000 *escudos*; and suppose that on Tuesday, without yet having paid off my debt to you, I borrow another 1000 *escudos*. I can then discharge both debts (or, if you like, the single cumulative debt) in a single act, perhaps by paying you 2000 *escudos* on Wednesday, or I can pay off the debt in installments, accomplishing the discharge of these obligations by a series of two or more actions. Contrast this with the case of my borrowing one book from you on Monday and a second on Tuesday. In this case the descriptions of the actions required to fulfill the two obligations will differ: one is the obligation to return *Os Lusíadas*, and the other the obligation to return *Don Quixote*. I may indeed return both at once, but this is easily handled in our formalism: the obligation to bring it about that A and the obligation to bring it about that B are both met by bringing it about that $A \wedge B$. But the problem with the case of monetary borrowing is that we would have an obligation to bring it about that C and an(other) obligation to bring it about that C, and must somehow avoid saying that these could be met by bringing it about that $C \wedge C$, i.e. simply by bring it about that C. If this were so, I could satisfy my cumulative debt of 2000 *escudos* by paying just 1000—a neat trick if I could do it! So the simple picture that applies to the case of borrowing books—that each obligation requires an identifiably corresponding discharging action—will not apply straightforwardly to standing or cumulative obligations. As a consequence, the task of providing a formal language suitable for discussing such obligations looks quite formidable, and it appears the better part of wisdom to bring other, more tractable, features of deontic diachronics under control before attempting to give a system intended to deal with such complications.

Mindful, therefore, that a formal system which aims to take in all these complexities at once may well be unmanageable, and in the interests of maintaining some hope of keeping the system under control, I propose to set aside for the moment several potential complications:

(1) we will leave aside conditional obligation;
(2) we will leave aside prohibitions;
(3) we will leave aside standing obligations;
(4) we will leave aside cumulative obligations;
(5) we will confine attention to a single agent, left tacit in the formalism.

What remain are what I shall call ***simply dischargeable obligations*** of a single agent. This may seem like very little but, as we shall see, providing a coherent formal language to describe the simply dischargeable obligations of a single agent, and giving a reasonable formal semantics for such a language, will be more than enough to keep us occupied for now. I shall have suggestions at the end about how we might go on at some later date to address these other complexities.

4 An Approach to Time

We might contemplate either of two basic strategies for incorporating time into our deliberations: using overt temporal indexing of some sort in our formulas, or using tense operators. In this paper, I want to explore the latter strategy, since it seems more in keeping with the generally modal character of the action, ability, and obligation operators we must consider. This basic strategy in turn offers (at least) two possibilities. In the first, which Thomason [2] calls the $\mathbf{T} \times \mathbf{W}$ approach, models provide a set of worlds and a set of times, and the satisfaction conditions for a formula are relative to a time at a world. In the second, called the branching time approach, models attach states to moments which are connected in a tree structure branching towards the future, and the satisfaction conditions for a formula are relative to a time in a history (a branch of the tree).

I shall adopt this last alternative, assuming as the appropriate background a picture of time of the general sort explored by Thomason [5], and used by Belnap and his various collaborators, as in [6], [7]. Given that we hope to discuss deontic diachronics, it is important not to make what might otherwise be reasonable simplifying assumptions about the structure of time. In particular, a picture of time which is finite, or even discrete, seems inappropriate. Accordingly, the formalism will, for the moment, be silent about whether any history has a first or last moment, and whether the moments in a history are discretely or densely ordered. But it will be well to keep the more complex possibilities in mind in our informal thinking, particularly as we introduce defined concepts. Ways in which such complexities might be handled are indicated in Burgess [8].

The basic treatment of the various notions we consider will to some extent follow the work of Belnap and Horty [7] on the deliberative *stit* (which in turn has antecedents in the work of Chellas [9] and [10]) but will relate that to work by Brown [11], [12] on the logic of ability and action and to work by van Fraassen [13] and later by Brown [14] on the logic of conflicting obligations. As a consequence the system presented here will have an unusually large number of distinguishable primitive modal operators, although the underlying frames used in our models will be only modestly more complex than those standardly considered in the logic of branching time.

5 Semantics

By a *branching temporal frame*, I mean a structure $\langle \mathbb{T}, < \rangle$ such that:

(1) $\mathbb{T}$ is a non-empty set (of *moments* of time);

(2) $<$ is a strict partial ordering (antisymmetric and transitive) on $\mathbb{T}$.

By a *forward-branching back-connected temporal frame*, I mean any branching temporal frame satisfying the following constraints, where we define $\leq$ in terms of $<$ in the obvious way:

(2.1) $(\forall m, m_1, m_2 \in \mathbb{T})[m_1 < m \wedge m_2 < m \Rightarrow m_1 \leq m_2 \vee m_2 \leq m_1]$, (no backward branching),

(2.2) $(\forall m, m' \in \mathbb{T})(\exists m_0 \in \mathbb{T})[m_0 < m \wedge m_0 < m']$ (backward connection).

On any branching temporal frame we can define *histories* as maximal linearly ordered chains of moments. For the special case of forward-branching back-connected temporal frames, this amounts to saying that if m is any element of $\mathbb{T}$, a history through m is any subset h of $\mathbb{T}$ satisfying the conditions

(i) $m \in h$ (through m),

(ii) $(\forall m_1, m_2 \in h)[m_1 \leq m_2 \vee m_2 \leq m_1]$ (connected),

(iii) $\neg(\exists m_\omega \in \mathbb{T})(\forall m \in h)[m < m_\omega]$ (complete forwards),

(iv) $(\forall m \in h)(\forall m_0 \in \mathbb{T})[m_0 < m \Rightarrow m_0 \in h]$ (complete backwards).

Given any moment m in such a frame, and any history h through m, we let

$\mathbb{H}(m)$	be the set of all histories through m, i.e. the set of all $h \subseteq \mathbb{T}$ satisfying constraints (i)-(iv) above;
$\mathbb{H}(m/h)$	be the set of all histories through m that branch off from h, if at all, only at some point after m, i.e. which include at least one point $m' \in h$ such that $m < m'$;
$\mathbb{m}$	be the set of all moment/history pairs m/h in which $m \in \mathbb{T}$ and $h \in \mathbb{H}(m)$;
$\mathbb{P}$	be the set of propositions, defined as $\wp\mathbb{m}$;

for m/h and $m'/h' \in \mathbb{m}$, we let

$m/h \lessdot m'/h'$ iff $m < m'$ and $h' \in \mathbb{H}(m/h)$.

By a *choice frame*, I mean a structure $\langle \mathbb{T}, <, \mathbb{C} \rangle$ in which $\langle \mathbb{T}, < \rangle$ is a forward-branching back-connected temporal frame, and

(3) $\mathbb{C}$ (the *choice function*) provides, for each moment m, a partition $\mathbb{C}(m)$ of $\mathbb{H}(m)$ such that:

(3.1) $(\forall m \in \mathbb{T})(\forall \mathbb{c} \in \mathbb{C}(m))(\exists h \in \mathbb{H}(m))[h \in \mathbb{c}]$ (non-vacuity),

(3.2) $(\forall m, m' \in \mathbb{T}: m < m')(\forall h, h' \in \mathbb{H}(m')(\forall \mathbb{c} \in \mathbb{C}(m))[h \in \mathbb{c} \Leftrightarrow h' \in \mathbb{c}]$ (no choice between undivided histories).

By a *diachronic deontic frame*, I mean a structure $\langle \mathbb{T}, <, \mathbb{C}, \mathbb{O} \rangle$ in which the structure $\langle \mathbb{T}, <, \mathbb{C} \rangle$ is a choice frame, and

(4) $\mathbb{O}$ (the *obligation function*) is a function from moment/history pairs to sets of propositions, i.e. from $\mathbb{m}$ to $\wp\mathbb{P}$, subject to the following constraints:

(4.1) $(\forall m/h \in \mathbb{m})(\forall o \in \mathbb{O}(m/h))(\exists m',h': m/h \lessdot m'/h')[m'/h' \notin o]$; (non-triviality)

(4.2) $(\forall m/h \in \mathbb{m})(\forall o \in \mathbb{O}(m/h))[m/h \notin o]$; (non-fulfillment)

(4.3) $(\forall m/h \in \mathbb{m})(\forall o \in \mathbb{O}(m/h))(\exists m',h': m/h \lessdot m'/h')[m'/h' \in o]$; (fulfillability)

(4.4) $(\forall m/h \in \mathbb{m})(\forall o \in \mathbb{P}: m/h \notin o)$
$[((\exists m' < m)(\forall m'': m' < m'' < m)[o \in \mathbb{O}(m''/h)] \wedge$
$(\exists m^*/h^*: m/h \lessdot m^*/h'^*)[m^*/h^* \in o]) \wedge$
$(\exists m^*/h^*: m/h \lessdot m^*/h'^*)[m^*/h^* \notin o]) \Rightarrow o \in \mathbb{O}(m/h)]$
(persistence).

Finally, by a *model* for diachronic deontic logic, I mean any quintuple $\mathbb{M} = \langle \mathbb{T}, <, \mathbb{C}, \mathbb{O}, \mathbb{V} \rangle$ in which $\langle \mathbb{T}, <, \mathbb{C}, \mathbb{O} \rangle$ is a diachronic deontic frame, and

(5) $\mathbb{V}$ is a *valuation* assigning, at each moment m, to each sentential constant S, a truth value $\mathbb{V}(m,S) \in \{\mathbf{t}, \mathbf{f}\}$.

Some discussion of the constraints imposed above may be helpful. First, we should note that, as Thomason [5] (elaborating on Prior [15]) shows, we cannot assess complex claims about the future, against a background of branching time, without sometimes taking account of *which* future is at stake. To put the matter simplistically: it may be that along one branch of the future, A will eventually become true, while along another branch, it will not. Without saying which branch is in question, we cannot simply assert, without being misleading, either that A will, or that it won't, at some future time become true. For such reasons Thomason, followed by Belnap and his various collaborators, have made all assessment of formulas in a model relative not merely to moments, but to moment/history pairs, in which the history in question runs through the moment in question. I shall follow in this tradition, as will appear more explicitly in the definition of satisfaction to be given shortly. The definition of models anticipates this feature of the satisfaction conditions in various ways, notably in the way in which propositions are construed as sets of moment/history pairs rather than as sets of moments, and the way histories are brought into play in the constraints on the obligation function.

The choice function partitions the histories which branch ahead from a given moment into mutually exclusive and jointly exhaustive non-empty sets of histories. Each such set is intended to correspond to a different action open to me at that moment. The sets are non-empty because it is not open to me to bring history to a halt. They are sets, rather than single histories, because normally my action at this moment does not determine the whole future course of the universe, or even the near portion thereof, in all its details. To say it did would be to deny the role of other agents, of chance, and indeed of my capacity for further choice in the future.

The most directly definable action operator, which will be designated as $[\delta]$, is based on the idea that the choice actually made at a given moment will be the one within which the actual future history falls, and that I can be said in some weak sense to bring it about that A at that moment, if A is true at all the histories within that choice. This operator corresponds closely to the action operator ⟐ found in Brown [12]. However, as Belnap has repeatedly pointed out, this operator is too weak to represent the fullest sense of *seeing to it that*. For that fuller sense, the positive condition which characterizes $[\delta]$ needs to be supplemented by a negative condition, as in Horty and Belnap [7]. We can accomplish that by using an operator $\Diamond$ from the logic of ability (analogous to the operator ◈ in Brown [11]) also definable using our choice function, and bearing the sense *there is a choice open to me which is compatible with the truth of* .... Combining the two, we can then use the formula $[\delta]A \wedge \Diamond\neg A$ to express what Horty and Belnap [7] express using their ***dstit*** operator. This defined composite operator will not support the claim that I see to the truth of tautologies, or that I see to (and therefore can take credit or receive blame for) all the logical consequences of the states I bring about. Instead, I genuinely bring about only those things I had some chance of avoiding, had I chosen to act differently.

The obligation function offered here differs from what might be expected—differs, certainly, from what *I* expected—in both its domain and its codomain. First, it is a function of moment/history pairs, not merely of moments. As a consequence, a formula of the form $\bigcirc A$ which is true at such a pair will not typically be settled true, i.e. true at all other pairs involving that same moment. But although this is not how the matter is treated elsewhere (e.g. in [2]), it seems nonetheless to be correct. If the pair *m/h* falls within, while the pair *m/h'* falls outside, some choice corresponding to an action of making a promise, for example, then at *m/h* I will be under an obligation which I am not under at *m/h'*.

The function's codomain is more surprising. By analogy with the choice function, we might expect the obligation function to take as its values families of sets of futures, with each set of futures representing an action I am obliged to take, and the various futures within the set representing various ways in which that action might turn out. The most conspicuous problem with that approach is that the naturally resulting deontic operator is of type 1 and—as nearly as I am able to see, at least—there would be no way of defining an obligation operator of type 2, i.e. one that displays necessary and sufficient conditions for the fulfillment of an obligation.

We could block some of the seemingly undesirable inferences that an operator of type 1 otherwise would permit by adding a negative condition, as we did with the action operator, but this time that device will not block enough of the undesirable consequences. For example, if $\langle[o]\rangle$ is our type 1 deontic operator, then we could use the compound $\langle[o]\rangle A \wedge \Diamond\neg A$ as an approximation to a type 2 deontic operator. Then even when B follows logically from A, the attempt to infer $\langle[o]\rangle B \wedge \Diamond\neg B$ from $\langle[o]\rangle A \wedge \Diamond\neg A$ will be blocked by the extra conjunct. That will mean that the schema associated with Ross's paradox will not be valid. Specifically, $\langle[o]\rangle(A \vee B) \wedge \Diamond\neg(A \vee B)$ will not in general follow from $\langle[o]\rangle A \wedge \Diamond\neg A$. So if A expresses the claim that I mail your letter, and B the claim that I burn it, and we use the formula $\langle[o]\rangle A \wedge \Diamond\neg A$ to express my obligation to mail your letter, it will not automatically follow from my having an obligation to mail your letter that I have an obligation to mail or burn it.

But this is false comfort, for if I still have at least one course of action open to me that does not involve either mailing or burning the letter you gave me to post, i.e. if $\Diamond\neg(A \vee B)$ should happen to be true (as it might easily be) then $\langle[o]\rangle(A \vee B) \wedge \Diamond\neg(A \vee B)$ will follow, after all, from $\langle[o]\rangle A \wedge \Diamond\neg A$ together with this additional fact. So if there happens to be something I can do, such as tearing up your letter and throwing it away, that would leave open the possibility that the letter will be neither mailed nor burnt, then I'll seem to be obliged to mail or burn it after all.

The only way I've been able to find to produce a genuinely type 2 operator is the way indicated here in the models, and followed up in the satisfaction conditions for the operator $\bigcirc$ presented below. This amounts to using the sort of minimal models introduced by Segerberg [16] and exploited by Chellas [17]. The resulting type 2 operator is very weak, and from a technical point of view not terribly interesting. It is, however, still possible to exploit this obligation function to define a type 1 operator $\langle[o]\rangle$, and such an operator also appears in the language. We are then enabled to explore some of the connections between the two operators.

Conditions 4.1-4.3 represent various aspects of the notion that *ought* implies *can*. The non-triviality constraint imposed in condition 4.1 corresponds in function to the negative condition for *stit* operators: it assures that there is no obligation to do what will assuredly happen anyway. This rules out obligations to bring about the truth of logical theorems, but also rules out any obligation to have the sun rise tomorrow. If we had such a negative condition for action, but not for obligation, it would be possible to be *obligated to do* things that might possibly *occur*, but that it was not possible to *do*. Similarly, I cannot be obligated to bring about what is already true, and this is reflected in the non-fulfillment constraint imposed by condition 4.2. Condition 4.3 reflects the fact that I cannot (reasonably) be obligated to do what is no longer possible at all, even if it once was possible. I may be guilty of having violated my obligation, or perhaps I am guilty of some lesser form of defaulting on it, but there is no point in saying that I still have the obligation.

The persistence constraint, given in condition 4.4, reflects the fact that obligations are not momentary states, but endure until something happens to bring them to an end. If I have been under an obligation for a while, right up until the present moment, and have not just this moment fulfilled it, and still have some chance of non-trivially fulfilling it, then I am still under that obligation.

6 Syntax

For our formal language, we assume the availability of denumerably many (non-logical) sentential constants. As logical constants, we include:

truth-functional operators:	$\neg$, $\wedge$, $\vee$, $\rightarrow$, $\leftrightarrow$
unary temporal operators:	$\langle\!\mathrm{P}\!\rangle$, $\boxed{\mathrm{P}}$, $\langle\!\mathrm{F}\!\rangle$, $\boxed{\mathrm{F}}$, $\Diamond$, $\Box$,
binary temporal connectives:	***since***, ***until***
unary action operators:	$\langle\delta\rangle$, $[\delta]$,
unary ability operators:	$\langle\!+\!\rangle$, $\boxed{\times}$,
unary deontic operators:	✧, $\bigcirc$, $\langle\langle o\rangle\rangle$, $\langle[o]\rangle$, $[\langle o\rangle]$, $[[o]]$.

This is a generous (some will think over-generous) collection of logical constants. On the whole, and at least for the moment, I prefer to err on the side of distinguishing too many concepts here rather than risk conflating subtly different notions. Despite the richness of the language, I believe it passes one important methodological test: merely technical operators should be avoided, i.e. operators which do not correspond fairly closely to comparatively simple expressions of careful discourse in natural language about actions and obligations.

We define formulas in the obvious way, and use italicized upper case letters as variables over formulas. As usual, some of the truth-functional operators could have been (or could be considered as being) defined in terms of others. In addition, some of the non-truth-functional operators are related as duals of one another. These dualities will be acknowledged in the set of valid schemata we shall put forward in a moment. Taking advantage of these facts, we could reduce the number of primitive logical constants approximately by half, if that were thought to be desirable.

Our intended interpretations for the logical constants can be indicated informally as follows:

- the truth-functional operators are interpreted as usual;
- $\langle\!\mathrm{P}\!\rangle A$ is true iff at some time in the past A has been true;
- $\boxed{\mathrm{P}} A$ is true iff at all times in the past A has been true;
- $\langle\!\mathrm{F}\!\rangle A$ is true relative to a certain currently possible future iff at some time in that future A would be true;
- $\boxed{\mathrm{F}} A$ is true relative to a certain currently possible future iff at all times throughout that future A would be true;
- $\Diamond A$ is true relative to a certain currently possible future iff there is some (possibly different) currently possible future relative to which A would be true;

- □*A* is true iff *A* would be true relative to each currently possibly future, i.e. iff *A* is, in Belnap's terminology, *settled true*;
- *A* ***since*** *B* is true iff at some time in the past *B* was true, and at each time from then until (but not necessarily including) the present moment *A* has been true;
- *A* ***until*** *B* is true relative to a certain currently possible future iff at every moment of that future which comes before all future moments at which *B* is true, *A* is true ;
- ⟨δ⟩*A* is true relative to a certain currently possible future iff I perform an action compatible with the truth of *A*;
- [δ]*A* is true relative to a certain currently possible future iff I perform an action which assures the truth of *A*;
- ⟐*A* is true iff some action currently possible for me would assure the truth of *A*, i.e. iff *A* would be true relative to each of the possible futures which might issue from that action;
- ⊠*A* is true iff every action currently possible for me is compatible with the truth of *A*;
- ✧*A* is true relative to a certain currently possible future iff none of my current obligations have as their objective the proposition that ¬*A*, i.e. the set of moment/history pairs at which *A* is not satisfied;
- ○*A* is true relative to a certain currently possible future iff the proposition that *A*, i.e. the set of moment/history pairs at which *A* is satisfied, is the objective of one of my current obligations;
- ⟨⟨o⟩⟩*A* is true iff I have an obligation which could still be fulfilled in a way compatible with the truth of *A*;
- ⟨[o]⟩*A* is true iff I have an obligation which hereafter could only be fulfilled in ways which would assure the truth of *A*;
- [⟨o⟩]*A* is true iff each obligation I have could still be fulfilled in a way compatible with the truth of *A*;
- [[o]]*A* is true iff, for each obligation I have, hereafter that obligation could only be fulfilled in ways which would assure the truth of *A*;

We might contemplate adding other operators:

- ◈*A* is true iff some action currently possible for me is compatible with the truth of *A*, in the sense that *A* would be true relative to at least one of the possible futures that might issue from that action;
- ▣*A* is true iff every action currently possible for me would assure the truth of *A*;

but these would be redundant, their jobs being already accomplished by ◇ and □.

If we fix attention on a single moment, and think of the histories through that moment as worlds, letting the choice function sort those "worlds" into sets corresponding to actions open to the agent at that moment, then the operators [δ] and its dual ⟨δ⟩ correspond closely to the action operators ⊠ and ⟐ of Brown [12], and the operators ◇, ⟐, ⊠, and □ given here correspond closely to the ability operators ◈, ⟐, ⊠, and ▣ in Brown [11].

We define some additional temporal operators which are occasionally useful:

$$\langle P\bullet\rangle A =_{df} A \vee \Diamond_{P} A \qquad [P\bullet]A =_{df} A \wedge \boxed{P}\, A$$

$$\langle F\bullet\rangle A =_{df} A \vee \Diamond_{F} A \qquad [F\bullet]A =_{df} A \wedge \boxed{F}\, A$$

$$\textit{recently}\ A =_{df} A\ \textit{since}\ A$$

We also define some additional agent-related operators as follows:

$$\textit{dstit}A =_{df} [\delta]A \wedge \Diamond\neg A \qquad \text{(cf. Horty and Belnap [7])}$$

$$\textit{incur}A =_{df} \bigcirc A \wedge \textit{recently}\ \neg\bigcirc A$$

$$\textit{discharge}A =_{df} (\bigcirc A\ \textit{since}\ \textit{incur}A) \wedge \textit{dstit}A$$

$$\textit{violate}A =_{df} (\bigcirc A\ \textit{since}\ \textit{incur}A) \wedge \textit{dstit}\Box\,[F\bullet]\,\Box\neg A$$

$$\textit{excused}A =_{df} (\bigcirc A\ \textit{since}\ \textit{incur}A) \wedge \Box\,[F\bullet]\,\Box\neg A \wedge \neg\textit{dstit}\Box\,[F\bullet]\,\Box\neg A\ .$$

To say that we incur an obligation at this moment is to say we now have one which we haven't had in the immediately recent past. We may have had an obligation with the same defining condition sometime in the past, but if we are incurring a new obligation of the same sort right now, that previous obligation must have been gone for a time by now. (Recall that we are excluding cumulative obligations from consideration in this system.)

To say that I now discharge an obligation to assure that A is to say that I have incurred such an obligation at some time in the past, that I have *had* such an obligation from that time up until now, and that I now see to it that A. This definition may be open to some question, however. Consider a case in which I owe you money, can't presently pay it, but give you my lottery ticket, with the understanding that if that should prove to be a winning ticket, the winnings will stand in satisfaction of my obligation. My action takes place at one time, but if (as I could not have guaranteed, and therefore cannot take credit for) the ticket actually does win, my debt is satisfied only at that later moment when the winning lottery ticket is announced (and perhaps even later, when payment is made). To do justice to such potential complications, we may need a still more sophisticated account of action, but it seems more likely to me that what is needed is an account of how we may negotiate changes in our obligations. Any account of such negotiations will have to wait for some future occasion. Meanwhile, the definition given here seems applicable to the vast majority of cases.

To say that I now violate my obligation to see to it that A is to say that I at some time incurred such an obligation, and have been under it from then until now, and that now I have taken some action which assures that there is no longer any chance of fulfilling that obligation now or in the future.

To say that I am excused from my obligation to see to it that A, at least in the sense being explored here, is to say that I at some time incurred such an obligation, and have been under it from then until now, but that now it has become, through no (current) action of mine, impossible ever to fulfill that obligation. There might well be some question about whether this definition captures precisely the right conditions for being excused from an obligation, but it at least gives a plausible first draft. We may hope that by exploring its consequences we can uncover and learn to correct any deficiencies from which it may suffer.

We might also contemplate attempting to characterize a notion of defaulting on an obligation, intermediate morally between violating it and being excused from it, but at the moment this seems a more subtle notion than I feel able to capture. It will have to await further thought.

7 Satisfaction Conditions

Given any formula A, we let $[\![A]\!]_{\mathbb{M}}$ (or just $[\![A]\!]$, if $\mathbb{M}$ is clear from context) be the set of all pairs m/h such that m is a moment in $\mathbb{M}$, $h \in \mathbb{H}(m)$, and in $\mathbb{M}$, at the moment m, relative to the history h through m, the formula A is satisfied. Satisfaction of formulas in a model $\mathbb{M}$, at a moment m, relative to a history h through m, is defined as follows, where A, B are any formulas and S is any sentential constant:

$m/h, \mathbb{M} \models S$	iff $\mathbb{V}(m,S) = \mathbf{t}$;
$m/h, \mathbb{M} \models \neg A$	iff $m/h, \mathbb{M} \not\models A$;
$m/h, \mathbb{M} \models (A \wedge B)$	iff $m/h, \mathbb{M} \models A$ & $m/h, \mathbb{M} \models B$;
$m/h, \mathbb{M} \models (A \vee B)$	iff $m/h, \mathbb{M} \models A$ or $m/h, \mathbb{M} \models B$;
$m/h, \mathbb{M} \models (A \rightarrow B)$	iff $m/h, \mathbb{M} \models A \Rightarrow m/h, \mathbb{M} \models B$;
$m/h, \mathbb{M} \models (A \leftrightarrow B)$	iff $m/h, \mathbb{M} \models A \Leftrightarrow m/h, \mathbb{M} \models B$;
$m/h, \mathbb{M} \models \langle P \rangle A$	iff $(\exists m_0 < m)[m_0/h, \mathbb{M} \models A\,]$;
$m/h, \mathbb{M} \models \boxed{P} A$	iff $(\forall m_0 < m)[m_0/h, \mathbb{M} \models A\,]$;
$m/h, \mathbb{M} \models \langle F \rangle A$	iff $(\exists m' \in h\colon m < m')[m'/h, \mathbb{M} \models A\,]$
$m/h, \mathbb{M} \models \boxed{F} A$	iff $(\forall m' \in h\colon m < m')[m'/h, \mathbb{M} \models A\,]$;
$m/h, \mathbb{M} \models \Diamond A$	iff $(\exists h' \in \mathbb{H}(m))[m/h', \mathbb{M} \models A\,]$
$m/h, \mathbb{M} \models \Box A$	iff $(\forall h' \in \mathbb{H}(m))[m/h', \mathbb{M} \models A\,]$
$m/h, \mathbb{M} \models A$ ***since*** B	iff $(\exists m_0 < m)[[m_0/h, \mathbb{M} \models B\,]$ & $(\forall m' < m\colon m_0 < m')[m'/h, \mathbb{M} \models A\,]]$;
$m/h, \mathbb{M} \models A$ ***until*** B	iff $(\forall m' \in h\colon m < m')[(\forall m_0 \in h\colon m < m_0)[[m_0/h, \mathbb{M} \models B\,] \Rightarrow m' < m_0] \Rightarrow m'/h, \mathbb{M} \models A\,]$;
$m/h, \mathbb{M} \models \lozenge A$	iff $[\![\neg A]\!] \notin \mathbb{O}(m/h)$;
$m/h, \mathbb{M} \models \bigcirc A$	iff $[\![A]\!] \in \mathbb{O}(m/h)$;
$m/h, \mathbb{M} \models \langle\langle o \rangle\rangle A$	iff $(\exists o \in \mathbb{O}(m/h))(\exists m'/h' \in o\colon m < m')[m'/h', \mathbb{M} \models A\,]$;
$m/h, \mathbb{M} \models \langle [o] \rangle A$	iff $(\exists o \in \mathbb{O}(m/h))(\forall m'/h' \in o\colon m < m')[m'/h', \mathbb{M} \models A\,]$;
$m/h, \mathbb{M} \models [\langle o \rangle] A$	iff $(\forall o \in \mathbb{O}(m/h))(\exists m'/h' \in o\colon m < m')[m'/h', \mathbb{M} \models A\,]$;
$m/h, \mathbb{M} \models [[o]] A$	iff $(\forall o \in \mathbb{O}(m/h))(\forall m'/h' \in o\colon m < m')[m'/h', \mathbb{M} \models A\,]$;
$m/h, \mathbb{M} \models \langle \delta \rangle A$	iff $(\exists c \in \mathbb{C}(m)\colon h \in c)(\exists h' \in c)[m/h', \mathbb{M} \models A\,]$;
$m/h, \mathbb{M} \models [\delta] A$	iff $(\exists c \in \mathbb{C}(m)\colon h \in c)(\forall h' \in c)[m/h', \mathbb{M} \models A\,]$;
$m/h, \mathbb{M} \models \langle + \rangle A$	iff $(\exists c \in \mathbb{C}(m))(\forall h' \in c)[m/h', \mathbb{M} \models A\,]$;
$m/h, \mathbb{M} \models \boxed{\Diamond} A$	iff $(\forall c \in \mathbb{C}(m))(\exists h' \in c)[m/h', \mathbb{M} \models A\,]$.

We take a formula to be ***valid*** iff it is satisfied at every moment, relative to every history through that moment, in every model.

8 Some Validities

With these interpretations in mind, we can begin to look for valid schemata and rules. The collection given below is no doubt not independent, and probably will not form the basis for a complete axiomatization. Indeed, observations in Thomason [18] make it appear extremely unlikely that any complete axiomatization can be found, even for just the temporal portion of the system. For convenience, the principles given below are grouped according to their primary subject matter. Where a principle is closely related to one of the constraints on models, or relates to some easily described feature of the models, I cite the relevant constraint or feature at the right. In some cases, I assign names to principles by analogy with existing naming conventions for analogous schemata and rules.

<u>Principles of non-modal logic:</u>

(PC) $\models A$, for A any substitution instance of any truth-functional tautology

(MP) $A, A \rightarrow B \models B$

<u>Principles of the logic of the past:</u>

(Df$\langle P\rangle$) $\models \langle P\rangle A \leftrightarrow \neg \boxed{P} \neg A$ (duality)

(K$\boxed{P}$) $\models \boxed{P}(A \rightarrow B) \rightarrow (\boxed{P}A \rightarrow \boxed{P}B)$ (normality)

(4$\boxed{P}$) $\models \boxed{P}A \rightarrow \boxed{P}\,\boxed{P}A$ (transitivity)

(G$\boxed{P}$) $\models \langle P\rangle\,\boxed{P}A \rightarrow \boxed{P}\,\langle P\rangle A$ (backward connection within histories)

$\models (\langle P\rangle A \wedge \langle P\rangle B) \rightarrow (\langle P\rangle(A \wedge \langle P\rangle B) \vee \langle P\rangle(A \wedge B) \vee \langle P\rangle(\langle P\rangle A \wedge B))$ (no backward branching within histories)

(RM$\boxed{P}$) If $\models A \rightarrow B$ then $\models \boxed{P}A \rightarrow \boxed{P}B$

(RN$\boxed{P}$) If $\models A$ then $\models \boxed{P}A$

<u>Principles of the logic of the future:</u>

(Df$\langle F\rangle$) $\models \langle F\rangle A \leftrightarrow \neg \boxed{F} \neg A$ (duality)

(K$\boxed{F}$) $\models \boxed{F}(A \rightarrow B) \rightarrow (\boxed{F}A \rightarrow \boxed{F}B)$ (normality)

(4$\boxed{F}$) $\models \boxed{F}A \rightarrow \boxed{F}\,\boxed{F}A$ (transitivity)

(G$\boxed{F}$) $\models \langle F\rangle\,\boxed{F}A \rightarrow \boxed{F}\,\langle F\rangle A$ (forward connection within histories)

$\models (\langle F\rangle A \wedge \langle F\rangle B) \rightarrow (\langle F\rangle(A \wedge \langle F\rangle B) \vee \langle F\rangle(A \wedge B) \vee \langle F\rangle(\langle F\rangle A \wedge B))$

(RM$\boxed{F}$) If $\models A \rightarrow B$ then $\models \boxed{F}A \rightarrow \boxed{F}B$

(RN$\boxed{F}$) If $\models A$ then $\models \boxed{F}A$

<u>Principles connecting past and future:</u>

$\models A \rightarrow \boxed{F}\,\langle P\rangle A$ (linearity of histories)

$\models A \rightarrow \boxed{P}\,\langle F\rangle A$ (linearity of histories)

<u>Principles of the logic of settledness:</u>

(Df$\Diamond$) $\models \Diamond A \leftrightarrow \neg \Box \neg A$ (duality)

(K$\Box$) $\models \Box(A \rightarrow B) \rightarrow (\Box A \rightarrow \Box B)$ (normality)

(T$\Box$) $\models \Box A \rightarrow A$ (reflexivity; there is a history)

(B$\Box$) $\models A \rightarrow \Box \Diamond A$ (symmetry)

(4$\Box$) $\models \Box A \rightarrow \Box\Box A$ (transitivity)
(5$\Box$) $\models \Diamond A \rightarrow \Box\Diamond A$ (euclideanness)
(RM$\Box$) If $\models A \rightarrow B$ then $\models \Box A \rightarrow \Box B$
(RN$\Box$) If $\models A$ then $\models \Box A$

Principles applying settledness to tense:

$\models \Diamond A \rightarrow \boxed{P}\, \Diamond\Diamond_{F} A$
$\models \Box \boxed{F} A \rightarrow \Box \boxed{F}\, \Box \boxed{F} A$
(C$\Diamond\Diamond_{F}$) $\models \Diamond\Diamond_{F}(A \vee B) \rightarrow (\Diamond\Diamond_{F} A \vee \Diamond\Diamond_{F} B)$
(V$\Diamond\Diamond_{F}$) $\models \Diamond\boxed{F}(A \vee B) \rightarrow (\Diamond\Diamond_{F} A \vee \Diamond\boxed{F} B)$

Principles connecting ***since*** and ***until***:

$\models A$ ***since*** $B \rightarrow \Diamond_{P} B$
$\models A$ ***since*** $B \rightarrow \Box$ (A ***since*** B) (settledness of the past)
$\models$ (A_1 ***since*** B) $\wedge$ (A_2 ***since*** B) $\rightarrow$ (($A_1 \wedge A_2$) ***since*** B)
$\models$ (A ***since*** $(B_1 \vee B_2)$) $\rightarrow$ ((A ***since*** B_1) $\vee$ (A ***since*** B_2))
If $\models A_1 \rightarrow A_2$ then $\models$ (A_1 ***since*** B) $\rightarrow$ (A_2 ***since*** B)
If $\models B_1 \rightarrow B_2$ then $\models$ (A ***since*** B_1) $\rightarrow$ (A ***since*** B_2)
$\models (\boxed{P} A \wedge \Diamond_{P} B) \rightarrow A$ ***since*** B
$\models \Diamond_{P}(B \wedge \Box \boxed{F} A) \rightarrow A$ ***since*** B
$\models A$ ***since*** $B \rightarrow \boxed{P}((A \wedge \Diamond_{P} B) \vee B \vee \Diamond_{F} B)$
$\models A$ ***until*** $B \rightarrow \boxed{F}(A \vee \langle \mathrm{P}\bullet \rangle B)$
$\models A$ ***until*** $B \rightarrow \boxed{F}$((A ***until*** B) $\vee \langle \mathrm{P}\bullet \rangle B$)
$\models A$ ***until*** $\neg A$
$\models$ ((A ***since*** B) $\wedge A \wedge \neg B \wedge$ (A ***until*** B)) $\rightarrow \Diamond_{P}$(A ***until*** B)
$\models$ ((A ***since*** B) $\wedge A \wedge \neg B \wedge$ (A ***until*** B)) $\rightarrow \Diamond_{F}$(A ***since*** B)

Principles of the logic of action:

(Df[δ]) $\models \langle\delta\rangle A \leftrightarrow \neg[\delta]\neg A$ (duality)
(K[δ]) $\models [\delta](A \rightarrow B) \rightarrow ([\delta]A \rightarrow [\delta]B)$ (normality)
(T[δ]) $\models [\delta]A \rightarrow A$ (reflexivity; efficacy of action)
(B[δ]) $\models A \rightarrow [\delta]\langle\delta\rangle A$ (symmetry)
(4[δ]) $\models [\delta]A \rightarrow [\delta][\delta]A$ (transitivity)
(RM[δ]) If $\models A \rightarrow B$ then $\models [\delta]A \rightarrow [\delta]B$
(RN[δ]) If $\models A$ then $\models [\delta]A$

Principles of the logic of ability:

(Df$\Diamond$) $\models \Diamond A \leftrightarrow \neg\Box\neg A$ (duality)
(Df$\Diamond_{+}$) $\models \Diamond_{+} A \leftrightarrow \neg\boxtimes\neg A$ (duality)
(C$\Diamond$) $\models \Diamond(A \vee B) \rightarrow (\Diamond A \vee \Diamond B)$
(V$\Diamond_{+}\Diamond$) $\models \Diamond_{+}(A \vee B) \rightarrow (\Diamond A \vee \Diamond_{+} B)$
(W$\Box\Diamond_{+}$) $\models \Diamond A \rightarrow (\Box B \rightarrow \Diamond_{+} B)$
(D$\Diamond_{+}\Diamond$) $\models \Diamond_{+} A \rightarrow \Diamond A$ (non-vacuity of actions)
(T$\Box$) $\models \Box A \rightarrow A$
(RM$\boxtimes$) If $\models A \rightarrow B$ then $\models \boxtimes A \rightarrow \boxtimes B$

(RM$\Box$)	If $\models A \rightarrow B$ then $\models \Box A \rightarrow \Box B$	
(RN$\Box$)	If $\models A$ then $\models \Box A$	

Principles of the logic of (potentially conflicting) type 1 obligation:

(Df$\langle\langle o\rangle\rangle$)	$\models \langle\langle o\rangle\rangle A \leftrightarrow \neg[[o]]\neg A$	(duality)
(Df$\langle[o]\rangle$)	$\models \langle[o]\rangle A \leftrightarrow \neg[\langle o\rangle]\neg A$	(duality)
(C$\langle\langle o\rangle\rangle$)	$\models \langle\langle o\rangle\rangle(A \vee B) \rightarrow (\langle\langle o\rangle\rangle A \vee \langle\langle o\rangle\rangle B)$	
(V$\langle[o]\rangle\langle\langle o\rangle\rangle$)	$\models \langle[o]\rangle(A \vee B) \rightarrow (\langle\langle o\rangle\rangle A \vee \langle[o]\rangle B)$	
(W$[[o]]\langle[o]\rangle$)	$\models \langle\langle o\rangle\rangle A \rightarrow ([[o]]B \rightarrow \langle[o]\rangle B)$	
(D$\langle[o]\rangle\langle\langle o\rangle\rangle$)	$\models \langle[o]\rangle A \rightarrow \langle\langle o\rangle\rangle A$	
(RM$[\langle o\rangle]$)	If $\models A \rightarrow B$ then $\models [\langle o\rangle]A \rightarrow [\langle o\rangle]B$	
(RM$[[o]]$)	If $\models A \rightarrow B$ then $\models [[o]]A \rightarrow [[o]]B$	
(RN$[[o]]$)	If $\models A$ then $\models [[o]]A$	

Principles of the logic of (potentially conflicting) non-trivial (type 2) obligation:

(Df$\diamondsuit$)	$\models \diamondsuit A \leftrightarrow \neg\bigcirc\neg A$	(duality)
	$\models \bigcirc A \rightarrow \langle[o]\rangle A$	
	$\models \bigcirc A \rightarrow \neg A$	(unfulfilledness)
(RM$\bigcirc$)	If $\models A \rightarrow B$ then $\models \bigcirc A \rightarrow \bigcirc B$	

More principles connecting the sublogics:

	$\models \Box A \rightarrow [\delta]A$	
	$\models \Box\boxed{F}A \rightarrow [[o]]A$	
	$\models \bigcirc A \rightarrow \Diamond\langle F\rangle A$	(fulfillability)
	$\models \bigcirc A \rightarrow \Diamond\langle F\rangle\neg A$	(nontriviality)
	$\models$ ***recently*** $\bigcirc A \wedge \neg A \wedge \Diamond\langle F\rangle A \rightarrow \bigcirc A$	(persistence)

9 Future Developments

To make the system here more manageable, it would help to have a complete axiomatization. However it seems unlikely this will be possible for the system as a whole, and particularly for the purely temporal portion. Nonetheless we may hope at least to give complete axiomatizations of some of the other principle sublogics: the sublogic of action, sublogic of ability, and sublogic of obligation, in particular. In light of the results in Brown [11], [12], and [14], it seems likely we have already incorporated principles sufficient to axiomatize the sublogics of action, ability and type 1 obligation. Results in Segerberg [16] suggest that the sublogic of type 2 obligation is also given here completely. The richer subsystem system combining the two types of obligation, however, remains to be investigated, as do various other combinations of subsystems.

At the same time, we may also hope to make the system even more rich and by that process make it more accurately applicable to real problems, by removing some of the restrictions we accepted in section 3. One major step forward would involve adding a characterization of conditional obligation. It seems likely to me that we can use conditional obligations to express most, perhaps all, standing obligations, spawning obligations of the sort considered here whenever the appropriate invoking conditions come to pass. It is plausible, for example, to suggest that my standing obligation to obey the speed limit does not come into play when I am dining at home, or asleep in bed, etc., but comes into play only when I am driving an automobile.

We may be able to handle some, perhaps even all, cumulative obligations by indexing each obligation with the moment at which it is incurred. This would differentiate the obligation created by my borrowing a thousand *escudos* from you yesterday from the additional obligation created by my borrowing a thousand more *escudos* today. It would not in any obvious way solve the problem of explaining what occurs when I pay off a debt in installments, however. In the end, it may be necessary to bring a "logic of accounting" into the picture more explicitly for such cases.

Bringing additional agents into the picture is in a way trivial, but we may expect that a full treatment of the possible interactions between agents will reveal subtleties every bit as important as those regarding time, and will require a fully dynamic, not merely a diachronic, deontic logic. I expect there will be real challenges there.[1]

REFERENCES

1. Åqvist L. and Hoepelman, J. Some theorems about a 'tree' system of deontic tense logic. In: R. Hilpinen (ed) New Studies in Deontic Logic. D. Reidel, Dordrecht, 1981, pp. 187-221 (Synthese library volume 152)

2. Thomason, R. H. Deontic logic as founded on tense logic. In: R. Hilpinen (ed) New Studies in Deontic Logic. D. Reidel, Dordrecht, 1981, pp 165-176 (Synthese library volume 152)

3. Thomason, R. H. Deontic logic and the role of freedom in moral deliberation. In: R. Hilpinen (ed) New Studies in Deontic Logic. D. Reidel, Dordrecht, 1981, pp 177-186 (Synthese library volume 152)

4. Ross, A. Imperatives and logic. Theoria 1941; 7: 53-71

5. Thomason, R. H. Indeterminist time and truth value gaps. Theoria 1970; 36: 264-281

6. Belnap, N. and Perloff, M. Seeing to it that: a canonical form for agentives. Theoria 1988; 54: 175-199

---

[1] I am indebted to the referees for several helpful suggestions and gentle corrections. An earlier version of this paper was given at the University of Lisbon on June 6, 1995. I am similarly indebted to the members of the audience present on that occasion. Thanks are due also to Robert Fudge, for discovering a number of typographic errors.

7. Horty, J. and Belnap, N. The deliberative stit: a study of action, omission, ability, and obligation. Journal of Philosophical Logic, forthcoming

8. Burgess, J. P. Basic tense logic. In: D. Gabbay and Guenther F. (eds) Handbook of philosophical logic, volume II: Extensions of classical logic. D. Reidel, Dordrecht, 1984, pp 89-133

9. Chellas, B. F. The logical form of imperatives. Perry Lane Press, Palo Alto, 1969

10. Chellas, B. F. Imperatives. Theoria 1971; 37: 114-129

11. Brown , M. A. On the logic of ability. Journal of Philosophical Logic 1988; 17: 1-26

12. Brown, M. A. Action and ability. Journal of Philosophical Logic 1990; 19: 95-114

13. van Fraassen, B. Values and the heart's command. Journal of Philosophy 1973; 70: 5-19

14. Brown, M. A. A logic of comparative obligation. Studia Logica, forthcoming.

15. Prior, A. Past, present, and future. Oxford University Press, Oxford, 1967

16. Segerberg, K. An essay in classical modal logic (3 vols). University of Uppsala, Uppsala, 1971

17. Chellas, B. F. Modal logic: an introduction. Cambridge University Press, Cambridge, 1980

18. Thomason, R. H. Combinations of tense and modality. In: D. Gabbay and Guenther F. (eds) Handbook of philosophical logic, volume II: Extensions of classical logic. D. Reidel, Dordrecht, 1984, pp 135-165

A deontic logic for reasoning about confidentiality

Frédéric Cuppens and Robert Demolombe
ONERA-CERT, 2 Av. E. Belin
31055, Toulouse Cedex, France
Fax: + 33 62 25 25 93
email: {cuppens,demolomb}@tls-cs.cert.fr

Abstract

This paper presents a deontic logic Σ for reasoning about permission or prohibition to know some parts of the datab¡ase content in the context of a multilevel confidentiality policy.

The most important logical features in the definition of a multilevel policy are that each confidentiality level is defined by a set of sentences and that, when the policy is designed, the permission to know is not necessarily the complement of the prohibition to know. These concepts are formalized in a modal logic where deontic modalities, doxastic modalities and confidentiality levels are interpreted by non-standard modal models. The corresponding axiomatics is also presented in the paper and its soundness and completeness have been proved. A limitation of the Σ logic is that sentences in the scope of modalities are sentences of Propositional Calculus.

Finally, it is shown how the logic can be used to express constraints to guarantee the consistency of a policy or to prevent the existence of inference channels. That is, the possibility to infer sentences that are not permitted to know from other sentences that are permited to know. Both deductive and abductive channels are considered.
Keywords: knowledge representation, deontic logic, database security.

1 Introduction

To guarantee confidentiality with regard to accesses to a database or, more generally, to an information system, a standard approach is to define multilevel security policies. A given multilevel policy assigns a clearance level to users and a classification level to sentences of the language used to represent the database content. This assignment is not necessarily complete for sentences. There may be sentences whose level is not defined.

With a multilevel policy a user is permitted to access sentences stored in the database whose classification is lower than or equal to his clearance and he is forbidden to access sentences whose classification is greater than his clearance. One may immediately notice that, as the permission and prohibition of a given user is derived from two independently defined sets – namely, the sentences wich are respectively lower than or equal to user's clearance and the sentences greater than user's clearance –, the permission is not necessarily the complement of prohibition. Moreover, there are sentences which are not assigned a classification. This means that the policy may be incomplete in the sense that there are sentences a given user may be neither permitted nor prohibited to know.

As a matter of simplification, we only consider here two security levels: the "confidential" level and the "secret" level. We consider that the assignment of a classification is defined independently of the database content and independently of the truth value value of sentences in a given situation. Therefore, it is, in one sense, syntactical. Also, the set of sentences to which a given classification is assigned should not be confused with the set of sentences actually stored in the database. For instance, the set of confidential sentences may be: $\{p, \neg p, q\}$, and the set of confidential sentences stored in the database may be: $\{p\}$ at a given time and $\{\neg p, q\}$ at another time.

In this context there are several issues related to confidentiality. The first is to find a formal definition for the "consistency" of a confidentiality policy. A confidentiality policy is consistent iff we cannot infer from this policy that there exists a sentence a given user is both permitted and prohibited to know. We can very briefly explain the reasons why the formalization of this definition is not a trivial problem. The two sets of sentences classified respectively at level confidential or secret, are intended to characterize the set of sentences a confidential user u is permitted to know or is forbidden to know. The user u is permitted to know all the sentences that are in the set of confidential sentences. He is also permitted to know the consequences of **each** consistent subset of this set but he does not necessarily have the permission to know, in a given situation, the conjunction of several sentences that are consequences of distinct subsets. For instance, if the set of confidential sentences is: $\{p, \neg p\}$, we can infer, for every sentence q, that user u has the permission to know $p \vee q$ and also to know $\neg p \vee q$. However, u does not have the permission to know $(p \vee q) \wedge (\neg p \vee q)$. Indeed, if this were the case, he would also be permitted to know the consequence q of $(p \vee q) \wedge (\neg p \vee q)$, whatever the sentence q may be. Reasoning about the consequences of the fact that a user is forbidden to know a sentence p is slightly less intuitive, as will be shown in the rest of the paper. Thus we can easily imagine that consistency checking of a confidentiality policy requires very clear definitions of the concepts of permission and prohibition to know and the associated inference rules have to be carefully defined.

The second issue is to characterize the "inference problem". There is an inference problem, for a given confidentiality policy and a given database state, if it is possible that a confidential user can infer from a set of sentences he has retrieved from the database a sentence he is not permitted to know. This problem can be refined into two sub-problems. In the first, we consider inferences based on deductive reasoning. In this case we say that there exists a "deductive channel". In the second, we consider abductive reasoning and say that there exists an "abductive channel". By abductive reasoning, we mean the kind of reasoning that allows the confidential user to deduce, by using an additional missing sentence that this user may abduce, a secret sentence. We want to be able to characterize, in our logical framework, the constraints that prevent the existence of deductive or abductive channels.

Most current research work (for instance [1, 2]) is restricted to the detection of deductive inference channels. There are few publications which are interested in abductive channels [3, 4]. Unfortunately, a formal definition of what is an abductive channel is not provided.

Moreover, in all these publications, the multilevel database is identified with a set of atomic facts and it is generally assumed that the classification procedure only applies to the database content, i.e., a given data may be assigned a classification only if it is a fact explicitly stored in the database. This leads to two important restrictions:

- first, it is not possible to specify, for a given sentence p_1 that both p_1 and $\neg p_1$ may be assigned a (possibly different) classification;
- second, it is not possible to specify that not only a given piece of information p_1 is classified as secret but also the disjunction $p_1 \vee p_2$ for some sentence p_2. Therefore, it is not possible to represent a situation where not only the "exact" disclosure of p_1 is forbidden but also a "partial" disclosure of p_1, represented by the formula $p_1 \vee p_2$, is to be prevented.

The next section presents the formal definition of the logic. The main reason we did not adopt a standard deontic logic and a standard doxastic logic is the necessity of dealing with complete representations of the regulation – in the sense that the regulation defines all the sentences and *only* the sentences that are permitted or prohibited to know – and complete sets of agent knowledge – i.e sets used to represent all the sentences and *only* the sentences that an agent know (see [5] for similar motivations). Another reason is that, the concept of permission is not defined as the complement of prohibition. This enables two different attitudes to be modeled [6]. In the first attitude, the multilevel database should only provide the user with data this user is explicitly permitted to know. We call **restrictive** this first attitude. With respect to the second attitude, the multilevel database should only provide the user with data he is not explicitly prohibited to know. We call **permissive** this second attitude. Finally, section 3 presents examples of practical application of the logic to regulation consistency and to inference channels.

2 Formal definition of the Σ logic

2.1 Basic Assumptions and Notations

We assume that the database content is represented by a consistent set db of sentences of a language L of Propositional Calculus. It is not assumed that sentences in db are true in the world. This means that db is considered a set of beliefs. By querying the database db, the users will know that the database believes some sentence. Therefore, the fact that a user u knows that db believes some sentence is represented by the modality KB_u, and the fact that u knows that db believes the sentence p is represented by $KB_u(p)$, $KB_u p$ for short. The modality KB_u can be viewed as the combination of an epistemic modality and a doxastic modality. This means that the modality KB_u can be formalized by the KD logic [7].

Let us now turn to the representation of a multilevel confidentiality policy. In this case, each user is associated with a clearance level and some sentences of the propositional language L are associated with a classification level. If we consider a user u cleared at a given level i, it is possible to associate this user with two sets of sentences: the set of sentences whose classification level l is such that $l \leq i$ and the set of sentences whose classification level l is such that $l > i$. These two sets are used to define the set of sentences a user cleared at level i is permitted or prohibited to know.

For the sake of simplicity, we only consider two security levels. Level 1 is called "confidential" and level 2 is called "secret". The set of sentences such that $l \leq 1$ is called C and the set of sentences such that $l > 1$ is called S. We also only consider the case of a user cleared at confidential (level 1). Generalization to more than two security levels is trivial; we have simply to replace level 1 by level i.

It is assumed that some sentences may not be classified, that is, we do not necessarily have: $C \cup S = L$. These two sets are formally represented in the language by two sentences of the form: $[C](p_1, \ldots, p_m)$, and $[S](p'_1, \ldots, p'_n)$, where $\{p_1, \ldots, p_m\}$ (resp. $\{p'_1, \ldots, p'_n\}$) is the set of **all** the sentences classified at the confidential (resp. secret) level.

It is worth noting that for sentences that are neither in C nor in S, it is not explicitly defined whether a user cleared at confidential is permitted or prohibited to know these sentences. In some sense, the regulation is not complete. However, when a confidential user is querying the multilevel database, we need to complete the regulation. This is because the multilevel database has to decide whether or not the answer to the query has to be delivered to the user. We can actually propose two different attitudes to complete the regulation.

The first attitude is to consider that a user cleared at confidential is only permitted to know sentences belonging to C. This attitude can be qualified as restrictive because it means that a confidential user, when he is querying a multilevel database should only be provided with an answer he is explicitly permitted to know.

The second attitude is to consider that this user is only forbidden to know sentences belonging to S. This second attitude can be qualified as permissive because it means that a confidential user, when he is querying a multilevel database, should only be provided with an answer he is not explicitly forbidden to know.

We consider that it is the responsability of the security administrator to choose one of these two attitudes. For this purpose, our model enables these two attitudes to be formally represented. That is why, in the model, the concepts of permission and prohibition are **independent** in the sense that what is permitted is not necessarily the complement of what is forbidden. These two concepts are formalized by the two modalities PKB_c and FKB_c. The fact that some confidential users are permitted to know that the database believes p is represented by $PKB_c(p)$, PKB_cp for short, and the fact that it is forbidden for some confidential users to know that the database believes p is represented by

$FKB_c(p)$, FKB_cp for short.[1]

2.1.1 Language Definition

Let L be a Propositional Calculus language defined on a finite set VAR of propositional variables.
Let L′ be a language defined from L as follows.

- If $p \in L$ then $p \in L'$.
- If $p_1 \in L, \ldots, p_n \in L$ then $[C](p_1, \ldots, p_n) \in L'$ and $[S](p_1, \ldots, p_n) \in L'$.
- If $p \in L$ then $\Box p \in L'$, $KB_u(p) \in L'$, $PKB_c(p) \in L'$ and $FKB_c(p) \in L'$.
- If $p \in L'$ then $\neg p \in L'$.
- If $p \in L'$ and $q \in L'$ then $p \vee q \in L'$.

There is no other sentence in L′ than the sentences defined by the above rules. As usual conjunction and implication are defined from negation and disjunction.

2.1.2 Semantics

Let Σ be the logic we are defining.
A structure M of Σ is a tuple $M =< W, R_u, D_{cr}, D_{cp}, R_{xcr}, R_{xcp}, C, S, P >$, where:

- W is a set of worlds.
- P is a function from VAR to 2^W.
- R_u, D_{cr}, D_{cp}, R_{xcr} and R_{xcp} are relations on $W \times W$.
- C and S are two functions from W to 2^{2^W}.

M is a model of Σ iff it satisfies the following constraints:

C1: if p is a satisfiable[2] sentence of L then $P(p) \neq \emptyset$.

C2: relations R_u, R_{xcr} and R_{xcp} are serial.

We shall use the notation $OB_{xcr}(p)$, $OB_{xcp}(p)$, $DB(C(w))$ and $DB(S(w))$ which are formally defined by:

- $M, w' \models OB_{xcr}(p)$ iff $\forall w''(w' R_{xcr} w''$ iff $M, w'' \models p)$
- $M, w' \models OB_{xcp}(p)$ iff $\forall w''(w' R_{xcp} w''$ iff $M, w'' \models p)$
- $\|p\|$ will denote the set of worlds $\{w \ : \ M, w \models p\}$

[1] In what follows, we consider only what a confidential user is permitted or prohibited to know. We shall ignore obligations with regard to a secret user.
[2] Here "satisfiable" means satisfiable in Classical Propositional Calculus (CPC).

- $DB(C(w)) = \{ X : \exists X_1 \ldots \exists X_m (X_1 \in C(w) \text{ and } \ldots \text{ and } X_m \in C(w) \text{ and } X = X_1 \cap \ldots \cap X_m \text{ and } X \neq \emptyset)\}$
- $DB(S(w)) = \{ X : X \subseteq W \text{ and } \forall Y(Y \in S(w) \text{ implies } X - Y \neq \emptyset)\}$

C3: $\forall p(\exists w'(M, w' \models OB_{xcr}(p)) \text{ if } \exists w(\|p\| \in DB(C(w)))$

C4: $\forall w \forall w'(wD_{cp}w' \text{ iff } \exists p(\|p\| \in DB(C(w)) \text{ and } M, w' \models OB_{xcr}(p))$

C5: $\forall p(\exists w' M, w' \models OB_{xcp}(p) \text{ if } \exists w(\|p\| \in DB(S(w)))$

C6: $\forall w \forall w'(wD_{cp}w' \text{ iff } \exists p(\|p\| \in DB(S(w)) \text{ and } M, w' \models OB_{xcp}(p))$

Comments:

These constraints may seem to be unnecessarily complicated. The following comments are intended to explain their intuitive meaning and their justifications.

Functions C and S define the regulation in each world, in other words the set of sentences which are respectively assigned the confidential level or the secret level. Each sentence p is represented by its truth set $\|p\|$.

A consequence of constraint C1 is that there is at least one world in W for each possible truth value assignment to propositional variables, in other words for each possible state of the world.

Relation R_u represents, for each state of the world, the confidential user's knowledge about the database beliefs.

Each set of worlds X_i in C(w) can be represented by a sentence p_i such that $\|p_i\| = X_i$, and each set of worlds X in DB(C(w)) can be represented by a sentence p such that $\|p\| = X$ and there exists a consistent subset of the p_is that implies p.

The elements of DB(C(w)), and the corresponding sentences p, can be interpreted as a set of mental states of a user who is only permitted to know the sentences defined by C(w) or their consequences. In this case permission is defined according to a restrictive attitude, since what is prohibited is the complement of what is explicitly permitted.

Each set of worlds X_j in S(w) can be represented by a sentence p_j such that $\|p_j\| = X_j$, and each set of worlds X in DB(S(w)) can be represented by a sentence p such that $\|p\| = X$ and p implies none of the p_js.

The elements of DB(S(w)), and the corresponding sentences p, can be interpreted as a set of mental states of a user who is **only** permitted to know any set of sentences that does not allow to infer any sentence defined by S(w). In this case permission is defined according to a permissive attitude, since what is permitted is the complement of what is explicitly prohibited.

Our deontic is then based on the concept of role [8, 9]. Intuitively, each individual is associated with a set of roles which represents the behavior the individual is playing in a given situation. Each role defines the permissions, obligations and prohibitions laid upon the role-holder. In our model of a multilevel confidentiality policy, a confidential user is associated with two different roles: the role cr (resp. cp) which represents the behavior of a user cleared at confidential with respect to the restrictive (resp. permissive) attitude.

The deontic accessibility relation D_{cr} (resp. D_{cp}) characterizes the set of ideal [10, 11] mental states of a user xcr (resp. xcp) who plays the role of a user cleared at confidential when a restrictive (resp. permissive) attitude is adopted. Accessibility relation R_{xcr} (resp. R_{xcp}) characterizes xcr's beliefs (resp. xcp's beliefs).

Constraint C3 imposes that for each element in DB(C(w)) there exists a world where xcr believes and **only** believes a sentence p that corresponds to this element. Constraint C4 imposes that the worlds accessible via D_{cr} are the worlds where xcr's mental states are exactly those ones represented by the elements in DB(C(w)) (see figure 1). That is the xcr's permitted mental states and elements in DB(C(w)) are in a one to one correspondence.

Constraint C5 imposes that for each element in DB(S(w)) there exists a world where xcp believes and only believes a sentence p that corresponds to this element. Constraint C6 imposes that the worlds accessible via D_{cp} are the worlds where xcp's mental states are exactly those ones represented by the elements in DB(S(w)) (see figure 2). That is the xcp's permitted mental states and elements in DB(C(w)) are in a one to one correspondence.

Satisfiability Conditions

$M, w \models p$ means that the sentence p of L′ is true in the world w of the model M. The satisfiability conditions are:

$$\begin{array}{l}
\text{If } p \in VAR : M, w \models p \text{ iff } w \in P(p). \\
M, w \models \neg p \text{ iff } M, w \not\models p. \\
M, w \models p \vee q \text{ iff } M, w \models p \text{ or } M, w \models q. \\
M, w \models \Box p \text{ iff } \forall w' \in W \;\; M, w' \models p. \\
M, w \models [C](p_1, \ldots, p_n) \text{ iff } C(w) = \{\|p_1\|, \ldots, \|p_n\|\}. \\
M, w \models [S](p_1, \ldots, p_n) \text{ iff } S(w) = \{\|p_1\|, \ldots, \|p_n\|\}. \\
M, w \models KB_u p \text{ iff } \forall w'(wR_u w' \Rightarrow M, w' \models p). \\
M, w \models PKB_c p \text{ iff } \exists w'(wD_{cr} w' \text{ and } M, w' \models KB_{xcr} p). \\
M, w \models FKB_c p \text{ iff } \forall w'(wD_{cp} w' \Rightarrow M, w' \models \neg KB_{xcp} p).
\end{array}$$

where:
$M, w' \models KB_{xcr}p$ is an abbreviation for: $\forall w''(w'R_{xcr}w'' \Rightarrow M, w'' \models p)$, and
$M, w' \models \neg KB_{xcp}p$ is an abbreviation for: $\exists w''(w'R_{xcp}w''$ and $M, w'' \models \neg p)$.

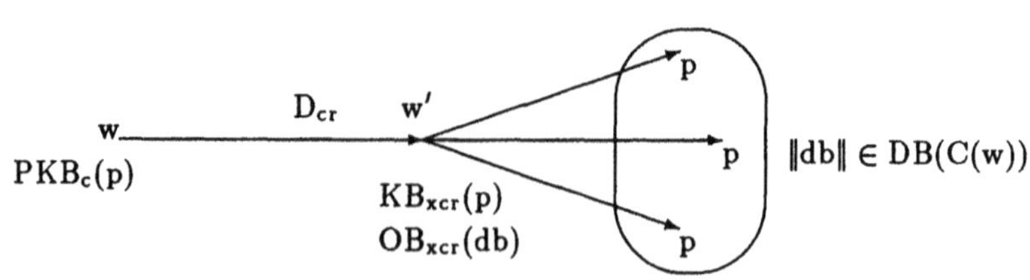

Figure 1: Permission to know.

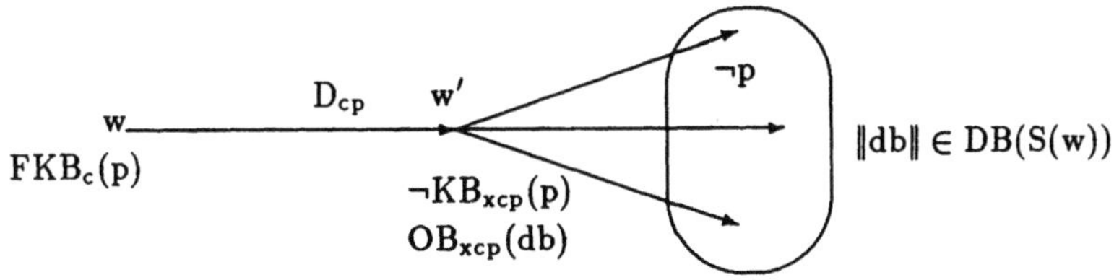

Figure 2: Prohibition to know.

Valid Sentences

$M \models p$ denotes the fact that p is true in the model M. We have:

$$M \models p \text{ iff } \forall w(M, w \models p).$$

$\models p$ denotes the fact that p is a valid sentence. We have:

$$\models p \text{ iff } \forall M(M \models p).$$

Properties of the Σ Models

Property 1: If $\|p\| \in C(w)$ then $M, w \models PKB_c p$.

Property 2: If $\|p\| \in S(w)$ then $M, w \models FKB_c p$.

Property 3: $\models p \rightarrow q \;\Rightarrow\; \models PKB_c p \rightarrow PKB_c q$.

Property 4: $\models p \rightarrow q \;\Rightarrow\; \models FKB_c q \rightarrow FKB_c p$.

Properties 1 and 2 show that the definitions of PKB_c and FKB_c work out as they should for sentences represented by C(w) and by S(w). Properties 3 and 4 show that the set of sentences a confidential user is permitted to know is extended by their logical consequences whereas the set of sentences that a confidential user is forbidden to know is extended by their logical implicants.

We have the following corollaries of properties 3 and 4:

Corollary 1: $\models PKB_c(p \wedge q) \rightarrow PKB_c p \wedge PKB_c q$.

Corollary 2: $\models PKB_c p \rightarrow PKB_c(p \vee q)$.

Corollary 3: $\models FKB_c p \rightarrow FKB_c(p \wedge q)$.

Corollary 4: $\models FKB_c(p \vee q) \rightarrow FKB_c p \wedge FKB_c q$.

Corollary 5: $\models p \leftrightarrow q \;\Rightarrow\; \models PKB_c p \leftrightarrow PKB_c q$.

Corollary 6: $\models p \leftrightarrow q \;\Rightarrow\; \models FKB_c p \leftrightarrow FKB_c q$.

Corollaries 5 and 6 show that in the definition of a regulation, formally represented by $[C](p_1, \ldots, p_n)$ and $[S](p'_1, \ldots, p'_n)$, each sentence p_i can be replaced by any logically equivalent sentence.

Finally, we have the following properties.

Property 5: $\not\models PKB_c p \wedge PKB_c q \rightarrow PKB_c(p \wedge q)$.

Property 6: $\not\models FKB_c(p \wedge q) \rightarrow (FKB_c p \vee FKB_c q)$.

The intuitive meaning of property 5 is that it may happen that p can be derived from a given permitted mental state of xcr and that q can be derived from another permitted mental state of xcr, whereas there is no permitted mental state of xcr that allows us to derive $p \wedge q$.

For instance, if, in a model M, we have: $C(w) = \{\|a\|, \|a \rightarrow p\|, \|\neg a\|, \|\neg a \rightarrow q\|\}$, then we have $M, w \models PKB_c p$, because there exists a world w′ accessible from w via D_{cr} where we have $M, w' \models OB_{xcr}(a \wedge (a \rightarrow p))$ and we have $M, w \models PKB_c q$, because there exists a world w′ accessible from w via D_{cr} where we have $M, w' \models OB_{xcr}(\neg a \wedge (\neg a \rightarrow p))$. However, we do not have $M, w \models PKB_c(p \wedge q)$ because there exists no world w′ accessible from w via D_{cr} where we have $M, w' \models OB_{xcr}(a \wedge (a \rightarrow p) \wedge \neg a \wedge (\neg a \rightarrow p))$ because $a \wedge \neg a$ is inconsistent.

The intuitive meaning of property 6 is that it is possible that a user u, cleared at level c, is forbidden to know $p \wedge q$, whereas u is neither forbidden to know p nor forbidden to know q. In fact, it may happen that neither p nor q independently allows the derivation of a secret sentence, whereas their conjunction allows to derive it.

For instance, if, in a model M, we have: $S(w) = \{\|a \wedge b\|\}$, then we have $M, w \models FKB_c(a \wedge b)$, whereas we have neither $M, w \models FKB_c(a)$ nor $M, w \models FKB_c(b)$. This corresponds to a "Chinese wall" type of regulation [12], where the conjunction of two sentences may have a classification level strictly greater than each term in the conjunction.

2.1.3 Axiomatics

The axiomatics of the Σ logic is defined by the following axiom schemas and inference rules.

- Axiom schemas of CPC plus Modus Ponens.
- For the modality KB_u we have:

  (K) $KB_u p \wedge KB_u(p \rightarrow q) \rightarrow KB_u q$

  (D) $\neg KB_u(\text{false})$

  $$\frac{\vdash p}{\vdash KB_u p}$$

- For the modality $PKB_c(p)$ we have:

  (P) $[C](p_1, \ldots, p_n) \rightarrow$

  $$\Big(\bigvee_{<i_1,\ldots,i_j> \text{ in } <1,\ldots,n>} \Diamond(p_{i_1} \wedge \ldots \wedge p_{i_j}) \wedge \Box(p_{i_1} \wedge \ldots \wedge p_{i_j} \rightarrow p)\Big) \leftrightarrow PKB_c(p)$$

where $< i_1, \ldots, i_j >$ in $< 1, \ldots, n >$ means that $i_1, \ldots, i_j$ is a subsequence of $1, \ldots, n$[3].

- For the modality $FKB_c(p)$ we have:
  (F) $[S](p_1, \ldots, p_n) \rightarrow$
  $(\bigvee_{i \in [1,n]} \Box(p \rightarrow p_i)) \leftrightarrow FKB_c(p)$

- for the modality $\Box$ we have:
  $\vdash \Box p$ iff $\vdash p$
  $\vdash \neg\Box p$ iff $\nvdash p$

The intuitive meaning of axiom schema (P) (resp. (F)) is that we can infer $PKB_c(p)$ (resp. $FKB_c(p)$) iff p is the consequence of a consistent subset $\{p_{i_1}, \ldots, p_{i_j}\}$ of $\{p_1, \ldots, p_n\}$ (resp. p is the implicant of a sentence p_i in $\{p_1, \ldots, p_n\}$).

Notice that the following sentences are theorems of the Σ logic:
$\Box(p \rightarrow q) \rightarrow (PKB_c p \rightarrow PKB_c q)$, and
$\Box(p \rightarrow q) \rightarrow (FKB_c q \rightarrow FKB_c p)$.

Theorem 1. The Σ logic is sound and complete.

The proof technique of Theorem 1 is based on a canonical model where the set of worlds is the set of maximal consisent sets of sentences of Σ (see [7]).

3 Security Constraints expressed in the Σ logic

This section defines, in the framework of the Σ logic, constraints which characterize acceptable states of a database in regard to a given confidentiality policy. These constraints depend on the choice of a security policy and we do not pretend that they have to be accepted in every context. Nevertheless, they show the expressive power and the flexibility of the Σ logic to formalize security policies.

A given state is defined by a tuple $< C, S, cdb >$, where C represents the set of confidential sentences, S represents the set of secret sentences, and cdb represents the set of sentences that user u has retrieved from the database.

A state $< U, S, cdb >$ is represented in the language L' by the three assumptions: $[C](p_1, \ldots, p_n)$, $[S](p'_1, \ldots, p'_m)$, $KB_u(cdb)$.

We shall use the notations: $R = [C](p_1, \ldots, p_n) \wedge [S](p'_1, \ldots, p'_m)$ and $CDB = KB_u(cdb)$. R represents a confidentiality policy called the "regulation" and CDB will be called the "confidential database".

We first present a security constraint to guarantee the consistency of a confidentiality policy. This constraint says that in the context of a given regulation there is no sentence p such that a confidential user is both permitted to know p and forbidden to know p. The formal representation of the constraint is:

[3]If n = 0, then conjunction should be interpreted to true and disjunction should be interpreted to false.

$$\text{(CR)} \quad \vdash R \rightarrow \neg(PKB_c p \wedge FKB_c p)$$

Property 7: From (CR) we can derive that $C(w) \cap S(w) = \emptyset$ for every world $w \in W$.

The proof of property 7 is trivial. Notice also that the converse of this property is generally not true.

3.1 Deductive Channels

The security constraint which has to be satisfied to prevent deductive channels means that for every sentence p there is no world where a confidential user u knows p and u is not permitted to know p. In other words, the set of sentences $\{R, CDB, KB_u p, \neg PKB_c p\}$ should never be satisfied in the Σ logic. Therefore this constraint is represented by:

$$\text{(SCD)} \quad \vdash R \wedge CDB \rightarrow (KB_u p \rightarrow PKB_c p)$$

The constraint (SCD) is violated for a sentence if we have: $\not\vdash R \wedge CDB \rightarrow (KB_u p \rightarrow PKB_c p)$. This constraint corresponds to the so-called restrictive attitude. Notice that from (CR) and (SCD) we can infer:

$$\vdash R \wedge CDB \rightarrow (KB_u p \rightarrow \neg FKB_c p)$$

In fact, from (CR) we have $\vdash R \rightarrow (PKB_c p \rightarrow \neg FKB_c p)$. This last constraint corresponds to the so-called permissive attitude. Notice that the converse does not hold. In the case where there exist sentences which are neither confidential nor secret, the (SCD) constraint imposes a stronger constraint. This means that the restrictive attitude formalizes a more conservative policy than the policy represented by the permissive attitude. For instance, let us assume that, in a model M we have: $C(w) = \emptyset$, $S(w) = \{\|a\|\}$ and $db = \{\|a\|\}$ and let us consider a confidential user who ask the query $a \vee b$. According to the restrictive attitude, the multilevel database should not answer this query, because in w we have $\neg PKB_c(a \vee b)$, whereas according to the permissive attitude, this user should be provided with true as an answer, because in w we have $\neg FKB_c(a \vee b)$.

The (SCD) constraint can be enforced in two different ways. The first is to allow user u to access only the sentences that do not lead to a violation of (SCD). That is, to control accesses to cdb to avoid violations. The second is to modify the regulation, in other words, to extend the set of sentences in C.

3.2 Abductive Channels

In this section, we investigate situations where abductive channels may exist. Informally, if sentence q cannot be deduced from p, it is said that it can be inferred by abductive reasoning if there exists a sentence h such that q can be deduced from {p,h}. However, this definition has to be refined to remove trivial abductions. There are two kinds of trivial abductions. The first is when {p,h} is an inconsistent set of sentences. In this case, any sentence can be derived from {p,h} and we can derive q even if there is neither a logical link between p and q nor a logical link between h and q. The second kind of trivial abduction is when h allows us to derive q without using p. Here again, there is no logical link between h and p. In formal terms, we say that q can be derived from p by abductive reasoning, using the assumption h, and this fact is denoted by ABD(p,h,q), if we have:

$$(1)\ \vdash h \rightarrow (p \rightarrow q),\ (2)\ \nvdash h \rightarrow \neg p \ \text{ and }\ (3)\ \nvdash h \rightarrow q$$

These properties are represented in the language L' by:

$$\mathrm{ABD}(p,h,q) \stackrel{\text{def}}{=} \Box(h \rightarrow (p \rightarrow q)) \wedge \neg\Box(h \rightarrow \neg p) \wedge \neg\Box(h \rightarrow q)$$

There exists an abductive channel for the sentence q if the set of sentences $\{R, CDB, ABD(p,h,q), KB_u p, \neg PKB_c q\}$ is satisfiable. From a practical point of view, we have to distinguish two cases to prevent the existence of an abductive channel.

In the first case, it is assumed that to know whether sentence h holds, it is not necessary for a user to access the database. In informal terms we say that h is part of common knowledge. In this case, it would be useless to define the regulation in order to have $FKB_c(h)$. Thus, the security constraint to be satisfied is:

$$(\mathrm{SCA}) \quad \vdash R \wedge CDB \wedge ABD(p,h,q) \rightarrow (KB_u p \rightarrow PKB_c q)$$

To enforce this constraint, we have either to restrict cdb in such a way that p cannot be deduced from cdb or to extend set C in order to have $PKB_c(q)$.

In the second case, user u has to access the database to know h. Then the security constraint to be satisfied is:

$$(\mathrm{SCA}') \quad \vdash R \wedge CDB \wedge ABD(p,h,q) \rightarrow (KB_u p \rightarrow (PKB_c q \vee FKB_c h))$$

To enforce the security constraint, we have to control access to cdb in such a way that p cannot be deduced from cdb, to extend C in order to have $PKB_c q$ or to extend S in order to have $FKB_c h$. A practical solution to extend S in this way is to add the sentence $p \rightarrow q$ to S.

For two given sentences p and q, the most general sentence h such that ABD(p,h,q) is $h = p \rightarrow q$, because condition (1) imposes that any h is an implicant of $p \rightarrow q$. Therefore, if $p \rightarrow q$ is in S, we have $FKB_c(p \rightarrow q)$. From properties 2 and 4, for all the implicants h of $p \rightarrow q$ we have $FKB_c h$. This means that if $p \rightarrow q$ is in S, for all the h such that ABD(p,h,q) a confidential user is forbidden to know that the database believes h. In other words, all the abductive channels from p to q are broken.

4 Conclusion

We have presented formal definitions of the concepts of permission and prohibition to know some parts of the content of a database. In the semantics they are interpreted by two accessibility relations D_{cr} and D_{cp} that indepedently characterize ideal worlds in regard to a restrictive and a permissive attitude. This enables us to consider that regulation is not necessarily complete, in the sense that some sentences are neither permitted nor prohibited. These relations are respectively combined with accessibility relations R_{xcr} and R_{xcp} in order to characterize mental states of abstract agents xcr and xcp that are compatible with the two different attitudes.

In the axiomatics, the combination of D_{cr} and R_{xcr} is represented by the modality PKB_c and the combination of D_{cp} and R_{xcp} is represented by the modality FKB_c.

It is worth noting that the logic allows us to represent the fact that we have a complete knowledge of the regulation even if the regulation itself is not complete, in the sense that if, in the context of a given regulation, we cannot infer $PKB_c p$ (resp. $FKB_c p$), then we can infer $\neg PKB_c p$ (resp. $\neg FKB_c p$).

We have also shown practical applications of the logic to define constraints to guarantee the consistency of a regulation or to prevent the existence of inference channels.

Further extensions of this work might be to extend the expressive power of the language to have two independent representations of the deontic concepts and doxastic concepts, for instance to represent facts of the form: $P_c(KB_c(p) \vee KB_c(q))$. Another interesting extension might be to apply deontic modalities to deontic modalities themselves, for instance to represent facts like: "it is permitted to a user cleared at the secret level to know that a user cleared at the confidential level is forbidden to know p".

Acknowledgement

This work was partially supported by the ESPRIT Basic Research Action MEDLAR 2. We are very grateful to Jose Cramo for his valuable comments.

References

[1] Thuraisingham B., Ford W., Collins M., and O'Keefe J. Design and implementation of a database inference controller. *Data & Knowledge Engineering*, 11(3), December 1993.

[2] Hinke T. H. Inference Aggregation Detection in Database Management Systems. In *IEEE Symposium on Security and Privacy*, Oakland, 1988.

[3] Garvey T. D., Lunt T. F., and Stickel M. E. Abductive and Approximate Reasoning Models for Characterizing Inference Channels. In *Proc. of the computer security foundations workshop*, Franconia, 1991.

[4] Garvey T. D., Lunt T. F., Qian X., and Stickel M. Toward a Tool to Detect and Eliminate Inference Problems in the Design of Multilevel Databases. In *Proc. of the Sixth IFIP WG 11.3 Working Conference on Database Security*, Vancouver, 1992.

[5] Chen J. The Generalized Logic of Only Knowing (GOL) that Covers the Notion of Epistemic Specifications. In Z. Ras and M. Zemankova, editors, *Proceedings of the 8th International Symposium, ISMIS' 94*, volume 869 of *Lecture Notes in Artificial Intelligence*, Charlotte, North Caralina, 1994. Springer Verlag.

[6] Cuppens F. A Logical Analysis of Authorized and Prohibited Information Flows. In *IEEE Symposium on Security and Privacy*, Oakland, 1993.

[7] Chellas B. F. *Modal Logic: An introduction*. Cambridge University Press, 1988.

[8] Cuppens F. Roles and Deontic Logic. In A. J. I. Jones and M. Sergot, editors, *Second International Workshop on Deontic Logic in Computer Science*, Oslo, Norway, 1994.

[9] Pörn I. *Action Theory and Social Science; Some Formal Models*, volume 120 of *Synthese Library*. D. Reidel, Dordrecht, 1977.

[10] Carmo J. and Jones A. J. I. Deontic database constraints and the characterization of recovery. In A. J. I. Jones and M. Sergot, editors, *Second International Workshop on Deontic Logic in Computer Science*, Oslo, Norway, 1994.

[11] Jones A. J. I. and Porn I. Ideality, Sub-ideality and Deontic Logic. *Synthese*, 65, 1985.

[12] Brewer D. and Nash M. The Chinese wall security policy. In *IEEE Symposium on Security and Privacy*, Oakland, 1989.

A Modal Approach to Intentions, Commitments and Obligations: Intention plus Commitment yields Obligation

F.Dignum[1] J.-J.Ch.Meyer[2] R.J.Wieringa[3] R. Kuiper[4]

1. Fac. of Maths. & Comp. Sc., Eindhoven University of Technology, Eindhoven, The Netherlands (dignum@win.tue.nl).
2. Dept. of Computer Science, Utrecht University, Utrecht, The Netherlands (email: jj@cs.ruu.nl).
3. Fac. of Maths. & Comp. Sc., Free University, Amsterdam, The Netherlands (email: roelw@cs.vu.nl). The research of J.-J.Ch.Meyer and R.J.Wieringa is partially supported by ESPRIT BRA 8319 ModelAge.
4. Fac. of Maths. & Comp. Sc., Eindhoven University of Technology, Eindhoven, The Netherlands (email: wsinruur@win.tue.nl). The research of R. Kuiper is partially supported by ESPRIT BRA P 6021 REACT.

Abstract

In this paper we introduce some new operators into our framework that make it possible to reason about decisions and commitments to do actions. In our framework, a decision leads to an intention to do an action. The decision in itself does not change the state of the world, but only the relation to possible future worls. A commitment to actually perform the intended action changes the deontic state of the world such that the intended action becomes obligated. Of course, the obligated action may never actually occur. In our semantic structure, we use static (ought-to-be) and dynamic (ought-to-do) obligation operators. The static operator resembles the classical conception of obligation as truth in ideal worlds, except that it takes the current state as well as the past history of the world into account. This is necessary because it allows us to compare the way a state is actually reached with the way we committed ourselves to reach it. We show that some situations that could formerly not be expressed easily in deontic logic can be described in a natural way using the extended logic described in this paper.

1 Introduction

In the fields of law as well of computer science there are frequently occurring situations in which intentions and obligations to perform actions arise depending on past events. In law, the main purpose of a court trial is to discover what was the *actual* sequence of events that led to a situation in which a crime had been committed. Usually there are many ways that the situation could have been reached. The task of the judge (and the jury) is to find out what the actual way was. On the basis of the findings a judgement is formed, which leads to an intended (i.e. preferred) action for the future. The intended action is either some punishment or a dismissal of the defendant. Sometimes the decision can still change a few times either through a ruling of the judge or through a ruling by a higher court. When a final judgement has been given and thus a commitment has been made there arises an obligation (for the public prosecutor) to fulfil the judgement.

In database systems the *transaction* concept is used to maintain the integrity of the database [1]. A transaction is a conceptually atomic state transition that, at any moment, either has occurred or has not yet occurred. A transaction is implemented as a process that is not atomic but guarantees atomicity by a "rollback/commit" protocol. (The database "commit" is a different concept from the commitment studied in this paper.) Before a transaction attempt is "committed", it can be rolled back to restore the state as if no transaction has occurred. After the transaction is "committed", it cannot be rolled back; however, a *compensating transaction* can be performed to undo some of the effects of the transaction. For example, a money transfer is a transaction, and it can be rolled back if it is not yet "committed". The history of the bank account will then not contain this money transfer. If after the transaction is "committed", it turns out that it should be undone, then the obligation arises to inform the account owner of this and to perform a compensating transfer, which returns the money (and lost interest). In this case, however, the account has both transactions in its history.

The purpose of this paper is to show how these examples can be modelled in a simple and elegant way, using deontic logic with actions, intentions and (a very limited set of) temporal operators. In [2, 4], the deontic operators are treated as a type of temporal operators. It should be noted that our approach is completely different. We do not define the deontic operators in terms of temporal objects, but use temporal operators in addition to the deontic aspects. Furthermore, the view taken in this paper is that the past is fixed, but that the future is open. Therefore no nexttime operator is present, as this would allow fixing which action occurs next, and also only pasttime and no future time temporal operators are employed. In this paper we consider *personal* or *private* intentions that lead to obligations via *individual* commitment. Thus we are not so much concerned with a theory of intention/commitment in general; rather we present an account of how an individual (agent) decides to do certain actions out of a range of possibilities open to him/her, yielding (personal) intentions to perform these actions. These intended actions can then become committed to by a (personal) commit action, which results in an (personal) obligation. To keep things as simple as possible, in this paper we have abstracted away from the notion of agency and the way other agents may be involved in making these choices, for which the notions of internal/external nondeterminism seem relevant. We also avoided to define intentions in terms of goals to be reached and epistemic concepts such as the agent's knowledge and belief, as in [3], and take intention to actions as a primitive modality in out approach. Further investigation of these important matters is deferred to a later occasion.

In the next section we will introduce the logic that we will use. In section 3 we will extend the logic with two types of meta-actions that express decisions and commitments. In section 4 we will show how the above examples can be modelled in our logic. Section 5 is used to draw some conclusions.

2 A logic of actions and norms

We now proceed with the definition of a set of *formulas* with which we can describe the behaviour of (interpreted) actions. This language is a variant of *dynamic logic* ([5]), and was first used for this purpose in [6]. In the present paper we add a few "new" formulas to this language. They are the ones defined in points (5) and (6) below. The formulas defined in (5) all involve some type of temporal operations on actions. The formulas defined in (6) define the "classical" deontic formulas as introduced by von Wright in [7, 8].

We assume a fixed set *Prop* of atomic propositions and a set *Act* of action expressions. The set *Form* of formulas is then the smallest set closed under:

(1). $Prop \subseteq Form$

(2). $\phi_1, \phi_2 \in Form \Longrightarrow \phi_1 \wedge \phi_2 \in Form$

(3). $\phi \in Form \Longrightarrow \neg\phi \in Form$

(4). $\alpha \in Act, \phi \in Form \Longrightarrow [\alpha]\phi \in Form$

(5). $\alpha \in Act \Longrightarrow PAST(\alpha), PREV(\alpha), INT(\alpha) \in Form$

(6). $\phi \in Form \Longrightarrow O(\phi) \in Form$

Note: Other propositional connectives such as $\vee$ and $\rightarrow$ are assumed to be introduced as the usual abbreviations. Also the special proposition $false$ is introduced as the abbreviation of $p \wedge \neg p$ for some $p \in Prop$. The informal meaning of $[\alpha]\phi$ is "doing α necessarily leads to a state where ϕ obtains". The meaning of $PAST(\alpha)$ is that α has actually been performed in the past. The meaning of $PREV(\alpha)$ is "the present state is actually reached by performing α". Note that we make a difference between the possible ways that the state can be reached and the way it actually *is* reached. In our semantics, we will therefore consider state-history pairs. The formula $INT(\alpha)$ intuitively means that we prefer or intend to perform the action expressed by α next. The informal meaning of $O(\phi)$ is that ϕ should be the case in the present state. Because we include histories in our semantic structure, we can express that the current state should, ideally, have been reached by another history. We introduce the other deontic operators using the usual abbreviations:

- $F(\phi)$ abbreviates $O(\neg\phi)$
- $P(\phi)$ abbreviates $\neg F(\phi)$

The semantics of the formulas in *Form* is given in two stages. First we will give the syntax and semantics of the action expressions, which we will subsequently use in section 2.2 to define the semantics of the formulas.

2.1 Action expressions and their semantics

We start out by giving a definition of *action expressions*, which we shall typically denote α, possibly with subscripts. To this end we assume a set At of *atomic action expressions* that are typically denoted by $\underline{a}, \underline{b}, \ldots$. Furthermore, we assume special action expressions **any** and **fail** denoting "don't care what happens" and "failure", respectively.

Definition 1 The set Act of action expressions is given as the smallest set closed under:

(i). $At \cup \{\textbf{any}, \textbf{fail}\} \subseteq Act$

(ii). $\alpha_1, \alpha_2 \in Act \Longrightarrow \alpha_1 + \alpha_2 \in Act$

(iii). $\alpha_1, \alpha_2 \in Act \Longrightarrow \alpha_1 \& \alpha_2 \in Act$

(iv). $\alpha \in Act \Longrightarrow \overline{\alpha} \in Act$

We further use the following terminology: atomic action expressions $\underline{a} \in Act$, as well as their negations $\overline{\underline{a}}$ ($\underline{a} \in At$) are called *action literals*. An action expression that is the conjunction (using the &-connective) of one or more action literals is called an *action term*. So, for example, $\underline{a} \& \overline{\underline{b}} \& \underline{c}$, for $\underline{a}, \underline{b}, \underline{c} \in At$, is an action term.

For simplicity, we do *not* have a notion of sequences of actions in the action expressions in this paper. We discuss the action sequence operator elsewhere [9, 10].

The semantics of action expressions is given in two stages. First we define an algebra of uninterpreted actions (called a *uniform* semantics elsewhere [11]), which allows us to interpret equalities between action expressions without taking their effect into account. In the algebraic semantics, each action expression will be interpreted as a choice over possible steps. Next, we give a state-transition semantics of action expressions in which we define the effect of steps on the state of the world.

Algebraic action semantics. With every atomic action expression $\underline{a} \in Act$, we associate an event a in a given class $\mathcal{A}$ of events, with typical elements $a, b, c....$ Different atomic action expressions are associated with different events in $\mathcal{A}$. Events are the semantical entities on which we shall base our interpretation of action expressions. We further assume a special event δ, which is not an element of $\mathcal{A}$, called failure (comparable to deadlock in process algebra ([12]). The relation between an action expression $\underline{a} \in Act$ and the associated event $a \in \mathcal{A}$ is more involved than just interpreting $\underline{a}$ as a. We shall interpret atomic action expressions $\underline{a} \in Act$ in a more sophisticated way, which we call "open": the meaning of an atomic action expression $\underline{a} \in Act$ will be the event $a \in \mathcal{A}$ corresponding with it, in combination with any other subset of the events in $\mathcal{A}$. Thus $\underline{a}$ expresses that a occurs, but it leaves open which other events occur simultaneously (in the same step) with a. The intuitive motivation for this is that if we say that an event a occurs, we do not mean that nothing else occurs in the world.

Definition 2

1. The set $\{\delta\}$ is a step.
2. Every non-empty finite subset of $\mathcal{A}$ is a step. The powerset of non-empty finite subsets of $\mathcal{A}$ will be denoted by $\mathcal{P}^+(\mathcal{A})$.
3. The domain $\mathcal{D}$ for our model for action expressions from *Act* is the collection of sets of steps, $\mathcal{D}=\wp(\wp^+(\mathcal{A}) \cup \{[\delta]\})$. An element of $\mathcal{D}$ is called a *choice set.*

Notation: In concrete cases we write the sets indicating steps with square brackets, in order to distinguish them easily from other sets that we will use. So, the step consisting of δ is written as $[\delta]$ and the step consisting of the events a and b is written as $\begin{bmatrix} a \\ b \end{bmatrix}$.

The above definition prevents the simultaneous execution of the special event δ with other events, because it is not in $\mathcal{A}$. This is necessary, because it is not possible to perform an event and at the same time have a deadlock. Below, we interpret action expressions as choice sets.

Definition 3 For $T, T' \in \mathcal{D}$:

$$T \mathbin{\dot{\cap}} T' = \begin{cases} T \cap T' & \text{if } T \cap T' \neq \emptyset \\ \{[\delta]\} & \text{otherwise} \end{cases}$$

Definition 4 Let T be a set of steps. Then

$$T^\delta = \begin{cases} T \backslash \{[\delta]\} & \text{if } \exists S \in T : S \neq [\delta] \\ \{[\delta]\} & \text{otherwise} \end{cases}$$

The operator T^δ is closely related to what is called "failure removal" in [11]. The idea is that failure is avoided when possible, i.e. when there is a non-failing alternative. In [13], this is called *angelic* nondeterminism. We can define the semantic operators on $\mathcal{D}$. For the parallel operator & we use a set-intersection $\dot{\cap}$, which is almost the same as the normal set-intersection, except that in case no joint action expression at all is present, deadlock ensues.

The semantical counterpart of the choice operator is defined as follows:

Definition 5 For $T, T' \in \mathcal{D}$:

$$T \mathbin{\dot{\cup}} T' = (T \cup T')^\delta$$

The above definition states that the choice between two sets of steps is the union of those two sets minus $[\delta]$, unless the union does not contain anything else. Finally, we define the semantic counterpart of the negation operator.

Definition 6 The definition of "$\sim$" is given as follows:

1. For a step S,

$$S^{\sim} = \wp^{+}(\mathcal{A})\backslash\{S\}$$

2. For a non-empty set $T \in \mathcal{D}$

$$T^{\sim} = \Cap_{S \in T} S^{\sim}$$

That is, for a step the negation just yields the set-theoretic complement of $\{S\}$ with respect to $\wp^{+}(\mathcal{A})$. The negation of a set of steps T is the set-theoretic complement of T with respect to $\wp(\wp^{+}(\mathcal{A}))$ if $T \neq \wp^{+}(\mathcal{A})$. Otherwise it is $\{[\delta]\}$. We can now give the algebraic semantics of action expressions:

Definition 7 The semantic function $\llbracket\rrbracket \in Act \rightarrow \mathcal{D}$ is given by:

$$\begin{array}{l}
\llbracket \underline{a} \rrbracket = \{S \in \wp^{+}(\mathcal{A}) | a \in S\} \\
\llbracket \alpha_1 + \alpha_2 \rrbracket = \llbracket \alpha_1 \rrbracket \Cup \llbracket \alpha_2 \rrbracket \\
\llbracket \alpha_1 \& \alpha_2 \rrbracket = \llbracket \alpha_1 \rrbracket \Cap \llbracket \alpha_2 \rrbracket \\
\llbracket \overline{\alpha} \rrbracket = \llbracket \alpha \rrbracket^{\sim} \\
\llbracket \mathbf{fail} \rrbracket = \{[\delta]\} \\
\llbracket \mathbf{any} \rrbracket = \wp^{+}(\mathcal{A})
\end{array}$$

The first clause of the above definition expresses that the meaning of the action expression $\underline{a}$ is exactly as we have described informally before: it is the set of steps that contain the event a, representing a choice between all (simultaneous) performances of sets of events which at least contain the event a, so that the performance of a is guaranteed but also other events may happen simultaneously. The meaning of the action expression **fail** is comparable to a deadlock. The only event that can be performed is δ. The action expression **any** is the complement of **fail**. It stands for a choice of any possible combination of events.

The following properties can be easily proven and will be used later on:

Proposition 1

1. $\llbracket \alpha \rrbracket \cap \llbracket \overline{\alpha} \rrbracket = \emptyset$

2. $\llbracket \overline{\overline{\alpha}} \rrbracket = \llbracket \alpha \rrbracket$

3. $\llbracket \overline{\mathbf{fail}} \rrbracket = \llbracket \mathbf{any} \rrbracket$

Finally we define equality and implication between action expressions.

Definition 8 Action expressions α_1 and α_2 are *equal*, written $\alpha_1 =_{\mathcal{D}} \alpha_2$, iff $\llbracket \alpha_1 \rrbracket = \llbracket \alpha_2 \rrbracket$. α_1 *involves* or *implies* α_2, written $\alpha_1 > \alpha_2$, iff $\llbracket \alpha_1 \rrbracket \subseteq \llbracket \alpha_2 \rrbracket$.

State-transition action semantics. To get a state-transition semantics, we postulate what effects events have in terms of state transformations (we do this relative to a set Σ of states). We assume that there is a function $eff : \mathcal{A} \rightarrow (\Sigma \rightarrow \Sigma)$, such that $eff(a)$ is a function from states to states. (For simplicity, we assume events to be deterministic. Elsewhere, we show how nondeterministic events can be incorporated [6, 14].) Two events are called *compatible* if their joint effect is independent of the order in which they occur.

Definition 9 Let $S = [a_1, ..., a_n] \subseteq \mathcal{A}$ be a step consisting of pairwise compatible events. The accessibility relation $R_S \subseteq \Sigma \times \Sigma$ is defined as follows:

$$R_S(\sigma, \sigma') \Longleftrightarrow_{def} eff(a_1)(...(eff(a_n)...)(\sigma) = \sigma'$$

We require that in every state, we can do at least one step: $\forall \sigma \in \Sigma \exists S \subseteq \mathcal{A} \exists \sigma' \in \Sigma : R_S(\sigma, \sigma')$

2.2 Semantics of formulas

Having defined the semantics of the action expressions within the formulas, we can now give the semantics of formulas in *Form* by means of the notion of a Kripke structure $\mathcal{M} = (\mathcal{A}, \Sigma, \pi, R_{\mathcal{A}}, R_O, I)$.
Σ is a set of states (worlds).
$\mathcal{A}$ is a finite set of events.
π is a truth assignment function to the atomic propositions relative to a state: π is a function $\Sigma \rightarrow (Prop \rightarrow \{tt, ff\})$, where tt and ff denote truth and falsehood, respectively. Thus, for $p \in Prop$, $\pi(\sigma)(p) = tt$ means that the atomic proposition p is true in state σ.

The accessibility relation $R_{\mathcal{A}}$ specifies how actions can change states. The relation $R_{\mathcal{A}}$ is defined as follows: $R_{\mathcal{A}} = \{R_S | S \subseteq \mathcal{A}\}$, reflecting that R_S is the relevant entity. We write, slightly abusing notation,

$$R_{\mathcal{A}}(\sigma, \sigma') \Longleftrightarrow \exists S \subseteq \mathcal{A} : R_S(\sigma, \sigma')$$

The relation R_O relates state/history pairs (σ,γ) to state/history pairs (σ',γ'). The history of a state is expressed by a *trace* of the actions that went before. In process algebra, temporal logic and semantics of parallel programs traces are widely used to record what actions have taken place. Traces are defined inductively as follows:

Definition 10 1. ϵ is a (empty) trace.

2. if γ is a trace and $S \subseteq \mathcal{A}$ then $\gamma \circ S$ is a trace.
   Furthermore $\gamma \circ \epsilon = \epsilon \circ \gamma = \gamma$. We use $\mathcal{A}^*$ to denote the set of all traces.

We would like to ensure that it is possible for a state to be reached by its history. Therefore we introduce the following relation between states and traces relative to a model $\mathcal{M}$:

Definition 11 For each model $\mathcal{M} = (\mathcal{A}, \Sigma, \pi, R_{\mathcal{A}}, R_O, I)$ and pair (σ, γ) with $\sigma \in \Sigma$ and $\gamma \in \mathcal{A}^*$ $(\sigma, \gamma) \in Comp(\mathcal{M})$ iff

1. $\gamma = \epsilon$ or

2. $\gamma = \gamma' \circ S \wedge \exists \sigma' \in \Sigma[R_S(\sigma', \sigma) \wedge (\sigma', \gamma') \in Comp(\mathcal{M})]$

From now on we assume that all pairs $(\sigma, \gamma) \in Comp(\mathcal{M})$ unless it is stated otherwise.

R_O is the deontic relation that with respect to a state σ reached by history γ indicates the ideal situation consisting of state σ' and history γ'. R_O resembles the classical deontic relation in modal interpretations, except that we do not consider only states, but pairs of states and traces. We assume the relation R_O to be serial. I.e. for every world σ and trace γ there exists at least one pair (σ', γ') such that $R_O((\sigma, \gamma)(\sigma', \gamma'))$ holds.

The last accessibility relation I specifies which of the actions that are possible in a certain state are "intended" to be performed. I.e. it indicates the actions that are not only possible but also intended to be done. I is defined in terms of the relation I_S.
The relation I_S can be given as primitive, but can also be given a deontic interpretation by relating it to deontically preferable actions. For now we will assume this relation to be primitive. In section 5 we will show how it can be related to a deontic ordering on states as defined in [15]. If $I_S(\sigma, \sigma')$ holds then the step S is intended to be done in σ and will lead to σ'. Because we intend this relation to be a subset of the accessibility relation for states, we require that $\forall S \subseteq \mathcal{A}, \sigma, \sigma' \in \Sigma : I_S(\sigma, \sigma') \Longrightarrow R_S(\sigma, \sigma')$.
Like $R_{\mathcal{A}}$, I is defined as $I = \{I_S | S \subseteq \mathcal{A}\}$. Again we write

$$I(\sigma, \sigma') \Longleftrightarrow \exists S \subseteq \mathcal{A} : I_S(\sigma, \sigma')$$

We now give the interpretation of formulas in *Form* in Kripke structures. We interpret formulas with respect to a structure $\mathcal{M}$ and a pair $(\sigma, \gamma) \in Comp(\mathcal{M})$

Definition 12 Given $\mathcal{M} = (\mathcal{A}, \Sigma, \pi, R_{\mathcal{A}}, R_O, I)$ as above and $(\sigma, \gamma) \in Comp(\mathcal{M})$, we define:

1. $(\mathcal{M}, (\sigma, \gamma)) \models p \Longleftrightarrow \pi(\sigma)(p) = tt$ (for $p \in Prop$)

2. $(\mathcal{M}, (\sigma, \gamma)) \models \phi_1 \wedge \phi_2 \Longleftrightarrow (\mathcal{M}, (\sigma, \gamma)) \models \phi_1$ and $(\mathcal{M}, (\sigma, \gamma)) \models \phi_2$

3. $(\mathcal{M}, (\sigma, \gamma)) \models \neg\phi \Longleftrightarrow$ not $(\mathcal{M}, (\sigma, \gamma)) \models \phi$

4. $(\mathcal{M}, (\sigma, \gamma)) \models [\alpha]\phi \Longleftrightarrow \forall S \in [\![\alpha]\!] \forall \sigma' \in \Sigma[R_S(\sigma, \sigma') \Rightarrow (\mathcal{M}, (\sigma', \gamma \circ S)) \models \phi]$

5. $(\mathcal{M}, (\sigma, \gamma)) \models O(\phi) \Longleftrightarrow \forall (\sigma', \gamma') \in Comp(\mathcal{M})[R_O((\sigma, \gamma), (\sigma', \gamma')) \Rightarrow (\mathcal{M}, (\sigma', \gamma')) \models \phi]$

6. $(\mathcal{M}, (\sigma, \gamma)) \models PREV(\alpha) \Longleftrightarrow \exists S \in [\![\alpha]\!], \gamma' \in \mathcal{A}^*[\gamma = \gamma' \circ S]$

7. $(\mathcal{M}, (\sigma, \gamma)) \models PAST(\alpha) \Longleftrightarrow \exists \sigma' \in \Sigma \exists \gamma', \gamma'' \in \mathcal{A}^*[\gamma = \gamma' \circ \gamma'' \wedge (\mathcal{M}, (\sigma', \gamma')) \models PREV(\alpha)]$

8. $(\mathcal{M}, (\sigma, \gamma)) \models INT(\alpha) \Longleftrightarrow \exists S \in [\![\alpha]\!], \sigma' \in \Sigma[I_S(\sigma, \sigma')]$

9. ϕ is *valid* w.r.t. model $\mathcal{M} = (\mathcal{A}, \Sigma, \pi, R_{\mathcal{A}}, R_O, I)$, notation $\mathcal{M} \models \phi$, if $(\mathcal{M}, (\sigma, \gamma)) \models \phi$ for all $\sigma \in \Sigma$ and $\gamma \in \mathcal{A}^*$.

10. ϕ is *valid*, notation $\models \phi$, if ϕ is valid w.r.t. all models $\mathcal{M}$ of the form considered above.

The first four definitions are quite standard and we will not explain them any further here. The definition of the static obligation involves both the state and the trace (and *not* just the state). In this way, we can express that the circumstances described by $PREV(\alpha)$, for example, are obligatory. For example, it might be obligatory to have just done the action indicated by α. This means that the history (i.e. the trace) of an ideal world might differ from the history of the present world. We will use this feature to define obligations on actions shortly.

It should be noted that using the semantic definition of $[\mathbf{any}]\phi$ we can express the usual temporal next-time operators over static formulas as given in e.g. [16]. Points (6), (7) and (8) define extra temporal operators reaching over action expressions! (6) and (7) are quite obvious. The definition of the semantics of INT indicates that an action α is intended to be performed (or preferred) whenever there is *some* way to perform α. Intuitively this seems evident. For example, if I intend to fly to Lisbon, then I intend to go to Lisbon. Conversely, if I intend to go to Lisbon, I do not intend to do this in every possible way (e.g. by walking) but just that there is one way in which I intend to do this.

Using the above definitions we can now introduce the deontic operators over actions as follows:

- $O(\alpha) \equiv [\mathbf{any}]O(PREV(\alpha))$
- $F(\alpha) \equiv O(\overline{\alpha})$
- $P(\alpha) \equiv \neg F(\alpha)$

So, α is obligated if, whatever I do now, it will be true immediately afterwards that I was just previously obligated to do α. This means that if I do $\overline{\alpha}$, I reach a state where a violation occurs, indicated by the fact that in that state both $O(PREV(\alpha))$ and $PREV(\overline{\alpha})$ hold true.
Note that by definition the following formula is valid for all actions $\alpha \in Act$:

$$[\alpha]PREV(\alpha)$$

Therefore

$$O(\alpha) \Longrightarrow [\overline{\alpha}](\neg PREV(\alpha \wedge O(PREV(\alpha)))$$

In earlier accounts this implication was used to define the obligation as:

$$O(\alpha) \equiv [\overline{\alpha}]Violation$$

using a kind of Andersonian reduction [17]. For an extensive account of the adequacy of these abbreviations we refer to [6, 18, 19]; here it suffices to note that e.g. the claim that it is forbidden to do the action denoted by α is equated with the fact that the resulting state is not ideal (i.e. is in violation), because $PREV(\alpha)$ holds in that state while $PREV(\overline{\alpha})$ holds in the ideal states alternative to that state.

An important difference from this earlier account is that now we do *not* have:

$$O(\alpha) \Longleftarrow [\overline{\alpha}](PREV(\overline{\alpha}) \wedge O(PREV(\alpha)))$$

A further difference is that whereas in the earlier account we did not have the D-axiom

$$\models \neg(O(\alpha) \wedge O(\overline{\alpha}))$$

for obligations (automatically), we do inherit this property now from the static deontic operator.

The following validities follow (easily) from the above definition of the semantics of formulas in *Form*:

Proposition 2

1. $\models [\alpha](\phi \rightarrow \psi) \rightarrow ([\alpha]\phi \rightarrow [\alpha]\psi)$
2. $\models \phi \Longrightarrow \models [\alpha]\phi$
3. $\models \neg[\mathbf{any}]false$
4. $\models O(\phi \rightarrow \psi) \rightarrow (O(\phi) \rightarrow O(\psi))$
5. $\models \phi \Longrightarrow \models O(\phi)$
6. $\models \neg O(false)$
7. $\models \neg O(\mathbf{fail})$
8. $\models O(\mathbf{any})$
9. $\models O(\overline{\alpha}) \rightarrow \neg O(\alpha)$
10. $\models O(\alpha_1) \vee O(\alpha_2) \rightarrow O(\alpha_1 + \alpha_2)$
11. $\models (O(\alpha_1) \wedge O(\alpha_2)) \leftrightarrow O(\alpha_1 \& \alpha_2)$
12. $\models [\alpha]PREV(\alpha)$
13. $\models PREV(\overline{\alpha}) \leftrightarrow \neg PREV(\alpha)$
14. $\models PREV(\alpha_1) \vee PREV(\alpha_2) \leftrightarrow PREV(\alpha_1 + \alpha_2)$
15. $\models (PREV(\alpha_1) \wedge PREV(\alpha_2)) \leftrightarrow PREV(\alpha_1 \& \alpha_2)$
16. $\models \neg PREV(\mathbf{fail})$
17. $\models PREV(\mathbf{any})$

18. $\models PREV(\alpha) \rightarrow PREV(\beta)$ *if* $\alpha > \beta$

19. $\models PAST(\alpha) \rightarrow PAST(\beta)$ *if* $\alpha > \beta$

20. $\models PREV(\alpha) \rightarrow PAST(\alpha)$

21. $\models [\alpha]\phi \Longleftrightarrow \models PREV(\alpha) \rightarrow \phi$

22. $\models INT(\alpha) \rightarrow INT(\beta)$ *if* $\alpha > \beta$

23. $\models INT(\alpha_1 \& \alpha_2) \rightarrow INT(\alpha_1) \wedge INT(\alpha_2)$

24. $\models INT(\alpha_1 + \alpha_2) \leftrightarrow (INT(\alpha_1) \vee INT(\alpha_2))$

25. $\models \neg INT(\mathbf{fail})$

Note that we do *not* have $\models INT(\alpha_1 \& \alpha_2) \leftarrow INT(\alpha_1) \wedge INT(\alpha_2)$. For example, if I intend to drink a beer and have a brandy, I do not intend to gulp them down simultaneously. Note that as a special case we have also:

$$\not\models INT(\alpha) \wedge INT(\overline{\alpha}) \rightarrow INT(\alpha \& \overline{\alpha}).$$

3 Decisions and Commitments

In the previous section we introduced a deontic logic that allows for the expression of both static as well as dynamic deontic formulas. It also contains several "temporal" operators that make it possible to state that an action has actually been performed or is intended to be performed next.

In the present section we will extend the logic with two types of meta-actions, the *DECIDE* and the *COMMIT* actions. The *DECIDE* actions establish which actions are intended to be performed in a certain state. Thus, it is possible to express within the logic the decision which action is intended to be performed next. These type of actions have to be distinguished from the other actions, because they change the accessibility relations for intentions and obligations. We first define the syntax of the meta-actions:

Definition 13

1. $\alpha \in Act \Longrightarrow DECIDE(\alpha) \in MAct$ and $COMMIT(\alpha) \in MAct$
2. $\mu \in MAct, \phi \in Form \Longrightarrow [\mu]\phi \in Form$

Next we define a set of intentions on the semantic domain that will serve as the semantics of the *DECIDE* function. On the basis of a state σ, an action expression α and the relation I it indicates all possible intention relations I' such that from state σ only the actions that can be done in parallel with the action expressed by α are intended to be performed in I'.

Definition 14

$$\begin{aligned} dec(\alpha,\sigma,I) = \{I' \mid\ & \forall\sigma' \neq \sigma[I(\sigma',\sigma'') \iff I'(\sigma',\sigma'')] \wedge \\ & \exists S \in [\![\alpha]\!] \exists\sigma' I'_S(\sigma,\sigma') \wedge \\ & \forall S \notin [\![\alpha]\!] \neg\exists\sigma' I'_S(\sigma,\sigma')\} \end{aligned}$$

So, the relation I does not change for any state except σ. In σ it is changed in a way that the action expressed by α is intended to be performed, while the action expressed by $\overline{\alpha}$ cannot be intended to be performed. The decision to (intend to) perform the action expressed by α does not fix the way the action will be performed. Because the function *dec* establishes a relation I'_S for all $S \in [\![\alpha]\!]$ from the state σ to some state σ' this means that every possible way to perform the action expressed by α is intended. This conforms to the idea that I may e.g. decide to go on holiday next week, leaving open where I will go to, how I will go there, etc. But if I intend to go on holiday next week I cannot, of course, also intend to stay at home and work next week. The semantics of $DECIDE$ is now defined as follows:

Definition 15

$$\begin{aligned} & ((\mathcal{A},\Sigma,\pi,R_{\mathcal{A}},R_O,I),(\sigma,\gamma)) \models [DECIDE(\alpha)]\phi \\ & \iff \\ & \forall I' \in dec(\alpha,\sigma,I) : ((\mathcal{A},\Sigma,\pi,R_{\mathcal{A}},R_O,I',(\sigma,\gamma)) \models \phi \end{aligned}$$

The fact that a certain action α is intended does not have any influence on the course of events. For α to influence (in a deontic way) the course of events, the agent must be *committed* to α. The act of commitment has been discussed in the literature at various places already. Notably, in the theory of speech acts it is one of the basic communication types (see [20, 21]). In [22] the commitment is related to a goal of an autonomous agent. In this paper we give the commitment a deontic connotation by defining the commitment to perform α as a (meta-)action with the result that it becomes obligatory to perform α. Moreover we restrict the commitment to only those actions that are already intended. The following definition gives a function that defines the exact changes of the model brought about by a commitment. It will serve as the semantics of the $COMMIT$ actions.

Definition 16
$com(\alpha,(\sigma,\gamma),I,R_O) = R'_O \iff$

(1) $[\exists S \in [\![\alpha]\!], \sigma' : I_S(\sigma,\sigma')] \implies$

(1a) $[\forall\sigma'\forall S \subseteq \mathcal{A} : [R_S(\sigma,\sigma') \implies$
$\forall(\sigma'',\gamma'')(R'_O((\sigma',\gamma \circ S),(\sigma'',\gamma'')) \iff (R_O((\sigma',\gamma \circ S),(\sigma'',\gamma'')) \wedge \exists T \in$
$[\![\alpha]\!] \exists\gamma' \in \mathcal{A}^* :\ \gamma'' = \gamma' \circ T)] \wedge$

(1b) $\forall\sigma'[(\neg\exists S \subseteq \mathcal{A} : R_S(\sigma,\sigma') \implies$
$\forall(\sigma'',\gamma'')R'_O((\sigma',\gamma'),(\sigma'',\gamma'')) = R_O((\sigma',\gamma'),(\sigma'',\gamma''))]] \wedge$

(2) $[\neg\exists S \in [\![\alpha]\!], \sigma' : I_S(\sigma, \sigma')] \Longrightarrow$
$\forall(\sigma', \gamma')(\sigma'', \gamma'') R'_O((\sigma', \gamma'), (\sigma'', \gamma'')) = R_O((\sigma', \gamma'), (\sigma'', \gamma''))$

Part (2) states that the relation R_O does not change if α does not denote an intended action. I.e. the commitment only has effect if the action in question is already intended.
Part (1) of the definition covers the case that α denotes an intended action. Even if α is intended only a limited part of the relation R_O changes. In part (1b) it is stated that the relation R_O does not change for any state σ' that is not reachable (by performing some action) from σ.
Part (1a) states the actual changes of the relation R_O. For all pairs of states and traces $(\sigma', \gamma \circ S)$ that can be reached from (σ, γ) (by performing any possible step S) it holds that for all their ideal alternatives the last step of their trace is a step from α (i.e. in σ' it is obligated that $PREV(\alpha)$) and also their ideal alternatives were already ideal according to R_O before taking this step. Together this states that the performance of the action denoted by α has become obligated in σ and all previous obligations are still in effect. If $COMMIT(\alpha)$ would bring us into a structure in which some state has no R'_O-successor, then $COMMIT(\alpha)$ is defined to be equivalent to **fail**.

The semantics of $COMMIT$ is now given as follows:

Definition 17
If $com(\alpha, (\sigma, \gamma), R_O)$ is serial, then
$((\mathcal{A}, \Sigma, \pi, R_{\mathcal{A}}, R_O, I), (\sigma, \gamma)) \models [COMMIT(\alpha)]\phi \Longleftrightarrow$
$((\mathcal{A}, \Sigma, \pi, R_{\mathcal{A}}, com(\alpha, (\sigma, \gamma), R_O), I)), (\sigma, \gamma)) \models \phi$;
if $com(\alpha, (\sigma, \gamma), R_O)$ is not serial, then $[COMMIT(\alpha)]\phi$ is by definition valid in $((\mathcal{A}, \Sigma, \pi, R_{\mathcal{A}}, R_O, I), (\sigma, \gamma))$. To explain this, if $com(\alpha, (\sigma, \gamma), R_O)$ is not serial, then $COMMIT(\alpha)$ is the **fail** action and $[\textbf{fail}]\phi$ is valid for any ϕ.

Note that we cannot use the semantics of the formulas of the form $[\alpha]\phi$ with $\alpha \in Act$ to provide the semantics of $[\beta]\phi$ with $\beta \in MAct$ because when α is in Act the formula $[\alpha]\phi$ and the formula ϕ are evaluated in the same model, while if α is in $MAct$ the formulas $[\alpha]\phi$ and ϕ are evaluated in different models!

The following validities follow directly from the above definitions:

Proposition 3

1. $\models [DECIDE(\alpha)]INT(\alpha)$
2. $\models [DECIDE(\alpha)]\neg INT(\beta)$ *if* $\alpha \& \beta =_{\mathcal{D}}$ **fail**
3. $\models INT(\alpha) \longrightarrow [COMMIT(\alpha)]O(\alpha)$
4. $\models [DECIDE(\alpha)][COMMIT(\alpha)]O(\alpha)$
5. $\models (\neg O(\alpha) \wedge \neg INT(\alpha)) \longrightarrow [COMMIT(\alpha)]\neg O(\alpha)$
6. $\models O(\beta) \longrightarrow [COMMIT(\alpha)]O(\beta)$ *if* $\alpha \& \beta =_{\mathcal{D}}$ **fail**

7. $\models \neg O(\beta) \longrightarrow [COMMIT(\alpha)]\neg O(\beta)$ *if* $\alpha \& \beta =_{\mathcal{D}}$ **fail**

8. $\models [\beta]INT(\alpha_2) \longrightarrow [DECIDE(\alpha_1)][\beta]INT(\alpha_2)$

9. $\models [\beta]\neg INT(\alpha_2) \longrightarrow [DECIDE(\alpha_1)][\beta]\neg INT(\alpha_2)$

10. $\models [\beta]O(\alpha_2) \longrightarrow [COMMIT(\alpha_1)][\beta]O(\alpha_2)$

11. $\models [\beta]\neg O(\alpha_2) \longrightarrow [COMMIT(\alpha_1)][\beta]\neg O(\alpha_2)$

(1) states that after deciding to do the action denoted by α I also intend to do α. (2) states the complement namely, that if I decide to perform the action denoted by α I do not intend to perform an action that is incompatible with α. Actually (3) states the property for which the definitions are devised. If I intend to perform the action denoted by α then I am obligated to perform that action after I have committed myself to it. (4) states basically the same in a different format. (5),(6) and (7) state that a commitment has no influence on actions that are not intended and on actions that have nothing to do with the action committed to. The last four items state that decisions and commitments only influence the state-history pair in which they are made.

Having introduced all the ingredients in our logic, in the next section we will show how they can be used to model the examples given in the introduction.

4 Examples modelled

The first example models the decisions taken in a court case. We have simplified the formulas a little bit to make it easier to follow the example. First we assume that if a crime has been committed in the past and it is not yet punished then it should be our intention to punish the crime:

$$PAST(crime) \wedge \neg PAST(punish) \longrightarrow O(INT(punish))$$
$$\neg PAST(crime) \longrightarrow O(INT(dismiss))$$

I.e. we have a moral obligation to punish crimes and to dismiss the defendant if no crime has been committed. However, we still might decide to dismiss the defendant (because of mitigating circumstances or not enough evidence to convince us of the fact that the crime actually was committed by the defendant). To model these possibilities we use the following formulas:

$$INT(punish) \longrightarrow [DECIDE(dismiss)]\neg INT(punish)$$
$$INT(dismiss) \longrightarrow [DECIDE(punish)]\neg INT(dismiss)$$
$$PAST(crime) \longrightarrow [DECIDE(punish)](INT(punish) \wedge O(INT(punish)))$$
$$PAST(crime) \longrightarrow [DECIDE(dismiss)](\neg INT(punish) \wedge O(INT(punish)))$$
$$\neg PAST(crime) \longrightarrow [DECIDE(punish)](\neg INT(dismiss) \wedge O(INT(dismiss)))$$
$$\neg PAST(crime) \longrightarrow [DECIDE(dismiss)](INT(dismiss) \wedge O(INT(dismiss)))$$

The above states that it is morally wrong to punish someone for a crime he did not commit or to dismiss someone while he committed a crime. The above formulas do not state yet what the course of actions should be. This depends on the commitment to the decision that is reached:

$$INT(punish) \longrightarrow [COMMIT(punish)]O(punish)$$
$$INT(dismiss) \longrightarrow [COMMIT(dismiss)]O(dismiss)$$

The above formulas can model a conflict between a moral obligation and a legal obligation. If the crime has been committed by the defendant there is a (moral) obligation to punish the defendant. However, if we decide to dismiss him and commit ourselves to that decision there also exists a (legal) obligation to set the defendant free. I.e. we have $O(INT(punish)) \wedge O(dismiss)$.

The second example does not make explicit use of the decision operator. If some money was subtracted from the account of a client of the bank by mistake and this mistake was rectified by a compensating money transfer later on then we want to inform the client and have an obligation to restore any interest that was payable because of a negative balance due to the mistaken money transfer. This is modelled by the following formulas:

$$PAST(mistake) \wedge PAST(compensate) \longrightarrow$$
$$INT(inform\_client) \wedge O(rest\_interest)$$
$$PAST(mistake) \wedge PAST(compensate) \longrightarrow$$
$$[COMMIT(inform\_client)]O(inform\_client)$$

The obligation to restore the interest arises directly from the mistake made in the past, while the obligation to inform the client arises from a (voluntary) commitment. We can contrast the above formulation with the situation in which the mistake was recovered by executing a roll-back operation. In this case the client does not notice the mistake at all and therefore does not have to be informed of it. The client also does not miss any interest. Therefore we get the following formula:

$$PAST(mistake) \wedge PAST(rollback) \longrightarrow$$
$$\neg INT(inform\_client) \wedge \neg O(rest\_interest)$$

The above simple examples show the power of the logic that is proposed in this paper. We simplified the examples to be able to concentrate on the central issues that we wanted to bring up. Therefore we did not model the fact that the compensating action should be executed after the mistake. We also assumed that only one mistake was made in the past. To model these details we would need to introduce sequences of actions and operators on them. This is left out because of the space limitations.

5 Discussion and Conclusions

In this paper we have given a modal approach to deontic logic that integrates static (Ought_to_be) and dynamic (Ought_to_do) deontic operators. Through

the inclusion of a few (simple) temporal operators it is possible to express the dependence of obligations, prohibitions and permissions on a past situation.
The introduction of the two (meta-)actions $COMMIT$ and $DECIDE$ makes it possible to model (in some way desired) future behaviour. Because, within the logic we cannot simulate the actual occurrence of actions, the effects of this (desired) future behaviour are expressed in obligations to perform certain actions.

In section 4 we have shown that with this integrated deontic logic it is possible to model the practical and frequently occurring examples as given in the introduction in a simple and elegant way.
Due to the limited space many aspects of the logic have not yet been discussed. For instance, we have not given an axiom and inference schema for this logic. We leave this for the full paper.

Another point that we have not discussed yet is the nature of the "intention" relation I. We have taken the semantic relation I as given. However, it is also possible to give this relation a deontic connotation by relating it to a deontic ordering of the states. Let $\leq_D$ denote a deontic ordering on the states such that $\sigma \leq_D \sigma'$ if σ is deontically worse than σ'. Then we can give the following restriction on the relation I_S:

Definition 18

$$I_S(\sigma,\sigma') \Longrightarrow R_S(\sigma,\sigma') \wedge \forall\sigma'' \in \Sigma(R_S(\sigma,\sigma'') \Longrightarrow \sigma'' \leq_D \sigma')$$

In order for this restriction to be sensible $<_D$ has to be reflexive and transitive. It does not have to be total over Σ, but it is plausible to require that it is total on every set $W \subseteq \Sigma$ such that $\exists\sigma \in \Sigma : \forall\sigma' \in W : R(\sigma,\sigma')$. I.e. the ordering is total on every set of states that is reachable from one state by performing some action. Some ideas about how to define this ordering can been found in [15].
The above restriction states that only deontically optimal action can be the actions that are intended to be done next. I.e. the restriction ensures us that the modelled agent will behave in a moral way. Of course, other types of restrictions can be given. E.g. one might take the more realistic option of demanding the opposite of the above restriction. I.e. intended actions are never deontically preferred. Here we do not want to make a choice in this respect, but just point out the possibilities of the approach introduced in this paper.

Acknowledgement. We are grateful for the very helpful comments of the anonymous referees on the submitted version of this paper.

References

[1] C.J. Date. An introduction to database systems. Addison-Wesley, Amsterdam, 1995.

[2] J. Fiadeiro and T. Maibaum. Temporal Reasoning over Deontic Specification. In *Journal of Logic and Computation*, 1 (3), 1991.

[3] Ph. R. Cohen and H. J. Levesque. Intention is Choice with Commitment. In *Artificial Intelligence*, 42, 1990, pages 213–261.

[4] F. Santos and J. Carmo. A Deontic Logic Representation of Contractual Obligations. In *Proceedings of the First International Workshop on Deontic Logic in Computer Science*, pp. 364-382, Amsterdam, 1991.

[5] D. Harel. First Order Dynamic Logic. LNCS 68 Springer, 1979.

[6] J.-J.Ch. Meyer. A different approach to deontic logic: Deontic logic viewed as a variant of dynamic logic. In *Notre Dame Journal of Formal Logic*, vol.29, pages 109–136, 1988.

[7] G.H. von Wright. Deontic logic. In *Mind*, vol.60, pages 1–15, 1951.

[8] L. Åqvist. Deontic logic. In D.M. Gabbay and F. Guenthner, editors, *Handbook of Philosophical Logic II*, pages 605–714. Reidel, 1984.

[9] F. Dignum and J.-J.Ch. Meyer. Negations of transactions and their use in the specification of dynamic and deontic integrity constraints. In M. Kwiatkowska, M.W. Shields, and R.M. Thomas, editors, *Semantics for Concurrency, Leicester 1990*, pages 61–80, Springer, Berlin, 1990.

[10] F. Dignum. Using transactions in integrity constraints. In *Workshop on Applied Logic*, Amsterdam, 1992.

[11] J.W. de Bakker, J.N Kok, J.-J.Ch. Meyer, E.-R. Olderog, and J.I. Zucker. Contrasting themes in the semantics of imperative concurrency. In J.W. de Bakker, W.P. de Roever, and G. Rozenberg, editors, *Current Trends in Concurrency: Overviews and Tutorials*, pages 51–121. LNCS 224 Springer, Berlin, 1986.

[12] J.C.M. Baeten and W.P. Weijland. Process Algebra. Cambridge University Press, 1990.

[13] M. Broy. A theory for nondeterminism, parallelism, communication and concurrency. In *Theoretical Computer Science*, vol.45, pages 1–62, 1986.

[14] F. Dignum and J.-J. Ch. Meyer and R. J. Wieringa. A solution to the free choice paradox by contextually permitted actions. in *Studia Logica*, to be published.

[15] F. Dignum, J.-J.Ch. Meyer, and R. Wieringa. A dynamic logic for reasoning about sub-ideal states. In J. Breuker, editor, *ECAI workshop on Artificial Normative Reasoning*, pages 79–92, Amsterdam, 1994.

[16] E.A. Emerson. Temporal and Modal Logic. In J. van Leeuwen, editor, *Handbook of Theoretical Computer Science*, pages 995-1072, North-Holland, Amsterdam, 1989.

[17] A.R. Anderson. Some nasty problems in the formalization of ethics. In *Noôs*, vol. 1, pages 345-360. 1967.

[18] J.-J.Ch. Meyer. A simple solution to the 'deepest' paradox in deontic logic. In *Logique et Analyse, Nouvelle Série*, vol.30, pages 81–90, 1987.

[19] R. Wieringa, H. Weigand, J.-J.Ch. Meyer, and F. Dignum. The inheritance of dynamic and deontic integrity constraints. In *Annals of Mathematics and Artificial Intelligence 3*, pages 393–428. Baltzer A.G., 1991.

[20] J.R. Searle. Speech Acts. Cambridge University Press. 1969.

[21] J.R. Searle and D. Vanderveken. Foundations of illocutionary logic. Cambridge University Press. 1985.

[22] B. van Linder, W. van der Hoek and J.-J.Ch. Meyer. How to motivate your agents. On making promises that you can keep. to appear as technical report of the RUU, Utrecht, The Netherlands, 1995.

Combining Agency and Obligation
(Preliminary Version)

John F. Horty
Philosophy Department and
Institute for Advanced Computer Studies
University of Maryland
College Park, MD 20901
USA

Abstract

The purpose of this paper is to explore a new deontic operator for representing what an agent ought to do; the operator is cast against the background of a modal treatment of action developed by Nuel Belnap and Michael Perloff, which itself relies on Arthur Prior's indeterministic tense logic. The analysis developed here of what an agent ought to do is based on a dominance ordering adapted from the decision theoretic study of choice under uncertainty to the present account of action. It is shown that this analysis gives rise to a normal deontic operator, and that the result is superior to an analysis that identifies what an agent ought to do with what it ought to be that the agent does.

1 Introduction

The purpose of this paper is to define and explore a new deontic operator for representing what an agent ought to do, a notion that must be distinguished from that of what ought to be. This new operator is cast against the background of a modal analysis of agency developed by Nuel Belnap and Michael Perloff in a series of papers beginning with [1]. The general approach to agency set out in these papers—which itself relies on a theory of indeterministic time due to Arthur Prior—is sometimes described as *stit semantics*, because it concentrates on a construction of the form 'α (an agent) sees to it that A', usually abbreviated simply as [α *stit*: A]. The goal is to provide a precise semantic account of this *stit* operator within the overall logical framework of indeterminism.

As it happens, Prior's indeterministic temporal framework allows also for the introduction of a standard deontic operator $\bigcirc$, meaning 'It ought to be that ...'. It is natural, therefore, to explore the interactions between this standard deontic operator and the *stit* operator representing agency; and it may seem reasonable to propose a logical complex of the form $\bigcirc$[α *stit*: A]—meaning 'It ought to be that α sees to it that A'—as an analysis of the notion that seeing to it that A is something α ought to do. The motive for this analysis, of course, is a philosophical thesis, advanced by some but disputed by others, according to which what an agent ought to do can be identified with what it ought to be that the agent does; a proposal based on this identification was investigated in [2], and defended there against certain objections found in the literature.

In the present paper, I set out a new and powerful objection to the general idea of identifying what an agent ought to do with what it ought to be that he does; and driven by this objection, I propose a new analysis of what an agent ought to do. This new analysis is based on a loose parallel between action in indeterministic time and choice under uncertainty, as it is studied in decision theory. Very roughly, a particular preference ordering—a kind of dominance ordering—is adapted from the study of choice under uncertainty to the present account of action; it is then proposed that an agent ought to see to it that A whenever the agent has available some action which guarantees the truth of A, and which is not dominated by another action that does not guarantee the truth of A. The primary technical point of the paper is the demonstration that this new analysis of what an agent ought to do gives rise to a normal deontic operator.

2 Branching time

The theory of indeterminism underlying the present work—introduced in Chapter 7 of Prior's [3], and developed in more detail by Richmond Thomason in [4] and [5]—is based on a picture of moments as ordered into a treelike structure, with forward branching representing the openness or indeterminacy of the future and the absence of backward branching representing the determinacy of the past.

Such a picture leads, formally, to a notion of branching temporal frames as structures of the form $\langle \mathit{Tree}, < \rangle$, in which *Tree* is a nonempty set of moments and $<$ is an ordering on *Tree* that is transitive, irreflexive, and that satisfies the treelike property according to which, for any m_1, m_2, and m_3 in *Tree*, if $m_1 < m_3$ and $m_2 < m_3$, then either $m_1 = m_2$ or $m_1 < m_2$ or $m_2 < m_1$. A maximal set of linearly ordered moments from *Tree* is a ***history***, representing some complete temporal evolution of the world. If m is a moment and h is a history, then the statement that $m \in h$ can be taken to mean that m occurs at some point in the course of the history h, or that h passes through m. Of course, because of indeterminism, a single moment might be contained in several distinct histories: we let $H_m = \{h : m \in h\}$ represent the set of histories passing through m, those histories in which m occurs.

These ideas can be illustrated as in Figure 1, where the upward direction represents the forward direction of time. This diagram depicts a branching temporal frame containing five histories, h_1 through h_5. The moments m_1 through m_4 are highlighted; and we have, for example, $m_2 \in h_3$ and $H_{m_4} = \{h_4, h_5\}$.

In evaluating formulas against the background of these branching temporal frames, it is a straightforward matter to define a notion of truth at a moment adequate for the truth functional connectives, and even for the operator P representing simple past tense: the definitions from standard (linear) tense logic suffice. Since these frames allow alternative possible futures, however, it is not so easy to understand the operator F, representing future tense. Returning again to Figure 1, suppose that, as depicted, the formula A is true at m_3 and at m_4, but nowhere else. In that case, what truth value should be assigned to $\mathsf{F}A$ at the moment m_1?

On the approach advocated by Prior and Thomason, there is just no way

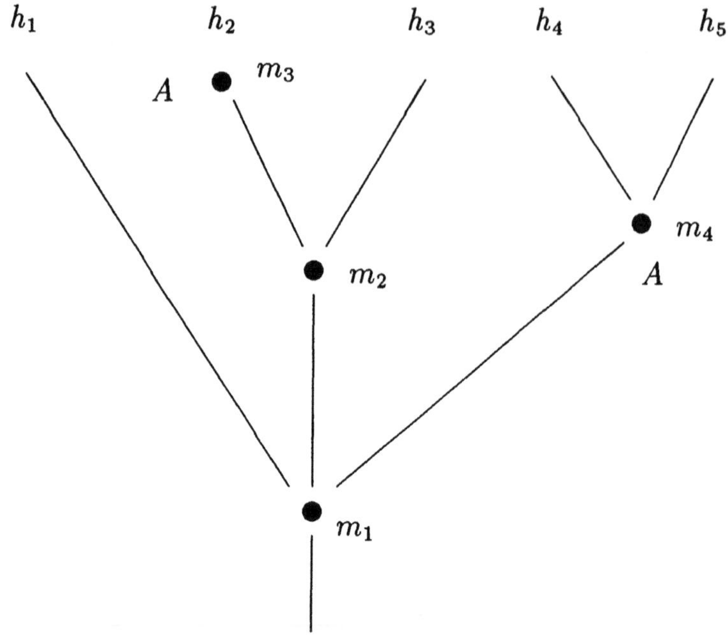

Figure 1: Branching time: moments and histories.

to answer this question. Evidently, FA is true at m_1—A really does lie in the future—if one of the histories h_2, h_4 or h_5 is realized; but it is false on the histories h_1 and h_3. And since, at m_1, each of these histories is still open as a possibility, that is simply all we can say about the situation. In general, in the context of branching time, a moment alone does not seem to provide enough information for evaluating a statement about the future; and what Prior and Thomason suggest instead is that a future tensed statement must be evaluated with respect to a more complicated index consisting of a moment together with a history through that moment. We let m/h represent such an index: a pair consisting of a moment m and a history h from H_m.

Since future tensed statements are to be evaluated at moments and histories together, semantic uniformity suggests that other formulas must be evaluated at these more complicated indices as well. We therefore define branching temporal models as structures of the form $\mathcal{M} = \langle \mathcal{F}, v \rangle$, in which $\mathcal{F}$ is a branching temporal frame and v is a valuation function mapping each propositional constant from the background language into the set of m/h pairs at which, intuitively, it is thought of as true. Where $\models$ represents, as usual, the relation between an index belonging to some model and the formulas true at that index, the base case of the truth definition for branching temporal models tells us simply that propositional constants are true where v says they are:

- $\mathcal{M}, m/h \models A$ if and only if $m/h \in v(A)$ for A a propositional constant.

And the definition extends to truth functions, past, and future as follows:

- $\mathcal{M}, m/h \models A \wedge B$ if and only if $\mathcal{M}, m/h \models A$ and $\mathcal{M}, m/h \models B$,

- $\mathcal{M}, m/h \models \neg A$ if and only if $\mathcal{M}, m/h \not\models A$,
- $\mathcal{M}, m/h \models \mathsf{P}A$ if and only if there is an $m' \in h$ such that $m' < m$ and $\mathcal{M}, m'/h \models A$,
- $\mathcal{M}, m/h \models \mathsf{F}A$ if and only if there is an $m' \in h$ such that $m < m'$ and $\mathcal{M}, m'/h \models A$.

As usual, we say that a formula is *valid* in some class of models if it is true at each index—in this case, each m/h pair—of every model belonging to that class.

It is easy to see that, as long as we confine ourselves to P, F, and truth functional connectives, the validities generated by this definition in branching temporal models coincide with those of ordinary linear tense logic, for the evaluation rules associated with these operators never look outside the (linear) history of evaluation. However, the framework of branching time allows us to supplement the usual temporal operators with an additional concept of settledness, or historical necessity, along with its dual concept of historical possibility. Here, $\Box A$ is taken to mean that A is settled, or historically necessary; $\Diamond A$, that A is still open as a possibility. The intuitive idea is that $\Box A$ should be true at some moment if A is true at that moment no matter how the future turns out, and that $\Diamond A$ should be true if there is still some way the future might evolve that would lead to the truth of A. The evaluation rule for historical necessity is straightforward:

- $\mathcal{M}, m/h \models \Box A$ if and only if $\mathcal{M}, m/h' \models A$ for all $h' \in H_m$;

and $\Diamond A$ can then be defined in the usual way, as $\neg\Box\neg A$.

It is convenient to incorporate this concept of settledness also into the metalanguage: we will say that A is *settled true* at a moment m in a model $\mathcal{M}$ just in case $\mathcal{M}, m/h \models A$ for each h in H_m, and that A is *settled false* at m just in case $\mathcal{M}, m/h \not\models A$ for each h in H_m.

Once the standard temporal operators are augmented with these concepts of historical necessity and possibility, the framework of branching time poses some technical challenges not associated with standard tense logics, but it is also directly applicable to a number of the philosophical presented by indeterminism. Details concerning both the technical issues surrounding branching time and its philosophical applications can be found in Thomason [5].

3 Agency

We now turn to the treatment of agency within this framework of branching time. Although we follow Belnap and Perloff [1] in its general approach, the particular account set out here differs in detail, resulting in a stit operator that is simpler than that of Belnap and Perloff and for certain purposes more natural. The present account derives most immediately from [2].

3.1 Agents and choices

The idea that an agent α sees to it that A is taken to mean that the truth of the proposition A is guaranteed by an action or choice of α. In order to represent this idea, then, we must be able to speak of individual agents, and also of their actions or choices; and so the basic framework of branching time is supplemented with two additional primitives, both drawn from [1].

The first is simply a set *Agent* of agents, individuals thought of as making choices, or acting, in time.

Now what is it for one of these agents to act, or choose, in this way? We idealize by ignoring any intentional components involved in the concept of action, by ignoring vagueness and probability, and also by treating actions as instantaneous. In this rarefied environment, the idea of acting or choosing can be thought of simply as constraining the course of events to lie within some definite subset of the possible histories still available. When Jones butters the toast, for example, the nature of his action, on this view, is to constrain the history to be realized so that it must lie among those in which the toast is buttered. Of course, such an action still leaves room for a good deal of variation in the future course of events, and so cannot determine a unique history; but it does rule out all those histories in which the toast is not buttered.

Our second additional primitive, then, is a device for representing the constraints that an agent is able to exercise upon the course of history at a given moment, the actions or choices open to him at that moment. Formally, these constraints are encoded through a choice function, mapping each agent α and moment m into a partition $Choice_{\alpha}^{m}$ of the histories H_m through m; and the idea is that, by acting at m, the agent α is able to determine a particular one of the equivalence classes from $Choice_{\alpha}^{m}$ within which the future course of history must then lie, but that this is the extent of his influence. Of course, in order for this choice information to make any sense, we must require that any two histories in H_m that have not yet divided at m must lie within the same choice partition; the choices available to an agent at m should not allow a distinction between two histories that do not divide until some later moment.

If K is an equivalence class belonging to $Choice_{\alpha}^{m}$, we speak of K as one of the *actions*, or choices, available to α at m; and we let $Choice_{\alpha}^{m}(h)$ (defined only when $h \in H_m$) represent the particular action or choice from $Choice_{\alpha}^{m}$ that contains the history h. If K is one of the actions available to α at m, we say that that α *performs* the action K at the index m/h just in case h is a history belonging to K. It is important to notice that, as in the evaluation of the future tense, all of the information provided by a full index is required in determining whether an agent performs an action: it makes no sense to say that an agent performs an action at a moment, but only at a moment/history pair. Finally, we speak of the histories belonging to an action K as the possible outcomes that might result from performing this action.

These concepts relating to choice functions can be illustrated as in Figure 2, which depicts a frame containing six histories, and in which the actions available to the agent α at three moments are highlighted. The cells at the highlighted moments represent the actions available to α at those moments. For example, there are three actions available to α at m_1—$Choice_{\alpha}^{m_1} = \{K_1, K_2, K_3\}$, with $K_1 = \{h_1, h_2\}$, $K_2 = \{h_3\}$, and $K_3 = \{h_4, h_5, h_6\}$. Because h_1 and h_2 are still undivided at m_1, these two histories must fall within the same partition there, and likewise for h_4 and h_5. The particular choice partition containing h_5, for example, is K_3, and so we have $Choice_{\alpha}^{m_1}(h_5) = K_3$.

The agent α faces two choices at m_2, but at at m_3 he effectively has no choice: histories divide, but there is nothing α can do to constrain the outcome.

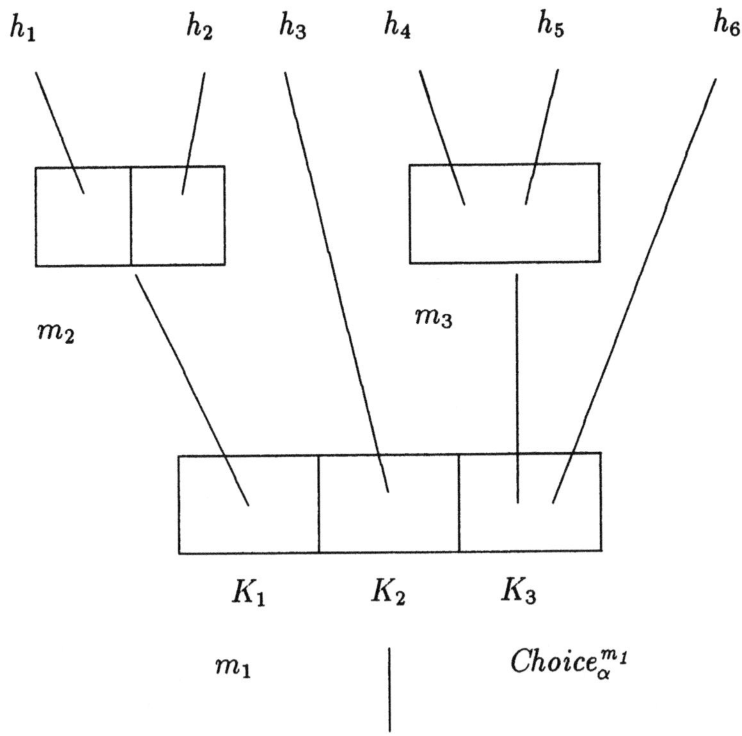

Figure 2: An agent's choices.

(It may be that the outcome can be influenced by some other agent whose choices are not depicted here; or perhaps it is something that just happens, one of nature's choices.) At such a moment, it would be possible to treat the choice function as undefined for α; but it is easier to treat it as defined but vacuous, placing the entire set of histories through the moment in a single equivalence class.

Returning to the moment m_1, we can say that α performs the action K_1 at the index m_1/h_2, for example, that he performs the action K_2 at m_1/h_3, and that he performs the action K_3 at m_1/h_6. Again: since the agent performs different actions along different histories through the moment m_1, it makes no sense to ask what action he performs at that moment. Finally, we can speak of h_4, h_5, and h_6 as the outcomes that might result from performing the action K_3, for example.

When the basic framework of branching time is supplemented with these additional primitives, the result is a *stit frame* of the form

$$\langle \mathit{Tree}, <, \mathit{Agent}, \mathit{Choice} \rangle,$$

with *Tree* and $<$ as before; and we can define a *stit model* as a structure of the form $\mathcal{M} = \langle \mathcal{F}, v \rangle$, in which $\mathcal{F}$ is a stit frame and v a valuation mapping each propositional constant, as before, into a set of m/h pairs. It is these structures that provide the backdrop for the current treatment of agency; the claim is that the structures are not just mathematical curiosities, but describe—up to a legitimate idealization—the world in which agents act.

3.2 The Chellas stit

The particular stit operator to be employed in this paper is described in [2] as the "Chellas stit"—and represented there as *cstit*—because it is an analog in the present framework of the agency operator first studied in Brian Chellas's [6].

The idea behind this *cstit* operator is simple: the statement $[\alpha\ \mathit{cstit}\colon A]$ is to hold at an index m/h just in case α performs an action at m/h that guarantees the truth of A; the action might result in a variety of possible outcomes, but the statement A must be true in each of them. This idea leads to a particularly straightforward evaluation rule:

- $\mathcal{M}, m/h \models [\alpha\ \mathit{cstit}\colon A]$ if and only if $\mathcal{M}, m/h' \models A$ for all $h' \in \mathit{Choice}^m_\alpha(h)$.

The rule is illustrated in Figure 3.[1] Here, the statement $[\alpha\ \mathit{cstit}\colon A]$ is true at m/h_1, because the truth of A is guaranteed by the action that α performs at that index: A holds at each m/h' for each h' belonging to $\mathit{Choice}^m_\alpha(h_1)$. But $[\alpha\ \mathit{cstit}\colon A]$ is not true at m/h_4, for example. Even though the statement A itself happens to hold at this index, the action that α performs at m/h_4 does not guarantee the truth of A.

In fact, this *cstit* operator is not the primary focus of [2]. Instead, that paper concentrates on another operator known as the "deliberative stit," represented

[1] A convention for interpreting figures: when a formula is written next to some history emanating from a moment, the formula should be taken as true at that moment/history pair. Thus, A should be taken as true at m/h_1 in Figure 3, for example, and $\neg A$ as true at m/h_3.

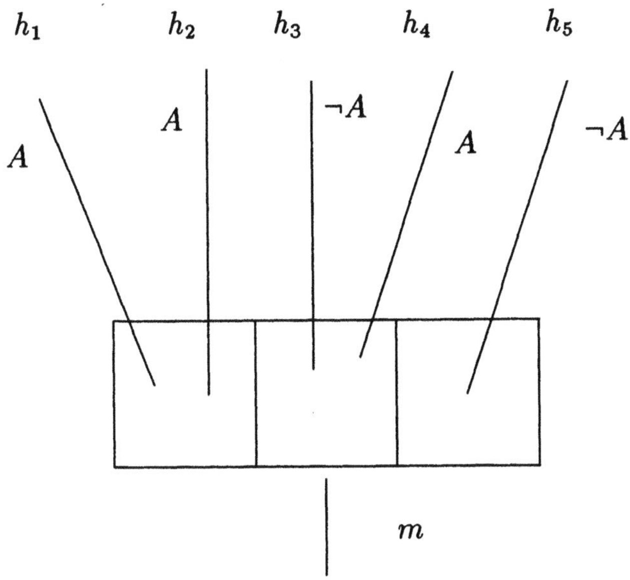

Figure 3: [α *cstit*: A] true at m/h_1.

as *dstit*, and definable through the equivalence

$$[\alpha\ dstit\colon A] \equiv [\alpha\ cstit\colon A] \wedge \neg\Box A.$$

The *dstit* operator has certain advantages over the *cstit* operator in the treatment of agency; for example, as shown in [2], it allows for an attractive analysis of the notion of refraining from an action. Nevertheless, the *cstit* operator, as an *S5* operator, is simpler and more transparent, and it will be best to concentrate on this operator in the present paper.

3.3 Ability

One benefit of employing either the *cstit* or the *dstit* operator in the analysis of agency is that, in either case, a natural treatment of ability lies close at hand. We can assume in either case that an agent's ability (personal can-do) can be represented through a simple combination of ordinary historical possibility (impersonal can) together with the appropriate stit operator (personal to-do). In the present context, the result is an analysis according to which the formula

$$\Diamond[\alpha\ cstit\colon A]$$

can be taken to expresses the claim that α is able to see to it that A.

This style of analysis runs contrary to a well-known thesis of Anthony Kenny's, who argues in [7] and [8] that the logic of ability cannot be formalized using the techniques of modal logic. Kenny follows G. H. von Wright in describing the 'can' of ability as a dynamic modality, and puts the point as

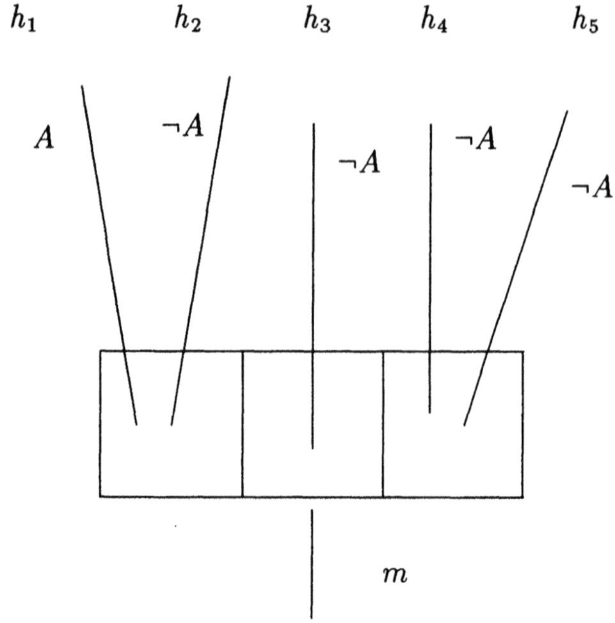

Figure 4: A without $\Diamond[\alpha\ cstit\text{: } A]$.

follows: "ability is not any kind of possibility; ...dynamic modality is not a modality" [8, p. 226].

The central thrust of Kenny's argument is directed against attempts to represent the 'can' of ability as a possibility operator in a modal system with the usual style of possible worlds semantics. Kenny claims that attempts along these lines are doomed to failure: any natural possibility operator, he says, must satisfy the two schemata

$$T\Diamond.\quad A \supset \Diamond A,$$
$$C\Diamond.\quad \Diamond(A \vee B) \supset .\Diamond A \vee \Diamond B;$$

and he argues persuasively that the 'can' of ability does not satisfy either of these. As a counterexample to the first, Kenny considers the case in which a poor darts player throws a dart and actually happens, by chance, to hit the bull's eye; although this shows that it is possible for the darts player to hit the bull's eye, it does not seem to establish his ability to do so. As a counterexample to the second, Kenny imagines a card player who, because he is able simply to draw a card, and all the cards are red or black, is able to draw either a red or a black card; it does not follow that he is able to draw a red card, or that he is able to draw a black card.

Our present analysis of ability escapes from this objection of Kenny's. The notion of historical possibility involved in our analysis, as an *S5* operator, does satisfy both $T\Diamond$ and $C\Diamond$. However, it is not this possibility operator alone that is taken to represent ability, but rather a combination of historical possibility and a stit operator; and the combination fails to satisfy the analogous schemata:

both

$$A \supset \Diamond[\alpha\ cstit\colon A],$$
$$\Diamond[\alpha\ cstit\colon A \vee B] \supset .\Diamond[\alpha\ cstit\colon A] \vee \Diamond[\alpha\ cstit\colon B],$$

are invalid. We provide a countermodel only to the first, based on Kenny's darts example, and depicted in Figure 4. Here, m is the moment at which α throws the dart; the cells belonging to $Choice^m_\alpha$ represent the possible actions or choices available to α at m; and the formula A means that the dart will hit the bull's eye. Evidently, if the player throws the dart and things evolve along the history h_1, then the dart will hit the bull's eye, but this is not a proposition whose truth the player has the ability to guarantee: although A is true at m/h_1, the formula $\Diamond[\alpha\ cstit\colon A]$ is not.[2]

4 Oughts in branching time

4.1 Standard deontic frames

We begin our treatment of deontic logic by considering a standard way of incorporating the deontic operator $\bigcirc$, representing 'It ought to be that ...', into the framework of branching time.[3] Typically in deontic logic, this ought operator is interpreted against a background set of possibilities, usually possible worlds. A number of these possibilities are classified as ideal, those in which things turn out as they ought to; and a sentence $\bigcirc A$ is then thought of as true just in case A holds in each of these ideal possibilities—just in case A is a necessary condition for things turning out as they ought to. In the context of branching time, the set of possibilities at a moment m is identified with H_m, the set of histories still available at m; and a nonempty subset of these is taken to represent the ideal histories. A sentence of the form $\bigcirc A$ is then defined as true at an index m/h just in case A is true at m/h' for each history h' from H_m that is classified as ideal.

This picture can be captured formally by supplementing the stit frames described earlier with a function *Ought* mapping each moment m into a nonempty subset $Ought(m)$ of H_m; the result is a *standard deontic* stit frame, a structure of the form

$$\langle Tree, <, Agent, Choice, Ought \rangle,$$

with *Tree*, $<$, *Agent*, and *Choice* as before. Where $\mathcal{M}$ is a standard deontic stit model—a model that results from interpreting our background language against a standard deontic stit frame—the evaluation rule for ought statements can be set out as follows:

- $\mathcal{M}, m/h \models \bigcirc A$ if and only if $\mathcal{M}, m/h' \models A$ for each $h' \in Ought(m)$.

Several logical features of the ought operator developed in this standard way are immediately apparent from the structure of its evaluation rule. First,

---

[2]Another response to Kenny's argument from the point of view of modal logic is found in Brown [9]; the relation between the present proposal and Brown's is discussed in [2].

[3]Our presentation follows the approach of Thomason [10]. Work along similar lines, but against the background of a slightly different temporal framework, had previously been carried out by Chellas [6], Montague [11], and Scott [12]; historical details can be found in Thomason [5].

it is clear that this ought is a normal modal operator—that is, an operator satisfying the principles

$$\begin{array}{ll} RE\bigcirc. & A \equiv B \ / \ \bigcirc A \equiv \bigcirc B, \\ N\bigcirc. & \bigcirc\top, \\ M\bigcirc. & \bigcirc(A \wedge B) \supset . \bigcirc A \wedge \bigcirc B, \\ C\bigcirc. & \bigcirc A \wedge \bigcirc B \supset \bigcirc(A \wedge B). \end{array}$$

Second, because the set *Ought* (m) is nonempty, it is easy to see that the formula $\bigcirc A \supset \Diamond A$ is valid; this formula expresses one version of the characteristic deontic idea that ought imples can—in this case: if it ought to be that A, then it can be that A. Finally, statements of the form $\bigcirc A$, like statements of the form $\Box A$, are always either settled true or settled false.

4.2 General deontic frames

Although the study of deontic logics has led the clarification of a number of problems involved in normative reasoning, the topic is often viewed with indifference by researchers interested in ethical theory more generally. Part of the reason for this, I believe, is the impression that these logics are able to model only very crude normative theories—theories that can do no more than classify situations, simply, as either ideal or non-ideal. However, while it is true that standard deontic logics have concentrated on this simple classification of situations, it turns out that the underlying semantic framework can be generalized in a natural way to accommodate a much broader range of normative theories.

In order to arrive at this generalization, in the present context of branching time, let us now imagine that each history through a moment, rather than being classified simply as ideal or non-ideal, is assigned a particular value at that moment. These values, chosen from some general space of values, are to represent the worth or desirability of the histories.

This change in perspective can be effected formally by replacing the primitive *Ought* in the frames described above with a function *Value* that associates each moment m with a mapping of the histories belonging to H_m into the set of values. Depending on the nature of the particular normative theory that is being modeled, the values themselves can be conceived of in different ways, and subjected to different ordering relations; but we will asuume that the space of values is always at least partially ordered by $\leq$, so that $Value_m(h) \leq Value_m(h')$ means that h' has a value at m greater than or equal to that of h. The result can be characterized as a *general deontic* stit frame, a structure of the form

$$\langle Tree, <, Agent, Choice, Value\rangle.$$

In the environment of these new frames, the evaluation rule set out above for ought statements must be abandoned, of course. But it is possible to define a coherent ought operator in this new environment by requiring that a statement of the form $\bigcirc A$ should be true at the index m/h whenever A is true along some history through m, and then true also at every history through m of equal or greater value. Where $\mathcal{M}$ is a general deontic stit model, resulting from the interpretation of our background language against a general deontic stit frame, this idea leads to the following evaluation rule:

- $\mathcal{M}, m/h \models \bigcirc A$ if and only if there is some history $h' \in H_m$ such that (1) $\mathcal{M}, m/h' \models A$, and (2) $\mathcal{M}, m/h'' \models A$ for all histories $h'' \in H_m$ such that $Value_m(h') \leq Value_m(h'')$.

The new rule is similar in spirit to the previous version. In the new environment, we can no longer think of $\bigcirc A$ as true whenever A is a necessary condition for achieving an ideal history, since we are no longer presented with a set of histories classified as ideal; instead, we think of $\bigcirc A$ as true whenever A is a necessary condition for achieving a history whose value is at least as great as some particular value.

It is easy to see that the general deontic framework presented here is, in fact, a conservative generalization of the standard deontic framework set out earlier: any standard deontic stit model can be coded into a general deontic stit model in such a way that the same set of ought statements is supported. Suppose that we allow only the two values 0 and 1, ordered so that $0 \leq 1$. We can then map each standard deontic stit model into a general deontic stit model just like the standard model, except that at each moment it assign the value 1 to those histories that the standard model classifies as ideal, and the value 0 to those histories that the standard model classifies as non-ideal. More exactly, where $\mathcal{M}$ is a standard deontic stit model, we let $\mathcal{M}'$ be a model just like $\mathcal{M}$, except that $Value_m(h) = 1$ in $\mathcal{M}'$ just in case $h \in Ought(m)$ in $\mathcal{M}$, and $Value_m(h) = 0$ otherwise. It is then a simple matter to verify that a statement $\bigcirc A$ will hold in the general deontic model $\mathcal{M}'$ at an index m/h according to our new evaluation rule just in case $\bigcirc A$ holds in the standard model $\mathcal{M}$ at the same index m/h according to the previous evaluation rule.

In addition to encoding the information provided by the standard deontic case, however, general deontic stit frames can be used also to represent normative theories that allow for more than two values, and in which the ordering among values is more complex. The most prominent of these, of course, are utilitarian theories, which take as their space of values a set of utilities usually thought of as isomorphic to the real numbers. In the present context, these theories can be represented through general deontic stit frames in which the function *Value* associates with each history passing through a moment, as its value, a real number representing the utility of the history at that moment, and in which the space of values, or real numbers, is subject to its usual ordering. Let us define structures of this kind—general deontic stit frames in which the values are real numbers under their usual ordering—as *utilitarian* stit frames, and the models based on these as utilitarian stit models. We will concentrate on utilitarian models throughout the remainder of the paper.

In depicting these utilitarian stit models, we mark each history through a moment with a number corresponding to its utility at that moment. Thus, Figure 5, for instance, represents a situation in which, at the moment m, the histories h_1, h_2, h_3, and h_4 are taken to possess the utilities 5, 0, 7, and 10, respectively. As a result, we can see that the formula $\bigcirc A$ is settled true at m in this situation, since A holds in h_3 and at each history at least as valuable as h_3. The formula $\bigcirc B$, however, is settled false, since for each history in which B is true, there is a history of equal or greater value in which it is false.

Although we concentrate in this paper on utilitarian models, I do wish simply to mention that the general deontic framework developed here is able to accommodate even more radical departures from standard deontic logic.

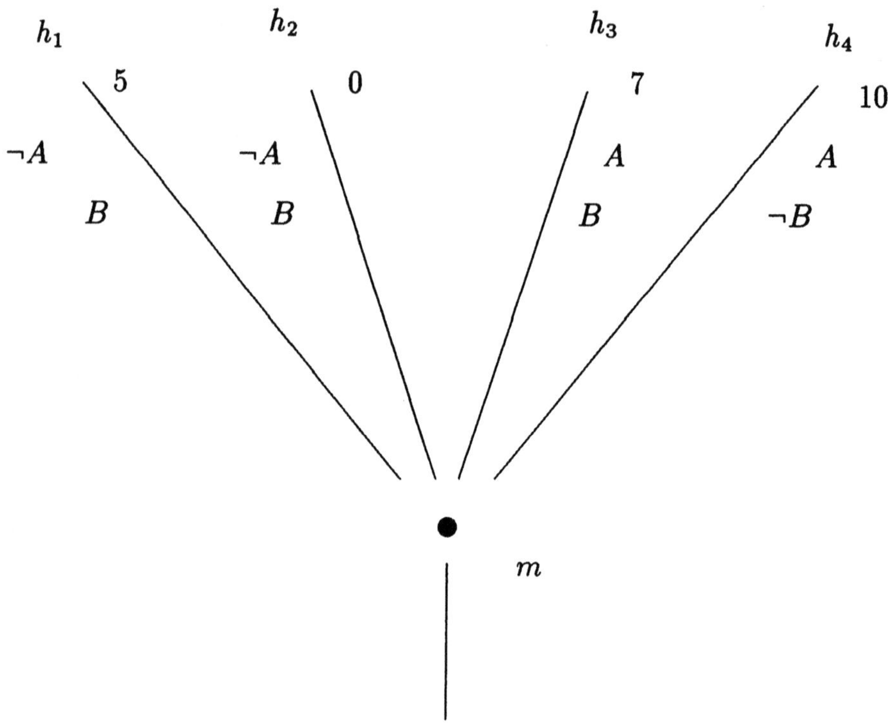

Figure 5: $\bigcirc A$ and $\neg \bigcirc B$.

In the utilitarian case, although there are a variety of different values, these values still stand in a linear ordering; but our general framework would allow us to represent theories in which even the assumption of a linear ordering among values is dropped. As an example, consider the approach to deontic logic described by Bas van Fraassen in [13]. Rather than defining oughts against a background set of ideal situations, van Fraassen postulates a background set of imperatives, possibly conflicting; an ought statement is then taken as true if it is is entailed by some maximal consistent subset of these imperatives. Of course, if the background set of imperatives does happen to contain conflicting but individually consistent statements—say, A and $\neg A$—then it will support the truth of conflicting oughts of the form $\bigcirc A$ and $\bigcirc\neg A$.

This idea could be incorporated into the present environment by supposing that each moment m is associated with a separate set $I(m)$ of imperatives—a set of formulas, possibly conflicting, each of which is taken to represent a statement that "ought" to hold at m. Let us now suppose that $Value_m(h)$ is defined as the set of imperatives from $I(m)$ that are true at the index m/h, so that the value assigned to a history at a moment represents the subset of those imperatives operative at that moment that are fulfilled in that history. Since it is better to satisfy more imperatives than fewer, we can take these values as ordered by subset inclusion, so that $Value_m(h) \leq Value_m(h')$ just in case $Value_m(h) \subseteq Value_m(h')$. It then turns out, as noted in [14], that the oughts generated by our new evaluation rule coincide with those supported by van Fraassen's own definition.

Returning to our new deontic evaluation rule, we can see that the ought operator it defines in general deontic stit models shares many of the logical properties of the operator defined by the previous evaluation rule in standard deontic stit models. It should be apparent, for example, that the characteristic deontic formula $\bigcirc A \supset \Diamond A$ is valid in the class of general deontic stit models, and also that any statement of the form $\bigcirc A$ is always either settled true or settled false. In addition, it is easy to verify that the principles $RE\bigcirc$, $N\bigcirc$, and $M\bigcirc$ listed earlier are valid in general deontic models. But if we consider the entire class of general deontic models, then it turns out that the ought operator defined by our new evaluation rule is not a normal modal operator, for the underlying space of values might be ordered in such a way that instances of $C\bigcirc$ are falsified. An example is provided by those general models mentioned above that are designed to represent the theory of van Fraassen [13]: the formula

$$\bigcirc A \wedge \bigcirc\neg A \supset \bigcirc(A \wedge \neg A)$$

will be false at any index at which both $\bigcirc A$ and $\bigcirc\neg A$ are true.

In the present paper, however, we focus on utilitarian stit models, and in these models, the underlying space of values is subject to a linear ordering: for any histories h and h' from H_m, we have either $Value_m(h) \leq Value_m(h')$ or $Value_m(h') \leq Value_m(h)$. It is easy to see that the schema $C\bigcirc$ is valid in any general deontic stit model in which the underlying space of values is subject to a linear ordering. Thus, as long as our attention is restricted to the class of utilitarian stit models, the ought operator defined by our new evaluation rule is a normal modal operator, satisfying $C\bigcirc$ as well as $RE\bigcirc$, $N\bigcirc$, and $M\bigcirc$.

5 Ought to do: the Meinong/Chisholm analysis

The utilitarian theory of oughts sketched so far is impersonal, an account of what ought to be. According to this theory, it makes perfect sense to say, for example, that it ought not to snow tomorrow; this means, simply, that there is some history in which it does not snow tomorrow, and that it fails to snow also in any history at least as valuable as that one. There is no implication that anyone ought to see to it that it does not snow, or that anyone can do this. However, just as we analyzed the idea of an agent's personal ability earlier through a combination of ordinary, impersonal possibility and a stit operator, we might hope to arrive at an account of what an agent ought to do in the same way: by combining a stit operator with our impersonal account of what ought to be, we might attempt to analyze what an agent ought to do as what it ought to be that he does.

The idea of analyzing what an agent ought to do as what it ought to be that he does was advanced by a number of Austrian and German writers toward the beginning of the century, notably Meinong and Nicolai Hartmann; and the strategy has been explicitly endorsed by at least one contemporary: Roderick Chisholm suggests in [15, p. 150] that "S ought to bring it about that *p*" can be defined as "It ought to be that S brings it about that *p*."[4] In developing this idea, Chisholm relies on his own treatment of what ought to be, in terms of requirement, and on a simple modal analysis of action that can be found already in the writings of St. Anselm. The same general strategy was studied in some detail in [2], which relied on the *dstit* operator for its treatment of agency, and on the account of what ought to be provided by standard deontic stit models; but in this paper, we will instead employ the *cstit* operator and the more general, utilitarian approach to what ought to be. The result is the proposal that a formula of the form

$$\bigcirc[\alpha\ cstit\colon A]$$

can be taken to express the claim that α ought to see to it that A.

This proposal gives us a picture according to which what an agent α ought to do at a particular moment m is determined by the way in which the histories of different value filter through the $Choice^m_\alpha$ partition. Consider, for example, the situation depicted in Figure 6. Here, $\bigcirc A$ is settled true at m. However, $\Diamond[\alpha\ cstit\colon A]$ is settled false: although A ought to hold, there is nothing that α can do about it. Since, as we have seen, any statement of the form $\bigcirc B \supset \Diamond B$ is valid in utilitarian models, we know that

$$\bigcirc[\alpha\ cstit\colon A] \supset \Diamond[\alpha\ cstit\colon A],$$

or that obligation implies ability: whenever it is true that α ought to see to it that A, he must be able to do so. Because α is unable at m to see to it that A, we can thus conclude that $\bigcirc[\alpha\ cstit\colon A]$ is settled false there as well. By contrast, Figure 7 depicts a situation in which $\bigcirc[\alpha\ cstit\colon A]$ is settled true: $[\alpha\ cstit\colon A]$ holds at m/h_1, and also at m/h'' for each history h'' from H_m whose utility is at least as great as that of h_1.

[4] Chisholm's paper contains a reference to Hartmann's work; a recent discussion of Meinong's proposal can be found in García [16].

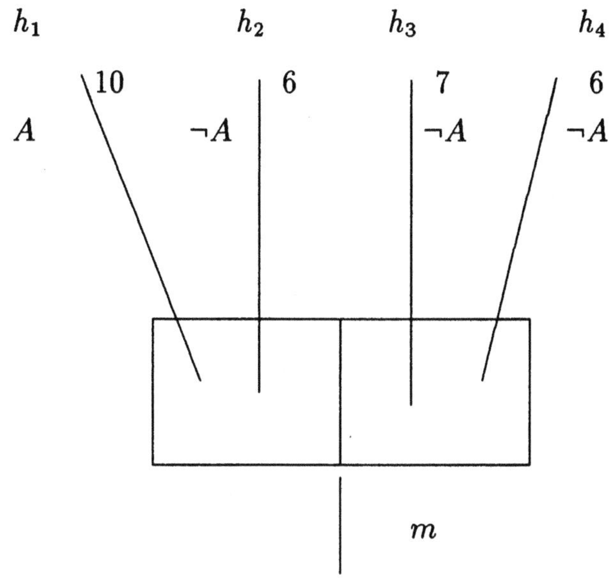

Figure 6: $\bigcirc A$ without $\bigcirc[\alpha$ *cstit*: $A]$.

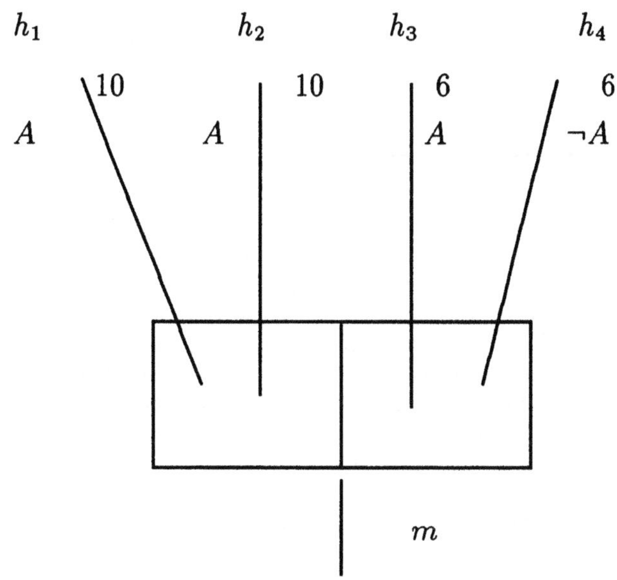

Figure 7: $\bigcirc[\alpha$ *cstit*: $A]$.

Although, according to this analysis, a statement of the form $\bigcirc[\alpha\ cstit\colon A]$ results from appending the generally applicable operator $\bigcirc$ to the formula $[\alpha\ cstit\colon A]$, it is still possible to focus on the logically complex connective $\bigcirc[\alpha\ cstit\colon \ldots]$, and to investigate its properties. It is then easy to see that this complex connective is a normal modal operator, satisfying the principles

$$\begin{array}{l} A \equiv B \ \ / \ \ \bigcirc[\alpha\ cstit\colon A] \equiv \bigcirc[\alpha\ cstit\colon B], \\ \bigcirc[\alpha\ cstit\colon \top], \\ \bigcirc[\alpha\ cstit\colon A \wedge B] \supset . \bigcirc[\alpha\ cstit\colon A] \wedge \bigcirc[\alpha\ cstit\colon B], \\ \bigcirc[\alpha\ cstit\colon A] \wedge \bigcirc[\alpha\ cstit\colon B] \supset \bigcirc[\alpha\ cstit\colon A \wedge B]. \end{array}$$

6 Ought to do: a dominance analysis

We have been considering a general approach, advocated by Meinong and Chisholm, according to which what an agent ought to do is determined by what it ought to be that he does. In the present framework, the formula $\bigcirc[\alpha\ cstit\colon A]$ represents the idea that it ought to be that α sees to it that A, and so according to the Meinong/Chisholm analysis, it is this formula also that represents the idea that α ought to see to it that A.

In fact, this particular version of the Meinong/Chisholm analysis is surprisingly robust: as shown in [2], it is able to withstand many of the objections advanced by Peter Geach in [17] and Gilbert Harman in [18] and [19, Appendix B] against the general strategy of identifying what an agent ought to do with what it ought to be that he does. Nevertheless, the proposal is vulnerable to another kind of objection. I refer to this objection as the gambler's problem, and use it in this section to motivate a different account of what an agent ought to do.

6.1 The gambler's problem

Imagine that an agent α is faced with two options at the moment m: to gamble the sum of five dollars, or to refrain from gambling. If α gambles, we suppose that there is a history in which he wins ten dollars, and another in which he loses and comes away with nothing; but of course, α cannot determine whether he wins or loses. If α does not gamble, we suppose that he preserves his original stake of five dollars no matter how things turn out. Finally, we suppose that the utility associated with each history at m is determined by the sum of money that α possesses in that history. The situation can thus be depicted as in Figure 8. Here, K_1 represents the option of engaging in the gamble, and K_2 the choice of refraining; A represents the statement that α gambles, and h_1 is the history along which α gambles and wins.

It turns out that $\bigcirc[\alpha\ cstit\colon A]$ is settled true at m: the formula $[\alpha\ cstit\colon A]$ is true at m/h_1, and also, trivially, at m/h'' for each history h'' at least as valuable as h_1. The Meinong/Chisholm analysis of what an agent ought to do thus tells us unambiguously that, in this situation, the agent ought to gamble: the most valuable history, with a utility of 10, is that in which he gambles and wins, and it is a necessary condition for achieving this utility that he should gamble. But this is a strange conclusion; for by gambling, the agent runs the real risk of achieving an outcome with the utility of 0, while he is able to

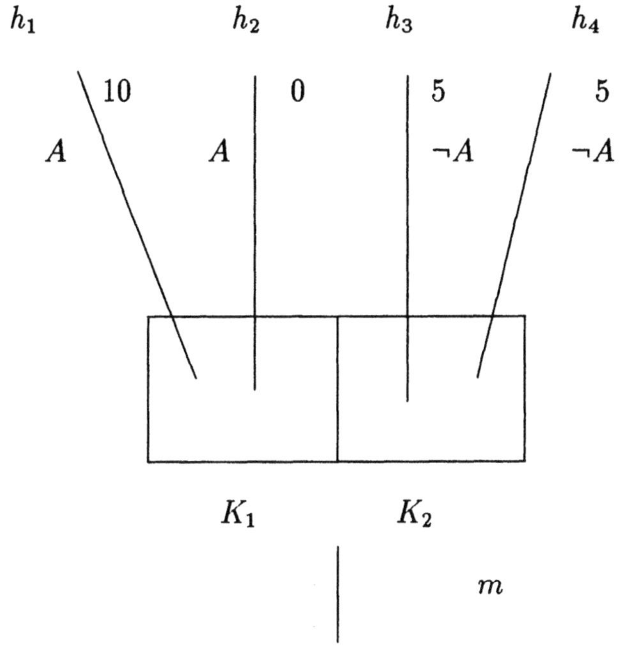

Figure 8: $\bigcirc[\alpha\ cstit\colon A]$, but not $\odot[\alpha\ cstit\colon A]$.

guarantee a utility of 5 by refraining from the gamble. From an intuitive point of view, it appears to be impossible to say whether the agent should gamble or not, at least without knowing the odds of winning; and we should be suspicious of any theory that makes a definite recommendation one way or the other.

The problem presented by this situation seems to reflect a real difficulty with the strategy of identifying what an agent ought to do with what it ought to be that he does. Perhaps it ought to be, in this situation, that the agent gambles; after all, this is what he does in the ideal outcome, the outcome of greatest utility. Still, it does not seem to follow that gambling is something the agent ought to do, since by doing so he risks attaining an outcome of less utility than he could otherwise guarantee.

It might appear that this kind of problem could arise only in a general utilitarian setting, with at least three different values, since it seems to rely upon the possibility that one choice might lead to outcomes both higher and lower in value than the intermediate outcomes resulting from another choice. But a related problem can be seen in a pure deontic setting, which represents outcomes as ideal or non-ideal through the assignment of only two values, 1 and 0. Consider the situation depicted in Figure 9, in which α again has two choices. Again, this situation can be thought of as one in which the agent is faced with accepting or refusing a gamble; and again, K_1 represents the option of engaging in the gamble, and K_2 the choice of refusing the gambling; A represents the statement that α gambles. In this case, however, the gamble is peculiar. If the agent accepts the gamble, we suppose he attains an ideal outcome if he wins,

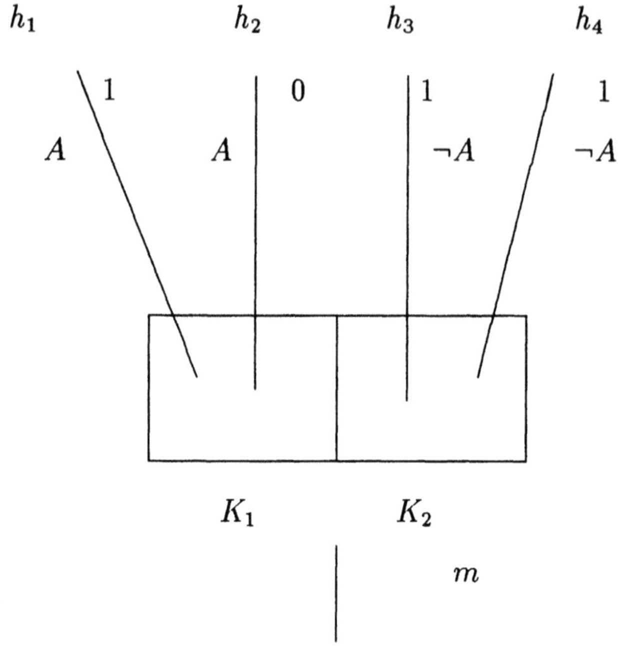

Figure 9: $\neg \bigcirc [\alpha\ cstit\colon \neg A]$, but $\odot[\alpha\ cstit\colon \neg A]$.

and a non-ideal outcome if he loses; what makes the case peculiar is that, here, the agent can guarantee an ideal outcome by declining the gamble.

It should be obvious that the gamble in this situation is not wise: why should the agent risk a non-ideal outcome simply for the chance of achieving an outcome no greater in value than one that he can guarantee by not gambling at all? Since the gamble is not wise, a correct account of what the agent ought to do should tell us that the agent ought not to gamble in this situation. But this is not the result of the theory that identifies what an agent ought to do with what it ought to be that he does: the statement $\bigcirc[\alpha\ cstit\colon \neg A]$ is settled false at m, since for each history in which the agent refrains from gambling, there is a history of equal value in which he gambles.

Let us return to the situation depicted in Figure 8. One natural way of reacting to situations like this is to ask for additional information—in particular, probabilistic information concerning the various outcomes that might result from the available actions. Suppose that for each action K open to the agent α at m—each K belonging to $Choice_\alpha^m$—we were provided with a probability distribution over the histories belonging to K, where the probability assigned to each history represented its chance of occurring should the agent choose K. We could then define the expected utility of an action K in the usual way, as the sum of the values of the various histories belonging to K, with each value weighted by the the probability assigned to its associated history. This introduction of expected utility would provide us with a linear ordering in value, not only of the histories through a moment, but of the actions themselves that are open to the agent at that moment. And it would then be natural to appeal to this new ordering in defining what an agent ought to do: we could suppose that in any given situation an agent ought to perform some one of those actions open to him whose expected utility is maximal.

The approach just sketched does seem like a promising way to proceed when one possesses the necessary probabilistic information concerning the various outcomes that might result from an agent's actions; but in many situations, this kind of probabilistic information is either unavailable or meaningless. In the literature on decision theory, a situation in which the available actions might lead to their various possible outcomes with known probability is characterized as a case of *risk*; a situation in which the probability with which actions might lead to their various possible outcomes is either unknown or meaningless is characterized as a case of *uncertainty*. Is there anything coherent to say about what an agent ought to do in these cases of uncertainty, when even probabilistic information concerning outcomes is absent?

6.2 Dominance

In fact, the decision theoretic treatment of choice under uncertainty describes a variety of ways in which preference orderings can be defined on actions as they are set out there; an introduction to this literature can be found in Chapter 13 of Luce and Raiffa [20]. We proceed in the present paper by adapting a particular one of these preference orderings—a dominance ordering sometimes referred to as "weak dominance"—to the current framework of action. The preference ordering to be defined here ignores the complexities that must be introduced for the proper the treatment of independent events, such as the actions of other agents. However, it seems reasonable to ignore these complexities, at least as a starting point; a more complete discussion can be found in [21] and [22].

Suppose, then, that K and K' are actions open to the agent α at m—members of $Choice_\alpha^m$—and that they are related as follows: each history belonging to K' is at least as valuable as any history belonging to K, and some history belonging to K' is more valuable than some history belonging to K. In these circumstances, a principle sometimes described as the "sure-thing" principle tells us that K' is a better set of outcomes, a better gamble, than K: by selecting an arbitrary outcome from K', the agent is sure to do at least as well as he would by selecting an arbitrary outcome from K, and he might do better.[5]

Let us now introduce the symbol $\prec$ to represent the preference ordering on actions given by the sure-thing principle. Where K and K' are actions open to an agent at m, we take

$$K \prec K'$$

to mean that: (1) $Value_m(h) \leq Value_m(h')$ for each history h in K and each history h' in K', and (2) $Value_m(h) < Value_m(h')$ for some history h in K and some history h' in K'. When $K \prec K'$, we say that the action K' *dominates* the action K, and we note for future reference that this dominance relation is transitive and asymmetric: if $K \prec K'$ and $K' \prec K''$, we can conclude that $K \prec K''$; and if $K \prec K'$, it is impossible to have $K' \prec K$.

[5]A discussion of the sure-thing principle from a different perspective can be found in Savage [23, Section 2.7].

The dominance ordering among actions can be illustrated through our gambling examples. In the situation depicted by Figure 8, we have neither $K_1 \prec K_2$ nor $K_2 \prec K_1$; neither of the actions open to the agent is preferable to the other. In the case of Figure 9, however, we do have $K_1 \prec K_2$, since in that situation it is better for the agent not to gamble.

6.3 A new deontic operator

As with expected utilities, this dominance ordering allows us to compare the actions themselves available to an agent at a moment, not merely the histories through that moment. This new ordering is weaker than the ordering derived from expected utilities; for example, it is not linear. Nevertheless, the dominance ordering is strong enough to support the definition of a reasonable deontic operator representing what an agent ought to do. Let us introduce the new, two-place operator $\odot[\ldots\ cstit\colon \_\_]$, allowing us to construct statements of the form

$$\odot[\alpha\ cstit\colon A],$$

with the intuitive meaning that α ought to see to it that A. The evaluation rule for this new operator is:

- $\mathcal{M}, m/h \models \odot[\alpha\ cstit\colon A]$ if and only if there is a history $h' \in H_m$ such that (1) $\mathcal{M}, m/h' \models [\alpha\ cstit\colon A]$, and (2) $Choice^m_\alpha(h'') \prec Choice^m_\alpha(h')$ for each history $h'' \in H_m$ such that $\mathcal{M}, m/h'' \not\models [\alpha\ cstit\colon A]$.

And the idea underlying this rule is as follows. The formula $\odot[\alpha\ cstit\colon A]$ is to be true at an index m/h whenever: there is a history h' through m along which $[\alpha\ cstit\colon A]$ is true, hence some action $Choice^m_\alpha(h')$ available to α that guarantees the truth of A, and which is such that, if α does not guarantee the truth of A, it must be that the action he performs is worse than $Choice^m_\alpha(h')$.

Having introduced this new deontic operator directly representing what an agent ought to do, we abandon the Meinong/Chisholm strategy of attempting to explicate what an agent ought to do as what it ought to be that he does. We continue to use the formula $\bigcirc[\alpha\ cstit\colon A]$ as a representation of the idea that it ought to be that α sees to it that A; but the distinct idea that α ought to see to it that A is now carried by the new formula $\odot[\alpha\ cstit\colon A]$.

It should be clear that the new analysis gives us the correct results in our two gambling examples. In the case of Figure 8, where it is appears to be impossible to conclude either that the agent should gamble or that he should not, both $\odot[\alpha\ cstit\colon A]$ and $\odot[\alpha\ cstit\colon \neg A]$ are settled false. In the case of Figure 9, where it seems that the agent should refrain from gambling, the statement $\odot[\alpha\ cstit\colon \neg A]$ is settled true, as desired.

We now turn to some observations concerning the logic of our new deontic operator.

Although perhaps apparent already, it is worth noting explicitly that the notion carried by this new operator of what an agent ought to do is logically neither weaker nor stronger than the notion of what it ought to be that he does, but incomparable: both the formulas

$$\bigcirc[\alpha\ cstit\colon A] \supset \odot[\alpha\ cstit\colon A],$$
$$\odot[\alpha\ cstit\colon A] \supset \bigcirc[\alpha\ cstit\colon A]$$

are invalid in the class of utilitarian stit models. A countermodel to the first is provided by Figure 8; a countermodel to (an instance of) the second is provided by Figure 9. It is interesting to note, however, that if we limit our attention to the class of those utilitarian stit models that can be taken to represent standard deontic stit models—those utilitarian models in which the space of values is limited to 1 and 0, representing the ideal and non-ideal histories—then the first of these two formulas is valid in this more restricted class. Thus, while the notion of what it ought to be that an agent does is incomparable in a general utilitarian setting to the notion of what an agent ought ought to do, it is a logically stronger notion in a pure deontic setting.

Even in a general utilitarian setting, however, although the notion of what an agent ought to do is incomparable to the notion of what it ought to be that he does, these two notions are at least guaranteed not to conflict: we will never come across a situation in which the agent ought to see to it that A, although it ought to be that he sees to it that $\neg A$. This guarantee is due to the validity of

$$\neg[\odot[\alpha\ cstit\colon A] \wedge \bigcirc[\alpha\ cstit\colon \neg A]].$$

It is clear from the structure of the evaluation rule for the new operator that any statement of the form $\odot[\alpha\ cstit\colon A]$ is always either settled true or settled false; and also that the characteristic deontic formula

$$\odot[\alpha\ cstit\colon A] \supset \Diamond[\alpha\ cstit\colon A]$$

is valid. The new operator is, moreover, a normal modal operator, satisfying the principles

$$\begin{array}{ll} RE\odot. & A \equiv B \ / \ \odot[\alpha\ cstit\colon A] \equiv \odot[\alpha\ cstit\colon B], \\ N\odot. & \odot[\alpha\ cstit\colon \top], \\ M\odot. & \odot[\alpha\ cstit\colon A \wedge B] \supset .\odot[\alpha\ cstit\colon A] \wedge \odot[\alpha\ cstit\colon B], \\ C\odot. & \odot[\alpha\ cstit\colon A] \wedge \odot[\alpha\ cstit\colon B] \supset \odot[\alpha\ cstit\colon A \wedge B]. \end{array}$$

Although, it is easy to establish $RE\odot$, $N\odot$, and $M\odot$, the verification of $C\odot$ is surprisingly difficult. The reason for this difficulty may become apparent if we recall that the proof validity of the analogous formula $C\bigcirc$ relied crucially on the assumption of linearity for the underlying ordering of values. Even though the dominance ordering on actions that figures in the definition of our new deontic operator is not linear, the validity of $C\odot$ can nevertheless be established.

7 A generalization

The analysis set out here presents, I believe, a coherent theory of what an agent ought to do, and one that improves on the Meinong/Chisholm idea of identifying what an agent ought to do with what it ought to be that he does. There are, however, a number of ways in which the theory as as it stands might be refined and generalized: I close simply by mentioning one of these.

As it stands, the theory focuses only on a moment. It specifies what an agent ought to do at a moment entirely on the basis of the actions available to the agent at that very moment, ignoring any actions that might be available

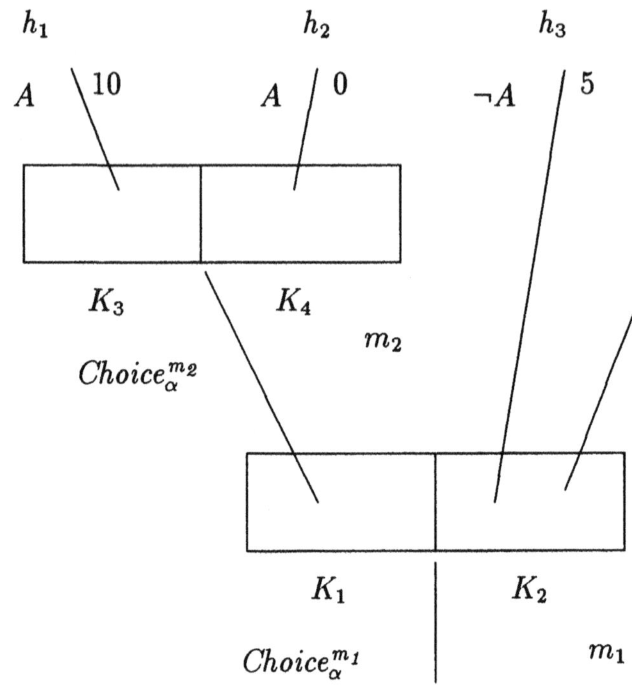

Figure 10: $\odot[\alpha$ *cstit*: $A]$ settled false.

later on. Of course, agents do not usually confine their attention to momentary actions; more often, they work out plans of action over intervals of time. Nevertheless, although a full account of what an agent ought to do over some period of time would ultimately have to be richer than the momentary theory presented here, it might seem that we could safely ignore the problems involved in developing such a full account if we were willing to settle, at first, only for an accurate momentary account of what an agent ought to do. Unfortunately, this is not so: there are situations in which, by concentrating only on a moment, ignoring the later actions available to an agent, we are left with a distorted picture of what the agent ought to do even at that very moment.

An example is provided by Figure 10, which depicts the options open to the agent α at the moment m_1, and then also at the later moment m_2. At m_1, the agent faces a choice between $K_1 = \{h_1, h_2\}$ and $K_2 = \{h_3, h_4\}$, and then at m_2 a choice between $K_3 = \{h_1\}$ and $K_4 = \{h_2\}$. The histories h_1 through h_4 possess the values indicated, relative to both m_1 and m_2; and the statement A is true at the indices m_1/h_1, m_1/h_2, m_2/h_1, and m_2/h_2.

Now what should the agent do at the moment m_1? Well, if we look at m_1 alone, the situation appears to be identical to that depicted in Figure 8, our first gambling example. Neither of the actions K_1 or K_2 dominates the other, and so the theory as it stands cannot recommend either action over the other. As a result, both $\odot[\alpha$ *cstit*: $A]$ and $\odot[\alpha$ *cstit*: $\neg A]$ are settled false at m_1.

From an intuitive point of view, however, this result is incorrect. The current situation is not like the earlier gambling example. In the present case, it is as if the agent could first gamble, and then later on choose whether or not he is to win. If the agent selects the action K_1 at m_1, he then faces at m_2 the further choice between K_3 and K_4. By adopting the strategy of first selecting K_1 and then selecting K_3, he can guarantee an outcome of value 10, the highest value that he can guarantee through any available strategy. Since K_1 is the action that α performs at m_1 in the best strategy available, it appears—from this more general perspective, which involves looking at later moments—that K_1 should be classified as a better action than K_2 even at m_1, and therefore, that at m_1 α ought to see to it that A.

Of course, generalizing the notion of what an agent ought to do in this way—evaluating present actions partly on the basis of later possibilities—would involve formulating precisely the notion of a strategy gestured at in the previous paragraph, and then working this notion into the semantics of a new deontic operator. An appropriate notion of strategy can be found in Belnap [24], but the detailed work involved in adapting this notion to the present deontic setting has not yet been carried out.

References

[1] Nuel Belnap and Michael Perloff. Seeing to it that: a canonical form for agentives. *Theoria*, 54:175–199, 1988.

[2] John Horty and Nuel Belnap. The deliberative stit: a study of action, omission, ability, and obligation. *Journal of Philosophical Logic*, forthcoming.

[3] Arthur Prior. *Past, Present, and Future.* Oxford University Press, 1967.

[4] Richmond Thomason. Indeterminist time and truth-value gaps. *Theoria*, 36:264–281, 1970.

[5] Richmond Thomason. Combinations of tense and modality. In Dov Gabbay and Franz Guethner, editors, *Handbook of Philosophical Logic, Volume II: Extensions of Classical Logic*, pages 135–165. D. Reidel Publishing Company, 1984.

[6] Brian Chellas. *The Logical Form of Imperatives.* PhD thesis, Philosophy Department, Stanford University, 1969.

[7] Anthony Kenny. *Will, Freedom, and Power.* Basil Blackwell, 1975.

[8] Anthony Kenny. Human abilities and dynamic modalities. In Juha Manninen and Raimo Tuomela, editors, *Essays on Explanation and Understanding: Studies in the Foundations of Humanities and Social Sciences*, pages 209–232. D. Reidel Publishing Company, 1976.

[9] Mark Brown. On the logic of ability. *Journal of Philosophical Logic*, 17:1–26, 1988.

[10] Richmond Thomason. Deontic logic as founded on tense logic. In Risto Hilpinen, editor, *New Studies in Deontic Logic*, pages 165–176. D. Reidel Publishing Company, 1981.

[11] Richard Montague. Pragmatics. In R. Klibansky, editor, *Contemporary Philosophy: A Survey.* Florence, 1968.

[12] Dana Scott. A logic of commands. Manuscript, Philosophy Department, Stanford University, 1967.

[13] Bas van Fraassen. Values and the heart's command. *The Journal of Philosophy*, 70:5–19, 1973.

[14] John Horty. Deontic logic as founded on nonmonotonic logic. *Annals of Mathematics and Artificial Intelligence*, 9:69–91, 1993.

[15] Roderick Chisholm. The ethics of requirement. *American Philosophical Quarterly*, 1:147–153, 1964.

[16] J. García. The *tunsollen*, the *seinsollen*, and the *soseinsollen*. *American Philosophical Quarterly*, 23:267–276, 1986.

[17] Peter Geach. Whatever happened to deontic logic? *Philosophia*, 11:1–12, 1982.

[18] Gilbert Harman. Human flourishing, ethics, and liberty. *Philosophy and Public Affairs*, 12:307–322, 1983.

[19] Gilbert Harman. *Change in View: Principles of Reasoning.* The MIT Press, 1986.

[20] R. Duncan Luce and Howard Raiffa. *Games and Decisions.* John Wiley and Sons, Inc., 1957.

[21] John Horty. Agency and obligation. *Synthese*, forthcoming.

[22] John Horty. *Agency, Deontic Logic, and Utilitarianism.* Book Manuscript, Philosophy Department and Institute for Advanced Computer Studies, University of Maryland, 1994.

[23] Leonard J. Savage. *The Foundations of Statistics.* John Wiley and Sons, 1954. Second revised edition published by Dover Publications, 1972.

[24] Nuel Belnap. An austere theory of strategies. Manuscript, Philosophy Department, University of Pittsburgh, 1994.

The Logic of Normative Systems

Arnold A. Johanson
Department of Mathematics, University of Toledo
Toledo, OH, USA

Abstract

This paper exploits the properties of adjointness to develop a theory of norms. The adjoints are a pair of mappings between two logics, one called IMP, a logic of imperatives, and another called PROP, which is ordinary logic. In particular, L is a mapping form Prop to IMP and it has a right adjoint R from IMP to PROP. Norms are defined, formally, as statements that contain occurrences of RU where U is in IMP. A sample theorem (reminiscent of Kant's "*ought* implies *can*"): Responsibility implies power and, moreover, incapacity implies immunity.

1 Introduction

The present paper may be viewed as an attempt to breathe new life into John Stuart Mill's long neglected theory of a link between science and art. Mill's theory suggests that we exploit the properties of mappings between the logic of imperatives and ordinary logic. In accordance with this suggestion we develop here a logical system that provides not only a theory of norms but a foundation for practical philosophy. A distinctive feature of this system is that "*ought*" disappears, to be replaced by modal concepts such as feasibility, tenability, avoidability, and revocability. A semantics for this system has been given along the lines laid down by Kanger [1], [2]. The relation of this system to deontic logic is discussed in the final remarks (Section 8).

According to Mill [3], the logic of practice is to be based on transformations (in both directions) between art, whose underlying framework is a logic of imperatives, and science, whose underlying framework is a logic of statements. Mathematics has long dealt with this kind of situation; for in mathematics one often finds that a problem which is intractable in one domain is easily solved when transformed into another domain. Transformations that perform this service occur so often in mathematics that mathematicians give them a special name: "functor", a name stolen from Carnap. But the mathematician's functors are not simply functions; on the contrary, they are, in essence, structure preserving operators from one mathematical category to another.

The most useful and interesting functors occur in pairs and have a property, described below, called "*adjointness*". Mill's idea suggests the possibility that there are pairs of adjoint functors between the categories represented, on one hand, by statements, and on the other, by imperatives or optatives; furthermore, it suggests that adjoint functors might be important—as they have proved to be in all branches of mathematics—for the foundations of practical philosophy. I hope that readers will agree that this expectation is borne out by the results contained in this paper.

As a concession to those who are not familiar with category theory, an effort is made to keep the treatment self-contained. Since our attention here is confined to logic and lattice theory, there is no need to define categorical terms such as "*functor*" and "*adjoint*" in their full generality. Though it has been pointed out, quite correctly, that the properties used in this paper as definitions are actually provable if the standard definitions are used, the present procedure is adopted in order to make the paper more accessible. Why the categorical terminology then? My excuse is that standard categorical results on adjoint functors are appealed to (see Section 4); moreover, since the theory is a simple application of category theory, it would not be proper to disguise the source from which the theory derives its concepts.

2 Lattices, Logic, and Functors

In this section, I outline a theory K in which to carry out the investigation. In order to avoid excessive pedantry, linguistic expressions are often treated autonymously, i.e. as names of themselves. In K there are two logics: one, PROP (a logic of statements); the other, IMP (a logic of imperatives). The logics are to be regarded as lattices ordered by entailment. A set theory, adequate for arithmetic, geometry, lattice theory, and real analysis (and, of course, category theory) is presupposed.

2.1 List of Symbols

(i)	parentheses:	(,)	
	brackets:	⟨, ⟩	
(ii)	connectives:	$\neg$	(negation)
		&	(conjunction)
		$\vee$	(disjunction)
		$\rightarrow$	(the conditional)
(iii)	statement letters:	$F, G, F(j),\ j = 1, 2, 3, \ldots$	
(iv)	imperative letters:	$U, V, U(j),\ j = 1, 2, 3, \ldots$	
(v)	individual variables:	a, b, x, y, z, t (time), $a(j),\ j = 1, 2, 3, \ldots$	
(vi)	time constant:	NOW	
(vii)	lattice constants:	0*, 1*	
(viii)	directive functor:	$(x,)$	(read: x sees to it that)
(ix)	primitive imperative operator:	!	(read: it shall be the case that)
(ix)	quantifier symbols:	$\exists, \forall$	
(x)	other operators:	L	(read: let it be [the case] that)
		S	(read: sustain that)
		R	(read: it is required that)
		A	(read: it is admissible that)

These operators are mappings as shown in the following diagram:

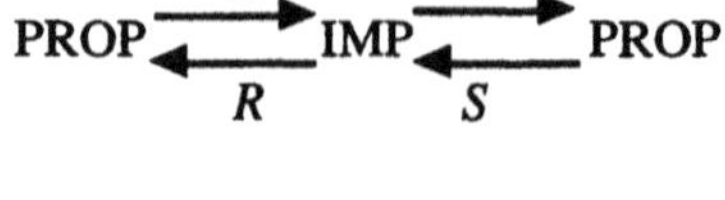

(xi) metalogical symbols: iff (if and only if)
$\Leftrightarrow$ (logical equivalence)
$\Rightarrow$ (entailment).

Ordered n-tuples are written, $\langle a(1), a(2), ..., a(n)\rangle$; ordered couples are written $\langle a, b\rangle$. Other symbols, including predicate letters, function symbols, and additional individual variables and constants, are introduced as needed. The sentential variables are metavariables, i.e. replaceable by *names* of sentences instead of sentences.

2.2 Definitions

In the following, let X and Y be lattices and let the order relation in each lattice be written "<".

D1. The lattice constants are defined as follows:
  0* for initial (smallest) element,
  1* for terminal (largest) element.

D2. In a lattice, $x = y$ provided $x < y$ and $y < x$.
  (In other contexts, $x = y$ means logical identity.)

D3. A mapping $f: X \longrightarrow Y$ is a (lattice) *functor* provided f preserves order, i.e. provided that if $a < b$ in X then $fa < fb$ in Y.

D4. Let $f: X \longrightarrow Y$ and $g: Y \longrightarrow X$ be a pair of lattice functors. Then $\langle f, g\rangle$ is a pair of *adjoint functors* provided $fa < b$ iff $a < gb$; the functor f is *left adjoint* to g and g is *right adjoint* to f. (See McLarty [4], MacLane [5] for the more general—and much more complex!—categorical definition.)

D5. PROP is the smallest complete, complemented, and distributive lattice that contains 0* and 1* together with the following:
  (i) $J\langle b(1), b(2),..., b(n)\rangle$ where J is an n-ary predicate letter and the $b(j), j = 1, 2,..., n$ are individual constants or variables;
  (ii) statement letters;
  (iii) $R\alpha$ where α is an element of IMP.

D6. Informally, a statement is a sentence in the indicative mood; formally, however, a ***statement*** is defined as a member of the lattice PROP. A statement is *closed* (or *t-closed*) if it does not contain free occurrences of the time variable.

D7. IMP is the smallest complemented distributive lattice that includes 0* and 1*, the imperative letters, and the elements of the form $!\alpha$ where α is a t-closed statement in PROP.

D8. Informally, imperatives are sentences in the optative or imperative mood; formally, however, an *imperative* is a member of IMP.

D9. The logical symbols internal to PROP and IMP are defined as follows:
  (i) $\neg a$ for the complement of a,
  $(a \,\&\, b)$ for the greatest lower bound of a and b, and
  $(a \vee b)$ for the least upper bound;
  (ii) $(a \rightarrow b)$ for $(\neg a \vee b)$, and
  $(a \leftrightarrow b)$ for $((a \rightarrow b) \,\&\, (b \rightarrow a))$;

D10. Given a complete lattice X (PROP, e.g.) and family Q of elements in X indexed by the parameter x, $(\exists x)Q$ is defined as the least upper bound of Q and $(\forall x)Q$ is defined as the greatest lower bound of Q.

D11. A *formula* is an element of PROP or IMP. A formula P is *valid* if $P = 1^*$.

D12. Operator symbols:

L for $!(\exists t)$.

S for $!(\forall t)$.

R is the *right adjoint* of L.

A for $\neg R \neg$.

The operator L is called "*let*", the operator S "*sustain*", the operator R "*requirement*", and the operator A "*admissibility*".

Note: There is no attempt in this theory to capture completely the intuitive meaning of the words "let", "sustain", "required", and "admissible". These words are merely meant to *suggest* the intended interpretation of the corresponding symbols. It is enough if the formal apparatus captures the intuitive meaning of these terms as much as but no more than the conditional in formal logic captures the intuitive meaning of implication.

D13. Some of the statements in PROP may be regarded as expressions of norms. Formally, *norms* are statements in PROP that have occurrences of RU (or a logical equivalent). Informally, a statement is normative if it expresses certain modal properties, e.g. feasibility, irrevocability, tenability, and avoidability.

D14. [PROP] is the sublattice of PROP consisting in t-closed statements, i.e. statements that do not contain the time variable free.

Note: The logics PROP and IMP, being complemented distributive lattices, are Boolean algebras—hence free use is made of the law of the excluded middle.

2.3 Axioms

A1. The imperative operator ! is a functor from [PROP] to IMP that preserves validity and commutes with negation, i.e. $!1^* = 1^*$ and if F does not contain t free, $!\neg F = \neg !F$. (Cf. Hofstader & McKinsey [6], Kanger [1].)

A2. The functor L has a right adjoint R.

Remark: Not all lattice functors preserve 1^*. As an example consider the functor H: PROP $\longrightarrow$ PROP defined as follows: let F be a fixed statement and set $Hp = (p \,\&\, F)$. Hp is never valid unless $F = 1^*$. On the other hand, if a lattice functor f, defined on a lattice that has a terminal element 1^*, is *onto*, then $f1^* = 1^*$. Note that unless F is a valid formula, the functor H is *not* onto.

The quantifiers $(\exists t)$ and $(\forall t)$ are mappings from PROP to [PROP]. They, like the imperative operator, are functors that preserve validity. Moreover, since composition of two functors is a functor, it follows that $(!(\exists t))$ and $(!(\forall t))$ are functors that preserve validity. Hence the result:

(0) L and S are functors from PROP to IMP that preserve validity.

Remarks: (1) The operators $L\neg$ and $\neg L$ (or $\neg S$ and $S\neg$) are *not* functors. They are, however, *contravariant functors* which may be characterized in the present context as operators that reverse the direction of entailment. (2) Note that if α does not contain t free, $!\alpha = !(\exists t)\alpha = L\alpha$. Moreover $(\alpha \vee \neg\alpha)$ is valid, but not necessarily $((x, \alpha) \vee (x, \neg\alpha)$.

3 The Semantics of the Functors *L* and *S*

Since history can be rewritten but not altered, it is useless to issue commands to alter the past. For this reason, in normal usage, an imperative of the form $!G$ (read: "It shall be the case that G") is relevant only to the future and, moreover, represents a tacit demand that G hold true *now or in the future*. Hence it is assumed that when the quantifiers $(\exists t)$ and $(\forall t)$ occur in an imperative, *t is restricted to present and future times*.

The time variable t ranges over real numbers. The time variable may be interpreted as a mapping from events to real numbers with the (moving) time frame so adjusted that an event occurring NOW receives the value 0 (Cf. Munroe [7]). In the quantifiers $(\forall t)$ and $(\exists t)$, t is restricted to events whose time coordinates are greater than or equal to 0; that is to say, "$(\forall t)$" may be read, "for all time t, greater than equal to zero", and "$(\exists t)$" may be read, "there exists a time t, t greater than equal to zero, such that"....

Logicians often assume that a statement F admits of only two possibilities: true or false. But quantified over time, a statement with a free time variable yields at least four possibilities: always true, sometimes true, sometimes false, and always false. (Of course these four outcomes are neither mutually exclusive nor exhaustive.) As a simple illustration, if F is the statement that the door is closed (or "The door is closed at time t") then there are four possibilities: sometimes closed, always closed, sometimes open, and always open. I conclude that the simplest formulas in PROP constructed out of a statement F and closed with respect to the time variable are the statements: $(\exists t)F$, $(\forall t)F$, $(\exists t)\neg F$, $(\forall t)\neg F$. Applying the imperative operator to these four forms we arrive at four possible formulas:

$$!(\exists t)F,\ !(\forall t)F,\ !(\exists t)\neg F,\ !(\forall t)\neg F.$$

As already noted, the first of these forms is abbreviated as LF—i.e. we set: L for $!(\exists t)$. Designations for the other three basic imperatives logically follow if we not only take note of the rules for the interchange of quantifiers but also note that (according to Axiom A1) the imperative functor commutes with negations; e.g., $\neg L\neg F = !(\forall t)F$. The operator $\neg L\neg$ is the "*associate functor*" of L and, according to definition D12, is designated by S; consequently, the four forms become simply LF, SF, $L\neg F$, $S\neg F$.

As examples, let F be the statement considered earlier: "The door is closed". Then LF is read "It shall be the case that sometime, now or in the future, the door is closed!" and SF is read "It shall be the case that from this moment onward the door is closed!" Informally, LF can also be read "Let it be that the door is closed!" and SF can also be read "Sustain that the door is closed!" If one wishes to express more timely compliance one can ensure this by including a time parameter in F, e.g. F might be "The door is shut immediately" or "The door is shut from two to four in the afternoon on the first of September 1996".

4 Functorial Semantics

According to D12, the functor R is the right adjoint of the functor L (and L is the left adjoint of R); that is to say, by definition R: IMP $\longrightarrow$ PROP is a functor such that

(1) $LF \Rightarrow U$ (*in* IMP) *iff* $F \Rightarrow RU$ (*in* PROP).

(By pre-established harmony, as it were, L is *left* adjoint and R is *right* adjoint.) The existence of R is assured by Axiom A2; uniqueness (up to logical equivalence) may also be shown. As already stated, RU is read: "It is required that U" and R is called "*requirement*".

As a consequence of (1), it can be shown that given $A = \neg R \neg$ and $S = \neg L \neg$, $\langle A, S \rangle$ is also an adjoint pair, i.e.

(2) $U \Rightarrow SF$ (*in* IMP) *iff* $AU \Rightarrow F$ (*in* PROP).

Thus A is a left adjoint to S and S is a right adjoint to A.

As already stated, AU is read: "It is admissible that U", and A is called "*admissibility*".

The role of R (and implicitly A) can be seen from the following results. According to (1), $L1^* \Rightarrow U$ iff $1^* \Rightarrow RU$, whence:

(3) *The imperative U is valid iff the statement RU is valid.*

As a corollary we have: *U entails V in* IMP *iff $R(U \rightarrow V)$ is provable in* PROP.

I conclude that requirement is analogous to "*truth*".

For an illustration of (1) and (2), consider traffic laws assumed as axioms. Assume as an axiom the law, "If the traffic light is red now, it is required that it shall be the case that no vehicle moves into the intersection". This law, according to (1), is valid iff "Let it be that the traffic light is red now!" entails "It shall be the case that no vehicle moves into the intersection!" For another example, consider the traffic law, "It is admissible that it shall be the case that x sees to it that a motor vehicle z moves into an intersection y at time a only if, if t is a, the traffic light at y is green at time t". Since, according to (2), the functor *sustain* is adjoint to the functor *admissibility*, the law holds iff "It shall be the case that x sees to it that a motor vehicle z moves into intersection y at time a!" entails "Sustain that if t is a then the traffic light at y is green at time a!"

In the latter example, the formal analysis may be facilitated by using the directive functor $(x,)$.

A further note on the semantics of R may be found in the final remarks (in Section 8).

5 The Logic of A, R, L, S, and $(x, \ )$

From the fact that L and A are left adjoints, one may infer (Cf. McLarty [1992, p.97], MacLane [1971, p. 114 ff.]) that L and A preserve disjunctions and therefore,
(4) $A(U \vee V) = AU \vee AV$, and $L(F \vee G) = LF \vee LG$.
Because of duality,
(5) $R(U \,\&\, V) = RU \,\&\, RV$, and $S(F \,\&\, G) = SF \,\&\, SG$.
Some further results are easy consequences of adjointness, e.g. $LF \Rightarrow LF$ and $SF \Rightarrow SF$ yield
(6) $(F \rightarrow RLF)$ and $(ASF \rightarrow F)$.
Finally, it follows from (4) together with the fact that A and L are functors that preserve validity that
(7) $(RU \rightarrow AU)$ and $(SF \rightarrow LF)$.
Remark: The purpose of the operator $(x, \ )$ is to bring directives within the scope of the theory. This primitive operator (which, it so happens, is a right adjoint) satisfies axioms in PROP similar to the properties of those listed by Kanger [2] for the "Do" operator. These are:

A3. $(x, F) \rightarrow F$.
A4. $\{(x, F) \,\&\, (x, (F \rightarrow G))\} \rightarrow (x, G)$.
A5. $(x, 1^*)$.

One may prove not only that $(x, \ )$ is a functor from PROP to PROP that preserves validity but that the formulas $((x, F) \rightarrow \neg(x, \neg F))$, $\neg(x, 0^*)$, $((x, F) \,\&\, (x, G)) \leftrightarrow (x, (F \,\&\, G))$ are valid.

6 Simple Norms

From a formal standpoint, a normative statement, abbreviated *norm*, is (cf. definition D13) a formula in PROP that contains occurrences of RU (or equivalents) where U is in IMP. The simplest norms based on a statement F may be constructed by applying the operators L and $\neg L$ to F and $\neg F$, then applying R and $\neg R$ to the result. The result is eight norms which may be written as follows: RSF, $RS\neg F$, $AS\neg F$, $RL\neg F$, ALF, $AL\neg F$, RLF, ASF. Let us call these norms "*simple norms*". The names chosen for the first four of these simple norms are as follows: *irrevocability* (RSF), *nonfeasibility* ($RS\neg F$), *avoidability* ($AS\neg F$), *nontenability* ($RL\neg F$). These we call the *primary simple norms*. The remaining four norms, being negations of the first four, are accordingly *revocability* ($AL\neg F$), *feasibility* (ALF), *nonavoidability* (RLF), *tenability* (ASF). Because of lack of space we cannot go into a justification of each of these names. Let it suffice to consider one of these: the norm RSF called "*irrevocability*".

As an example of irrevocability, take the marriage norm: "It is irrevocable that x is married to y". I claim it is reasonable to translate this norm as RSF, i.e. "It is required that it shall be the case that from now on x is married to y". Note that the entailment $RSF \Rightarrow F$ reflects perfectly the fact that it is irrevocable that x is married to y only if x is married to y.

7 The Atomic Structure of Norms

Remark: It is commonplace to say that the material world is made out of atoms. What is less appreciated is that the moral world also has an atomic structure. In this section we explore the structure of the Boolean algebra of norms constructed out of simple norms.

Let F be a fixed statement and denote by $e^*(F)$ the smallest Boolean algebra that contains the primary simple norms based on F. The elements of $e^*(F)$ are called "*elementary norms*". The algebra $e^*(F)$ may be viewed as a finite-dimensional vector space over the integers modulo 2 in which addition is interpreted as the exclusive or; i.e. $(p + q) = ((p \,\&\, \neg q) \vee (p \,\&\, \neg q))$. The vector space $e^*(F)$ is a quotient space of the free module generated by the elements in Boole's law of development reduced modulo the logical relations among the simple norms. The free module has as a basis the summands in the disjunctive normal form generated from the primary simple norms $\{RSF, ASF, RLF, ALF\}$. There are sixteen of these summands since each of them is a conjunction of four factors in which the nth factor is either the nth primary simple norm or its negation, $n = 1, 2, 3, 4$. It turns out that eleven of the summands are contradictions; when these are eliminated (and the remainder simplified), five norms remain: RSF, $RS\neg F$, $R(LF \,\&\, L\neg F)$, $ASF \,\&\, AL\neg F$, $AS\neg F \,\&\, ALF$. I conclude:

(8) *$e^*(F)$ is a five-dimensional vector space over the integers modulo* 2, and *each elementary norm based on F is expressible uniquely in what may be called "disjunctive normal form"*:

$a(RSF) + b(RS\neg F) + c(R(LF \,\&\, L\neg F)) + d(AL\neg F \,\&\, ASF) + e(ALF \,\&\, AS\neg F)$,

where the letters a, b, c, d, e are replaceable by 0 or 1. Note that $e^*(F)$ has an "orthogonal" basis in the sense that the conjunction of any two of the basis elements in the above form is 0^*. As a corollary, we have:

(9) *The number of elementary norms (excluding the trivial norms* 0^* *and* 1^**) is thirty.*

If the *co-norm* of a given norm based on F is defined as the norm obtained from a given norm by replacing F by $\neg F$, the elements in the foregoing basis for $e^*(F)$ consists of two pairs of co-norms $(\{RSF, RS\neg F\}, \{(ALF \,\&\, AL\neg F), (AL\neg F \,\&\, ALF)\})$ together with the fifth norm $R(LF \,\&\, L\neg F)$, which is the co-norm of itself. The elements in this basis are called "*atomic norms*". So far we have names for only two of these: irrevocability (RSF) and nonfeasibility ($RS\neg F$). The other three are named as follows: the norm $(RLF \,\&\, RL\neg F)$, which is the conjunction of nonavoidability and nontenability, is called "*incapacity*"; the conjunction $(AL\neg F \,\&\, ASF)$ of revocability and tenability is called "*proper tenability*"; finally, $(ALF \,\&\, AS\neg F)$, the conjunction of feasibility and avoidability, is called "*proper avoidability*".

Because of the duality between "and" and "or", there is also a *conjunctive normal form*:

$$\{a \vee AL\neg F\} \,\&\, \{b \vee (ALF)\} \,\&\, \{c \vee (AS\neg F \vee ASF)\}$$
$$\&\{d \vee (ASF \rightarrow RSF)\} \,\&\, \{e \vee (AS\neg F \rightarrow RS\neg F)\},$$

where the letters a, b, c, d, e are replaceable by 0^* or 1^*. Note that the factors in this form, $AL\neg F$, ALF, $(AS\neg F \vee ASF)$, $(ASF \rightarrow RSF)$, $(AS\neg F \rightarrow RS\neg F)$, are all negations of atomic norms. For example, the factor $(AS\neg F \vee ASF)$ is the negation of incapacity and hence is called "*power*" (or "*capability*"). The factor $(AL\neg F \rightarrow RL\neg F)$, the negation of proper tenability, is called "*relative nontenability*" and its co-norm $(ALF \rightarrow RLF)$, the negation of proper avoidability, is called "*relative nonavoidability*".

Remark: A norm (based on F) is called "*relative*" if it (or its co-norm) has the form $(ALF \rightarrow \alpha)$.

As a consequence of the disjunctive normal form, each elementary norm in $e^*(F)$ has an associated (binary) number $(abcde)$. Thus (11111) represents the logically true 1^*, (00000) (or (0)) represents the logically false 0^*, (01001) (or (1001))—which is the number of $(RS\neg F \vee (ALF \& AS\neg F))$—represents avoidability, i.e. $AS\neg F$, etc. (Note that, as usual, (00000) may be written (0) and in the remaining cases all initial zeroes may be dropped.)

The disjunctive normal form (or the dual conjunctive normal form) enables us to take an inventory of all elementary norms in $e^*(F)$. The thirty two norms (including 0^* and 1^*) may be broken down into three groups: the first group consists in those in which $a = b = 1$; the second, those in which $a = b = 0$; and the third group those in which $a = 1$ and $b = 0$ or $a = 0$ and $b = 1$. Space does not permit an examination of each of these groups. Here, I simply make a note of some of the more important norms.

For obvious reasons, the norm (11000), which has the form $(RSF \vee RS\neg F)$, is called "*responsibility*". The negation of responsibility, i.e. the norm (111), which has the form $(AL\neg F \& ALF)$, is therefore called "*immunity*". The conjunction of power and immunity, i.e. (11), is called "*license*", and its negation (11100), the disjunction of incapacity and responsibility, is called "*constraint*". The conjunction of feasibility and power (i.e. (10011)) is called "*freedom-to*" and the conjunction of revocability and power is called "*freedom-from*".

Two more norms remain to be mentioned: one (10100), which is of the form $\{RLF \& (ASF \rightarrow RSF)\}$, is the negation of freedom-from and therefore called "*bondage*"; the other (1100) is the negation of freedom-to and is called "*co-bondage*". The norm bondage, whose name perhaps suggests servitude; has the alternative form (power $\rightarrow$ irrevocability) and hence may be called "the principle of *noblesse oblige*".

Non-elementary norms may also be considered. Perhaps the most important non-elementary norm is the norm $\{RLF \& (ASG \rightarrow RSF)\}$ called "*conditional commitment*". If $F = G$, conditional commitment reduces to bondage. If one defines the mapping $(/G)$: PROP $\longrightarrow$ PROP by $(F/G) = RLF \& (ASG \rightarrow RSF)$ one obtains an operator analogous to "conditional obligation" (Cf. Hansson [8]). $(/G)$ is a functor that preserves validity.

A few obvious results on elementary norms are as follows:

1) Responsibility implies power since $(RSF \vee RS\neg F) \Rightarrow (ASF \vee AS\neg F)$; therefore, incapacity implies immunity. This result is reminiscent of Kant's "*Ought* implies *can*".
2) Irrevocability implies responsibility since $RSF \Rightarrow (RSF \vee RS\neg F)$.
3) Freedom-to is, by definition, feasibility in the presence of power.

4) Responsibility is constraint in the presence of power, and license is power in the presence of immunity.
5) Irrevocability, which implies both power and freedom-to, is bondage in the presence of power.
6) Avoidability is power together with nontenability, and tenability is power together with nonavoidability.
7) Tenability implies freedom-to, and avoidability implies freedom-from.
8) Constraint is the disjunction of responsibility and incapacity, and moreover, is also the disjunction of bondage and co-bondage.
9) Power is the disjunction of license and responsibility, while freedom-to is the disjunction of license and irrevocability.
10) The disjunction of freedom-to and freedom-from is power, but the conjunction of freedom-to and freedom-from is license.
11) Incapacity is logically equivalent not only to constraint together with immunity but to bondage together with co-bondage.
12) Incapacity is (by definition) nonavoidability together with nontenability, and similarly, constraint is relative nonavoidability together with relative nontenability.
13) Bondage is the conjunction of constraint and nonavoidability.
14) Proper avoidability implies proper nontenability together with relative proper avoidability, relative proper avoidability is freedom-to together with relative avoidability, and proper nontenability implies weak bondage together with immunity. A corollary of 4) is:
15) Responsibility implies constraint, and license implies immunity.

8 Final Remarks

Though the present system is in a sense a variety of deontic logic, there are significant differences from, as well as resemblances to, *classical* deontic logic. The present system depends basically on a *pair* of adjoint functors $\langle L, R \rangle$, while classical deontic logic depends basically on a *single* functor O (or on its associate functor P). One significant difference, which I hope is not ephemeral, is that the present system is free from serious paradoxes such as those that beset classical deontic logic (Cf. Johanson [8]). Though the functor RS has properties similar to O (and therefore the associate functor AL has properties similar to P), the resemblance is not absolute. RS is better interpreted as irrevocability rather than obligation since it satisfies the condition ($RSF \rightarrow F$). (For something to be irrevocable it must first exist and hence the fact that RSF entails F is a reasonable one.) It can be shown that the functor RS satisfies a modal semantics of the Brouwer type (Cf. Lewis and Langford [10].) Finally, let it be noted that unlike classical deontic logic, there is a functor in the present system that may be interpreted as a "conditional obligation operator"; namely, the operator ($/G$). Space does not permit further comment on this and other matters.

References

1. Kanger, S. New foundations for ethical theory. Privately distributed, Stockholm, 1957. Reprinted in Hilpinen [11]

2. Kanger, S. Law and logic. Theoria 1972. 38.3: 105-132

3. Mill, J.S. A system of logic. Longmans, Green, and Co., London, 1898 ed.

4. McLarty, C. Elementary categories, elementary toposes. Clarendon, Oxford, 1993

5. MacLane, S. Categories for the working mathematician. Springer, New York, 1971

6. Hofstader, A. & McKinsey, J. On the logic of imperatives. Philosophy of Science 1939. 6: 446-457.

7. Munroe, M. E. The language of mathematics. University of Michigan Press, Ann Arbor, 1963

8. Hansson, B. An analysis of some deontic logics. Noûs 1969. 3: 373-398. Reprinted in Hilpinen [11]

9. Johanson, A. Imperative logic as based on a Galois connection. Theoria 1988. 54.1: 1-24

10. Lewis, C.I. & Langford, C.H. Symbolic logic. Dover, New York, 1959

11. Hilpinen, R. (ed) Deontic logic: introductory and systematic readings. Reidel, Dordrecht, 1971

12. Birkhoff, G. Lattice theory. American Mathematical Society, New York, 1948

13. Kanger, S. & Kanger, H. Rights and parliamentarianism. Theoria 1966. 32.6: 85-115.

14. Snyder, P. Modal logic. Van Nostrand Reinhold, New York, 1971

Getting Personal
Some Notes on the Relationship Between Personal and Impersonal Obligation

Christen Krogh*
NRCCL and Department of Philosophy, University of Oslo
P.O.Box 6702 St. Olavs Plass, 0130 Oslo, NORWAY
http://www.uio.no/~krogh christen.krogh@filosofi.uio.no

Henning Herrestad
NRCCL, University of Oslo
P.O.Box 6702 St. Olavs Plass, 0130 Oslo, NORWAY
henning.herrestad@jus.uio.no

Abstract

The contribution of this paper lies in the analysis of the relationship between impersonal and personal deontic notions, keeping them as close to standard deontic logic as possible. In doing so, we offer arguments against reducing both personal notions to impersonal notions, and impersonal notions to personal notions. We discover that we need to introduce two new notions of impersonal obligation and permission. Finally we argue in favour of one particular representation of ought–to–do and permitted-to–do statements.

1 Introduction

Debates on deontic logic are largely carried out without revealing the identity of the persons involved. Obligations are commonly expressed by sentences such as OA, reading 'it is obligatory that A', even though they are usually thought of as obliging *some particular person* (c.f. Jaakko Hintikka's comment on this in [1, page 60].) The same consideration holds for permissions. In this paper, we 'get personal' when analysing obligations and permissions. In particular, we argue in favour of the need to be specific about who bears the obligation (permission, prohibition) by analyzing the relationship between personal and impersonal notions. We use one common reduction of personal to impersonal notions exemplified by Risto Hilpinen in [2] for opening up these issues. We end our considerations by arguing in favour of one particular representation of personal ought–to–do and permitted–to–do statements.

One motivation for getting personal can be found in applications of deontic logic in the analysis of rights. Here, there seems to be a particular need

*On sabbatical leave from SINTEF Information Technology. This work has been carried out within the the terms of reference of ESPRIT Basic Research Project No. 6471, MEDLAR II, and ESPRIT basic research working group no. 8319, ModelAge.

to be precise about whom the norms concern. In earlier publications (e.g. in [3]) we have argued that notions of personal obligation and permission[1] play a substantial part in the logical analysis of rights–relations between two parties. Another motivation can be found in the rather more pragmatically oriented area of multiagent systems (e.g., [4], and [5]). Here, we have argued the feasibility of applying impersonal and personal deontic notions when characterizing multiagent behaviour. However, we have not supplied a satisfactory treatment of the inter–relationship between personal and impersonal notions in these publications. The present paper is an attempt to correct this omission.

Below, we will refer to rules and principles for various modal operators by first giving the name of the operator and then the name of the rule/principle, e.g., we will refer to the rule of necessitation for the O–operator as 'O.RN', and the T-principle for the ${}_iE$–operator as '${}_iE$.T'. We name the rules and principles as Brian Chellas (c.f. [7]). The logics are classified by Lemmon codes (c.f. [6, page 51]), still following the practice of Chellas.

Throughout the paper we will seek to keep as close to standard deontic logic (SDL) as possible, e.g. as described in [8]. Unless otherwise stated, the logic of any operator O may be assumed to be of type KD. Operators ${}_iO$ may – unless otherwise stated – be considered as relativised KD operators. Throughout, we assume standard models for our deontic logics. Reasons for wanting to stay as close to SDL as possible are threefold. First, it reflects a 'minimal change' approach which enables us to build on existing intuitions and insights embedded in the received view, while also making it easier to point out exactly how the standard approach must be modified to accommodate new insights. Second, it makes our findings accessible to the wider audience who are already familiar with the standard approach. Third, it makes it easier to combine our findings with other, differently motivated deviations from the standard approach.

When necessary, we will assume a minimal account of agency in the tradition of Kanger [9], Pörn [10], and Elgesem [11]. We read sentences such as ${}_iEA$ as 'the person i sees to it that A (is the case)'. This minimal action logic may be axiomatised by the inference rule ${}_iE$.RE and the axiom schemata ${}_iE$.T. For our purposes, this minimal account will suffice. We argue that a full treatment of agency should at least have these properties. Semantics for a minimal account of agency may be supplied by minimal models in a straightforward manner.

2 Deontic notions, personal and impersonal

Sentences such as 'it is obligatory that ...', 'it is forbidden that', 'it is permitted that', are conventionally analysed by deontic logic. The operators of such logics are usually denoted by O (for 'obligatory'), P (for 'permitted'), and F (for 'forbidden'). Deontic notions expressed by non–agentive sentences qualified by this operator are conventionally considered impersonal. Examples of impersonal notions may be found in, e.g., normative systems such as national legislation or moral commandments.

Impersonal notions may be contrasted with personal notions. We will analyze sentences such as 'it is obligatory for (the person) i that ...', and 'i is under an obligation that ...', in terms of a personal deontic operator ${}_iO$. We read ${}_iOA$ as 'it is obligatory for i that A'. Similarly, sentences such as 'it is

[1]Previously called *special* obligation and permission.

permitted for i that ...', and 'i is permitted that ...' will be analysed in terms of the personal deontic operator ${}_iPA$. We read ${}_iPA$ as 'i is permitted that A'. A similar analysis for ${}_iF$ expressing personal prohibitions may be offered. Personal deontic notions typically result from acts such as making contracts or promises.

A simple argument in favour of the distinctiveness of the personal and impersonal notions may be found in considering the case where something is obligatory for one person, but not for another. A married man, for instance, has certain obligations an unmarried man does not have. Let m be a married man, and let u be an unmarried man, and let A express what m ought to do.[2] The sentence ${}_mOA$ expresses that it is obligatory for the married man that A, and $\neg_u OA$ expresses that it is not obligatory for the unmarried man that A. Having a collapse of impersonal and personal obligation (i.e., $OA \equiv {}_xOA, \forall x$), would cause an inconsistency here.

Having introduced personal and impersonal deontic obligations as expressing separate notions, we believe it pertinent to raise the question of how they relate to each other.

3 Relating impersonal and personal notions: the received view

The idea that there is a relation between what simply 'ought to be the case' and what is 'obligatory for someone (to bring about)' is not new. In [12, page 149], Chisholm quotes Samuel Clark (1706) on the idea, and in [13, page 6]), van Fraasen quotes G.E. Moore (1922). In [14], John Horty and Nuel Belnap mention that Alexius Meinong and Nicolai Hartmann put forward proposals concerning this at the beginning of the 20th century. In [15], J.L.A. Garcia discusses Alexius Meinong's views. In [16], Stig Kanger offers a treatment on the relationship in the context of formalising rights. Basically, the same ideas is expressed by Anderson in [17, page 143], as well as by Chisholm in [12, page 150].

In [1, p. 60], Jaakko Hintikka poses the question of whether or not, and if so, how, *personal* obligations of a particular bearer can be defined in terms of *impersonal* obligations. We are aware of several attempts to answer this question in the positive; e.g. Carl Kordig in [18], and Belnap and Horty in [14]. In the context of standard deontic logic and a minimal account of actions, the question is perhaps treated most clearly by Risto Hilpinen in the papers [19] and [2]. Here, he proposes to equate a sentence expressing the personal obligations of an agent by a sentence expressing the impersonal obligation that the agent does something:[3]

$$ {}_iOA =_{def} O({}_iEA) \qquad (1) $$

The equation (1) states that 'A is obligatory for i' is defined by 'it is obligatory that i sees to it that A'. We have here reflected Hilpinen's own reading of the various formulae (c.f. [2, Page 167]).

[2] We note that whether A is an action sentence or not seems to play no role here.

[3] Note that both Hilpinen's notation as well as his action logic differ slightly from ours. This does not, however, affect the arguments below.

Hilpinen offers a similar reduction of personal permission:

$${}_iPA =_{def} P\neg_i E\neg A \tag{2}$$

The equation (2) states that 'the person i is permitted that A' is defined by 'it is (impersonally) permitted that it is compatible with everything i does that A'. Our reading of $\neg_i E\neg$ as 'it is compatible with everything i does that ...' is in line with, e.g., Ingmar Pörn's treatment of action in [20, Page 3].

There are several problems with these definitions as we see them. In the following five sub–sections we will discuss that (i) we are unable to represent conflicts of personal obligation within Hilpinen's system; (ii) there is a problem concerning the bearer of the obligation; (iii) the definition of personal permission is counterintuitive; (iv) the system is unable to express private obligations; and (v) that the impersonal notions are asymmetric. The notion of asymmetry will be defined below.

By going through these problems, it is our intention to shed more light on both impersonal and personal notions. Subsequent to this discussion, we shall be constructive in offering a positive proposal for how to treat both impersonal and personal notions.

3.1 The inability to express conflicts of personal obligation

We hold that the proposed inter–relation of personal and impersonal obligation is unable to express interesting nuances related to conflicts of obligation.

We regard a conflict between two obligations to mean that it is impossible to fulfill both obligations simultaneously. Hence, if i has two conflicting obligations, he will violate one of them by fulfilling the other. Similarly, if two persons have mutually conflicting obligations, one of them will be violated by the other fulfilling his obligations.

It is not clear whether impersonal obligations may conflict at all. Some authors have claimed that systems of norms must be consistent (i.e., without conflicting obligations). Hilpinen, as well as Kanger (c.f. [16]), accepts that the O.D schema holds (i.e., $OA \supset PA$); hence, they both regard it as intuitively acceptable that a set of impersonal norms must be consistent, that is, without *conflicting obligations.* This is in accordance with well–established traditions in moral philosophy. In [21, pp. 196–198], Kant and W.D. Ross are described as authors who attempt to argue how a set of actual moral obligations must be a consistent set in spite of the appearance that moral obligations may be conflicting.

We are aware, though, that there is also a tradition for rejecting this hypothesis. When a set of norms is the result of multiple authorities issuing them, possibly at different times and sometimes without sufficient knowledge of what norms are already in force, we have what may be termed a *coordination problem.* In such situations it seems plausible that conflicting norms may be issued.

We regard the coordination problem as particularly acute with respect to personal obligations and permissions. This is because we hold that personal obligations and permissions typically result from acts such as making contracts or promises, acts by which particular individuals independently exercise a power to change their own or other people's normative status.

Consider the following examples, the first of which concerns obligations resulting from contracts:

(i) Due to fraud or bureaucratic mess, a house has two, new, legally entitled owners, each hiring a man to move furniture. The first man i is under an obligation to see to it that the house is empty of all furniture, the second man j is under an obligation to see to it that the house is not empty (perhaps full) of furniture.

The second example concerns obligations resulting from promises:

(ii) A man promises his friend that he and his wife will come to his friend's party, and his wife promises her mother that she and her husband will not go to the party (both women disliking the husband's friend).

Impersonal obligations expressed by operators adhering to SDL do not allow conflicts. On the reduction (1) above, we see that this is inherited by the personal obligations. This means that we would be at a loss for representing the examples given above by the formalisations of personal obligations offered by Hilpinen.

It is tempting to attempt to avoid this problem by allowing impersonal conflicts and thus personal conflicts of obligation, i.e., to reject the principle O.D for O. This, however, leads us to unwanted consequences for the personal notions. Consider the following argument:

${}_iOA \wedge {}_jO\neg A$	assumption	(3)
$O({}_iEA) \wedge O({}_jE\neg A)$	by 3, 1	(4)
$OA \wedge O\neg A$	by 4, ${}_iE.T, O.RM$	(5)
$O\bot$	by 5, O.C	(6)
$O({}_kEB)$	by 6, O.RM	(7)
$O({}_kE\neg B)$	by 6, O.RM	(8)
$O({}_kEB) \wedge O({}_kE\neg B)$	by 7, 8	(9)

In other words, we see that the following implication would be valid on rejecting the principle O.D:

$$({}_iOA \wedge {}_jO\neg A) \supset (O({}_kEB) \wedge O({}_kE\neg B)) \tag{10}$$

The principle (10) states that if two persons i and j have conflicting obligations, then a person k (possibly different from i and j) holds arbitrary, conflicting obligations. We thus see that giving up the consistency principle O.D for the impersonal obligation seems to have unfortunate consequences. We take the demonstrated inability of the current logical apparatus to express conflicting personal obligation to be one reason for rejecting the reduction (1).

3.2 A problem concerning the bearer of the obligation

A simpler, more direct problem with the reduction (1) is that we may infer obligations for seemingly unconcerned persons, i.e., that the personal obligation 'leaks' to other persons.[4]

[4] We owe this argument to José Carmo.

Let us assume that the chairman i of the board of a company is under an obligation that an accountant j perform a particular act, e.g., report on the financial status of the organisation to the Norwegian authorities. This is an actual obligation under Norwegian regulations. The chairman bears the obligation, not the accountant performing the task. The following formalisation offers itself:

$$_iO(_jEA) \tag{11}$$

We read (11) as 'it is obligatory for i that j sees to it that the financial status is reported'. Now, consider the following argument:

$$_iE(_jEA) \supset {_jEA} \qquad \text{from } {_iE.T} \tag{12}$$

$$O(_iE(_jEA)) \supset O(_jEA) \qquad \text{from 12, O.RM} \tag{13}$$

$$_iO(_jEA) \supset {_jOA} \qquad \text{from 13, 1} \tag{14}$$

From $_iO(_jEA)$, we may infer $_jOA$. In other words, from a personal obligation on i that j see to it that A, we may infer that j is under a personal obligation that A. Contrary to this, it may be argued that the responsibility as well as the obligation rests with i only, not j. This is the case under Norwegian regulations.

The following, perhaps simpler, example highlights the inadequacy of the implication (14). Consider that, as children are not legal subjects, their parents bears the obligation that their children keep the law. Hence, parents have personal obligations to influence their children to act in certain ways, e.g., to see to it that 'they attend school'. Again, we formalise this as $_iO(_jEA)$, where i is a parent, j the child of i, and A expresses that j attends school. By the formal argument above, we may infer a personal obligation on the child j. This is contrary to our assumptions.

We hold these two examples to illustrate another problem with the proposed reduction (1).

3.3 The definition of personal permission is counterintuitive

The definition of personal permission (2) states that 'the person i is permitted that A' is defined as 'it is (impersonally) permitted that it is compatible with everything i does that A'; i.e., $_iPA =_{def} P\neg_iE\neg A$. We note that this facilitates interdefinability of obligation and permission for the personal as well as the impersonal notions; i.e., def.$_iP$: $_iPA \equiv \neg_iO\neg A$ and def.P: $PA \equiv \neg O\neg A$.

There are, however, several problems with this definition. First, we observe that it is praxiologically quite weak. The personal permission that A does not state anything positive about what the person i is permitted *to do*. It only states that it is permitted that the person i does *not* see to it that the opposite is the case. We hold this to conflict with common intuitions about personal permissions as being permissions to bring about something.

In order to overcome this praxiological weakness, we may be tempted to offer the following, seemingly more intuitive, definition of personal permission:

$$_iPA =_{def} (P_iEA) \tag{15}$$

We immediately note that, with this definition, we will have to let go of either the interdefinability of personal permission and personal obligation (i.e., reject def.$_iP$), or the same relationship between impersonal obligation and impersonal permission (i.e., reject def.P). Upholding both the definition (15) as well as def.$_iP$ and def.P will lead to quite serious problems. Consider the following argument:

$$_iOA \equiv O(_iEA) \quad \text{from 1} \tag{16}$$

$$\neg_iP\neg A \equiv \neg P\neg(_iEA) \quad \text{from 16, def.P, } def._iP \tag{17}$$

$$_iP\neg A \equiv P\neg(_iEA) \quad \text{from 17} \tag{18}$$

$$_iPA \equiv P(_iEA) \quad \text{from 18, subst A for } \neg A \tag{19}$$

$$P(_iEA) \equiv P\neg_iE\neg A \quad \text{from 19, 15} \tag{20}$$

From this, we may (amongst other things) show that all the personal notions collapse (e.g., $\forall i.\forall j._iOA \equiv {_jOA}$). This is too high a price to pay. We will not yield to the temptation of accepting (15).

A second problem with the definition (2) is similar in structure to the problem we discussed in section 3.2. Here, personal permissions rather than personal obligations seem to 'leak'. Consider the following argument:

$$_jOB \supset O(_jEB) \quad \text{from 1} \tag{21}$$

$$_jOB \supset OB \quad \text{from 21, } {_jE.T}, O.RM \tag{22}$$

$$PB \supset {_jPB} \quad \text{from 22, contraposition, } def.P, def._jP \tag{23}$$

$$_iPA \supset P\neg_iE\neg A \quad \text{from 2} \tag{24}$$

$$_iPA \supset {_jP}\neg_iE\neg A \quad \text{from 23, 24, subst } \neg_iE\neg A \text{ for } B \tag{25}$$

From (25), we see that a personal permission for i (that A) lets us infer a personal permission for an arbitrary individual j (that it is compatible with everything i does that A). We hold this to be yet another problem with Hilpinen's reductions.

3.4 The inability to express private obligations

One may raise the question whether one should want to express *private* obligations – obligations that may not (should not) be fulfilled by any other person than the one who is subject to the obligation. Consider the following example: a man, i, is married to a woman. Being married, i enjoys certain rights and is subject to certain duties. Let us assume that one of these duties concerns the pregnancy of the wife (such a duty may be derived from catholic canonical law). We let A express the state of affairs that i's wife is pregnant. The expression $_iOA$ states that it is obligatory for i that i's wife is pregnant. One may now raise the issue that it makes sense to claim that no other person than i (and i's wife) should be permitted that A.[5] We see, however, that the following implication holds: $_iOA \supset {_jPA}$. In other words: that it is obligatory for a person i that A implies that an arbitrary person j is permitted that A. This relationship rejects the possibility of representing obligations that are private to one particular person.

[5] We owe this example to Dagfinn Føllesdal.

3.5 The impersonal notions are asymmetric

Trying to diagnose why we are unable to represent private obligations, we find that the two following principles are central:

$$_iOA \supset OA \quad (26)$$

$$PA \supset {_iPA} \quad (27)$$

(26) states that a personal obligation of i that A implies an impersonal obligation that A. (27) states that an impersonal permission that A implies a personal permission for an arbitrary person i. These two principles are direct consequences of the reduction schemes (1) and (2). Keeping in mind that the principle D holds for the impersonal notions, we may infer the problematic principle $_iOA \supset {_jPA}$ which blocks the possibility of representing private obligations.

Let the *scope* of an operator, denote the number of individuals concerned by the notion the operator expresses. Taking the impersonal permission operator, we see that, since $PA \supset \forall x.{_xPA}$ holds (by generalisation on (27)), the scope of P is the number of all individuals. By contrast, we see that the scope of the impersonal obligation operator is nil, since it need not concern any individuals.

When coupled with personal notions, we observe that impersonal obligation and impersonal permission are impersonal in different senses – strong and weak – according to their scope (how many persons they concern). They are not 'symmetric' with respect to their scope. *They are asymmetric.* It is this asymmetry which is the cause of the inability of representing private obligations within the current framework.

It may appear that the problem of asymmetry may be *identified* with the inability of representing private obligations. This is not so. We hold asymmetry to be problematic even though one rejects the importance of private obligations. The reason for this being that we find it intuitively unclear why impersonal obligation and impersonal permission should be different with respect to their scope. We also argue that it seems somewhat arbitrary that the scope of permission is larger than the scope of obligation.

In summary, we hold that we have pointed out yet another problematic point with Hilpinen's reduction of personal deontic notions to impersonal deontic notions.

3.6 Other relationships

Above, we have argued that a common reduction of personal to impersonal notions is problematic. The five points we have criticised may be largely divided into two groups: (i) the inability of the notions resulting from the reduction to express certain interesting nuances, and (ii) the leaking of norms to normatively unrelated individuals, contrary to common intuitions.

Below, we shall consider some further issues regarding the relationship between personal and impersonal notions. In particular, we shall spend some efforts discussing whether impersonal notions may be reduced to personal notions.

4 Reducing impersonal to personal notions: bridging the gap

Given that we accept that personal obligations and permissions may be represented by a deontic logic relativised to specific individuals (e.g., ${}_iO$ and ${}_iP$), we might ask whether this deontic notion can figure as a primitive notion, such that the notion of impersonal obligation and permission is defined in terms of personal obligations and permissions. We might consider the possibility of quantifying over different individuals, allowing expressions such as $\forall x.{}_xOA$ and $\exists x.{}_xOA$, and asking whether one of these expressions captures what is meant by an impersonal obligation. For instance, Bengt Hansson argues that:

> Just as "it is obligatory to do *p*" often means "it is obligatory for everyone to do *p*" the phrase "it is permitted to do *p*" often contains a tacit "for everyone".[22, page 246]

This suggests the following pair of definitions of impersonal obligation and impersonal permission:

$$(\forall x.{}_xOA) \equiv OA \tag{28}$$
$$(\forall x.{}_xPA) \equiv PA \tag{29}$$

However, Hansson points out that, as a consequence, we must give up def.P (i.e., the interdefinability of non–relativised obligation and permission). One might also give up interdefinability of bearer–relativised obligation and permission, or some other yet more basic relationship.[6] Given what we have said about our minimal change approach, we shall hold this to be a reason for rejecting the pair of definitions (28) and (29): in [23] von Wright wanted def.P for (an impersonal) deontic logic as it conveyed that obligation and permission are related to each other in the same way as necessity and possibility: A is necessary if and only if its negation, $\neg A$, is not possible, and similarly, A is obligatory if and only if $\neg A$ is not permitted. According to Simon Knuuttila (cf. [24]), this intuition was stated as far back as to the 14th century.

Jaakko Hintikka points out that obligations are usually thought of as obliging *some particular person*, and that permissions are similarly 'personal' (cf. [1, page 60]). This suggests the following pair of definitions of impersonal obligation and impersonal permission:

$$(\exists x.{}_xOA) \equiv OA \tag{30}$$
$$(\exists x.{}_xPA) \equiv PA \tag{31}$$

However, here we meet the same problem as with Hansson's suggestion; we have the same choice: either to give up ${}_iOA \equiv \neg{}_iP\neg A$, or $OA \equiv \neg P\neg A$, or to give up the definitions (30) and (31). Again, we argue in favour of the latter.

We may also consider choosing one definition from each of the above pairs, in order to uphold interdefinability of both the impersonal and the personal

[6] Assume (29). By def.P we get the equivalence $OA \equiv \neg(\forall x.{}_xP\neg A)$. By elementary quantification theory, we get the equivalence $\neg(\forall x.{}_xP\neg A) \equiv \exists x.(\neg_x P\neg A)$. By assuming def.${}_iP$, we get the equivalence $\exists x.(\neg_x P\neg A) \equiv \exists x.{}_xOA$. Hence, we have $OA \equiv \exists x.{}_xOA$. But, by (28), we have $OA \equiv (\forall x.{}_xOA)$, and hence we get the principle $(\forall x.{}_xOA) \equiv (\exists x.{}_xOA)$ which is clearly problematic for systems where there are more than one individual.

deontic operators; i.e., we may consider as definitions of impersonal obligation and impersonal permission the following pair (29) and (30):

$$(\forall x._x PA) \equiv PA$$
$$(\exists x._x OA) \equiv OA$$

Note that this pair of equivalences does not preclude upholding def.P as well as def.$_i P$. However, if both kinds of operators (relativised and non-relativised) adhere to the principles of SDL, we may prove from (30) a problematic instance of the axiom schema O.C (where we have replaced OA by $\exists x._x OA$ according to (30)):

$$((\exists x._x OA) \wedge (\exists x._x OB)) \supset (\exists x._x O(A \wedge B)) \qquad (32)$$

(32) may be read as 'if there is an individual for whom A is obligatory, and there is an individual (possibly different from the first individual) for whom B is obligatory, then there is an individual for whom both A and B are obligatory'. This is counter–intuitive: a janitor might be under an obligations to the effect that the floor in a building is swept clean every morning, and a minister of finance might be under an obligation that the rate of inflation is as low as possible. That there is a person for whom it is obligatory both that the floor is swept and that the inflation rate is as low as possible seems strange. This suggestion also has the problem of asymmetry described below. We thus reject the pair of definitions (29) and (30).

Finally, consider the opposite choice of definition from the first two pairs above; i.e., consider that (28) and (31) are chosen as definitions:

$$(\forall x._x OA) \equiv OA$$
$$(\exists x._x PA) \equiv PA$$

This pair of equivalences does not preclude upholding def.P as well as def.$_i P$, and does not allow the derivation of (32).

However, we still feel that there is something intuitively wrong about choosing this pair as definitions of, respectively, impersonal obligation and impersonal permission. The reason for this being that we have the problem of asymmetry as discussed above: if one says "it is obligatory that A", this implies that all individuals have a personal obligation that A, but if one says "it is permitted that A", is is only implied that there exists some individual who has a personal permission that A. If there is only one notion of impersonal obligation, and one notion of impersonal permission interdefinable with impersonal obligation, then it is intuitively more plausible that the scope of how many individuals are thereby personally obliged and permitted should be the same. In summary, we shall reject also this pair of definitions.

5 Conceptual explosion

In [22, page 246], Bengt Hansson gave up the interdefinability of impersonal obligation and permission to obtain symmetric inter–relations of his operators. He suggests the following arrangement (where the arrows indicate provable consequences):

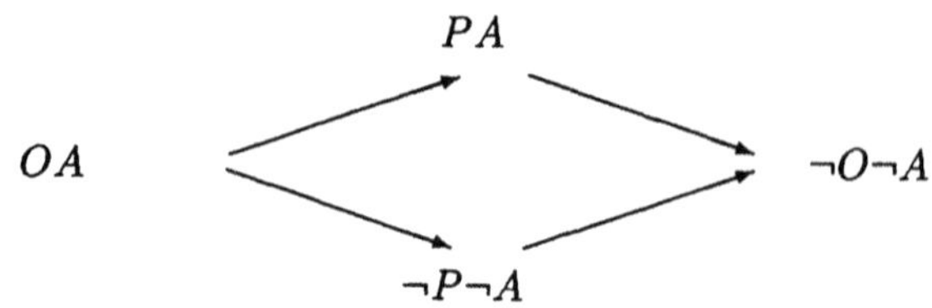

In order to solve the problem of asymmetry, there is a different path open to us; we may contend that there is *more than one* notion of impersonal obligation and permission that we are trying to capture. We may 'explode' the impersonal concepts into two inter–related pairs, one pair consisting of a strong obligation and a weak permission, the other pair consisting of a weak obligation and a strong permission. Let $O^{\oplus}$ and $P^{\oplus}$ represent the pair of strong obligation and weak permission. Let $O^{\otimes}$ and $P^{\otimes}$ represent the pair of weak obligation and strong permission. Consider the following intuitive motivation for holding that there may be two distinct notions of impersonal obligation in circulation.

The sentence "It is obligatory to wear clothes" appears to express an impersonal obligation that allows us to infer that all individuals have an obligation to wear clothes. Hence this sentence appears to express a notion of strong obligation. We believe that many general commands such as "Smoking is prohibited" or "Do not disturb" may be interpreted as expressing strong obligations. These sentences seems to confirm Bengt Hansson's argument that "It is obligatory to do p" often means "It is obligatory for everyone to do p", or the stronger relation that an impersonal obligation may be held to imply a personal obligation for everyone.

Now, consider the sentence "The door should be locked". Does this sentence also express an impersonal obligation, by which it may be inferred that each and every person has a personal obligation to lock the door? If so, every person who does not lock the door will have violated his or her personal obligation, regardless of the fact that some other person may have locked the door. Next, consider section 2 of Norway's Personal Data Registers Act of 1978 which states: "There shall be established a Data Inspectorate"[7]. Does this sentence allow us to infer that all individuals are under an obligation to establish a Data Inspectorate? In fact, does this sentence, in isolation, allow us to infer anything about certain individuals having a personal obligation to establish a Data Inspectorate?

There are certain ways of interpreting this law which we want to exclude. To abbreviate, we shall represent 'a Data Inspectorate is established' as A. First, one might say that what is expressed is not an obligation that A, but simply a preference for states of affairs where A is true over states of affairs where $\neg A$ is true. This is not the type of interpretation we are discussing here, though we admit that this might be one admissible interpretation, given the ambiguity of such sentences. Secondly, one might want to interpret such a sentence as referring to (or creating) an obligation for some public body, like the ministry of justice, to establish a Data Inspectorate. One might think that for sentences such as this and the previous example to be effective instruments in achieving the state of affairs described, there must exist institutional arrangements for

[7] An English translation of the Act is printed in [25, pp. 584 – 595]

fixing responsibilities. We do not accept, however, that facts about such arrangements must be taken into account in order to interpret the meaning of these sentences. We want to distinguish the analysis of the *meaning* of such sentences from the analysis of how they may be *used* to effect that someone locks the door or establishes a Data Inspectorate. These sentences may be regarded as meaningful sentences, even if they are entirely ineffective in fixing responsibilities. We may hold these sentences to express a *weak* notion of impersonal obligation which does not entail a personal obligation for any specific individuals.

The following bridge (reduction) principles (28) and (30) suggests themselves:

$$(\forall x._x OA) \equiv O^{\oplus} A \tag{33}$$

$$(\exists x._x OA) \equiv O^{\otimes} A \tag{34}$$

Conferring with our earlier argumentation we maintain that def.$P^{\oplus}$ and def.$P^{\otimes}$ should hold for these two impersonal obligation operators. Having $P^{\oplus} A \equiv \neg O^{\oplus} \neg A$, as well as $P^{\otimes} A \equiv \neg O^{\otimes} \neg A$ we see that the following relationships hold:

$$(\exists x._x PA) \equiv P^{\oplus} A \tag{35}$$

$$(\forall x._x PA) \equiv P^{\otimes} A \tag{36}$$

Going back to Hansson's diagram (above) we see that our notion of strong obligation ($O^{\oplus}$) corresponds to Hansson's operator O, our weak obligation ($O^{\otimes}$) corresponds to Hansson's expression $\neg P \neg$, our strong permission ($P^{\otimes}$) corresponds to Hansson's operator P, and our weak permission corresponds to Hansson's expression $\neg O \neg$. We note that the impersonal notions thus defined satisfy the relations of Hansson's scheme, above: the dual of the strong notion $O^{\oplus}$ is the weak notion $P^{\oplus}$, while the dual of the weak notion $O^{\otimes}$ is the strong notion $P^{\otimes}$.

The question arises whether we shall conceive the impersonal notions defined above as having their own logic, or whether we shall consider them completely determined by the given relationships with the personal notions. If we prefer the former, and we opt for a normal logic we will have the problem described in (32) reappearing for the weak notion for obligation. If we prefer the latter, there is a break with tradition (i.e., SDL) and we will be at loss for representing some interesting nuances described below.

6 A bridge too far

A direct argument in favour of rejecting the current bridge principles may be found when considering obligations which seem not to be associated with any one individual. Recall the example from section 5 which states that 'there shall be established a Data Inspectorate'. We offered an impersonal formalisation of this sentence, as OA. We now compliment the formalisation by holding it to be possible that both 'there shall be (established) a data inspectorate' and 'there does not exist a person for whom it is obligatory that there is (established) a data inspectorate' could hold at the same time; i.e., $OA \wedge \neg \exists i._i OA$. We consider this a counter–example to the reduction of some notion of impersonal obligation to either of the current renderings of impersonal obligation (strong or weak).

Two further, indirect, but in our view suggestive, examples may be found by considering a short diversion into investigating the (highly interesting, but difficult) notion of collective, or group, norms. A natural formalization of group norms, given the available machinery, would be to consider defining ${}^{g}O^{\oplus}A$ (to be read as 'the group g is under a (strong) obligation that A'), by the following expression:

$$ {}^{g}O^{\oplus}A \equiv (\forall x \in g.{}_{x}OA) \tag{37} $$

Of course, this will be a *strong* notion of group obligation, and its dual ${}^{g}P^{\oplus}A$ will be a *weak* notion, satisfying the relationship ${}^{g}P^{\oplus}A \equiv (\exists x \in g.{}_{x}PA)$. In order to avoid the problem of asymmetry for group obligations, we will duplicate our efforts above, and introduce a weak notion of group obligation which, by interdefinability, is coupled with a strong notion of group permission:

$$ {}^{g}O^{\otimes}A \equiv (\exists x \in g.{}_{x}OA) \tag{38} $$

The notion of strong group permission (${}^{g}P^{\otimes}$) is defined as the dual of weak group obligation.

We note that, when g is the unit set (i.e., contains only one person), the notion of strong and weak group obligations collapses. The same holds for strong and weak group permission. When g is the universal set (i.e., it contains all persons), strong *group* obligations collapse into strong impersonal obligation, weak *group* permissions collapse into weak impersonal permission, and so on. Obviously, a strong impersonal obligation implies a group obligation for any group g, and similarly a strong impersonal permission implies a group permission for any group g. The converse relationships do not hold. We also note that any group obligation implies a weak impersonal obligation, and that any group permission implies a weak impersonal permission.

In what follows, we shall argue that the notions of group obligation and permission defined above is inadequate for explicating some interesting examples. Analogously, we shall argue that the notions of impersonal obligation and permission defined above is inadequate for explicating similar examples.

Consider the following situation: at the Munch Museum in Oslo, school classes may enter for a special entrance fee. We say that there is a permission pertaining to such a group that they may enter for a special fare. We hold that from this (collective) permission, it should *not* be possible to infer that there is a permission for any *particular* individual in that group that (s)he may enter for a special entrance fee – individually. Note that in order to appreciate this example, it is necessary to appreciate the distinction between entering *as* part of a group, and entering *as* an individual.

Analysing the situation in terms of the set of group notions defined above, we find that both weak and strong group permission let us infer that there is someone in the group that has that permission (since ${}^{g}P^{\otimes}A \supset (\forall x \in g.{}_{x}PA)$ and ${}^{g}P^{\oplus}A \supset (\exists x \in g.{}_{x}PA)$). We claim this to be counterintuitive. The point we are making is that interesting (or genuine) group permissions do not pertain to one person in particular (or all the persons (in particular)) in the group. It is *the group* that 'carries' the permission, not the persons.[8] We note,

[8]It is no counter–argument to this example to offer a formalization of group permission

in passing, that there seems to be a controversy about whether group norms pertains to 'collective individuals', or whether they direct 'collective actions'. As is apparent, we have chosen the first approach. A similar argument for the latter approach seems readily devised.

An argument of similar structure as to one above, but of slightly different content, may be offered for group *obligation*. Reconsider the example from section 5, where some institution has issued the rule that "It is obligatory that the door is locked after five". This expression is vague in that no procedure is mentioned to fix responsibilities; i.e., we would neither want to infer that all members of the organization have an obligation to lock the door after five (as it is sufficient that one person locks the door), nor does the rule express which person is under an obligation to lock the door. This rule may be ineffective without a procedure to fix responsibilities, but it is not vacuous or meaningless. As the expression is meaningful, it should be possible to give a formal representation. However, any deontic operator implying or expressing an obligation on any specific individual would be too strong.

We view these group considerations as instrumental in contrasting the difference between notions pertaining to several persons, and the notions pertaining to one person. *The limit case of group notions is where the group consists of all individuals. Obligations for this group may, by analogy, be considered as different from an obligation carried by each and every individual person. The same holds for permission.* In summary, we argue that we have found arguments against the current bridge (actually reduction) principles between impersonal and personal obligation. In other words, we do not accept any of the discussed reductions of impersonal notions to personal notions.

7 Weaker bridges

In the above section, we have argued that we need an impersonal obligation–operator from which we *cannot* infer that there is an obligation for any one person, and a permission operator from which we *cannot* infer that there is a permission for any one person. This cannot be all, however. We need also other kinds of obligation and permission. The general prohibition to kill in most legislative bodies seems to be one such obligation. Freedom of speech, is commonly understood to be a permission which is generalisable for everyone. It seems that we need both strong and weak impersonal notions with weaker relations to the personal notions.

Hilpinen's reduction implies such a weaker bridge principle. We superscript the following impersonal deontic notions with an asterix to distinguish them from previously discussed notions.

$$\neg((\exists x. {}_xOA) \wedge P^*\neg A) \tag{39}$$

---

as $Mx \supset {}_xPA$, where Mx denotes membership in the group and where ${}_xP$ is the notion of personal permission from above. If x is a member of the group, then by modus ponens x will be personally permitted that A. This is exactly what we would like to avoid. Another argument against our formalisation that has been raised is that one may consider expressing group obligations by introducing special group 'persons' and relativising deontic operators to these. We find this proposal both intuitively unclear as well as ontologically taxing since it implies that there will be two kinds of persons, one which will be member of the other.

We see that if we assume interdefinability of obligation and permission for both the impersonal and the personal notions, we can infer the following principles: $(\exists i._i OA) \supset O^* A$, and $P^* A \supset (\forall i._i PA)$. This establishes O^* as expressing a weak impersonal notion, and P^* as expressing a strong impersonal notion.

In [18], Carl R. Kordig suggested the bridge–principle given below. Note that in order to differentiate between the different notions, we have renamed the impersonal notions stemming from Kordig's considerations O^+, and P^+:

$$\neg(O^+ A \wedge (\exists x._x P \neg A)) \tag{40}$$

If we (again) assume interdefinability of obligation and permission for both the impersonal and the personal notions, we can infer the following principles: $O^+ A \supset (\forall i._i OA)$, and $(\exists i._i PA) \supset P^+ A$. This establishes O^+ as expressing a strong impersonal notion, and P^+ as expressing a weak impersonal notion.

From the above naming, and discussion, we would expect strong obligation to imply weak permission (i.e. $O^+ A \supset P^+ A$), but we would *not* expect weak obligation to imply strong permission (i.e., *not* $(O^* A \supset P^* A)$). Thus we would want the principle D for O^+, but not for O^*. Another, perhaps more pragmatic, reason for rejecting the principle D for O^* is that adopting it would prevent us from representing conflicts of obligation between different persons:

$O^* A \supset P^* A$	Assumption	(41)
$_j OA \supset O^* A$	From 39	(42)
$_j OA \supset P^* A$	From 41, 42	(43)
$O^* A \supset {}_j PA$	From 43, contraposition, df.$_i P$, df.P,	(44)
$_i OA \supset O^* A$	From 39	(45)
$_i OA \supset {}_j PA$	From 44, 45	(46)
$\neg(_i OA \wedge {}_j O \neg A)$	From df.$_i P$, 46	(47)

In other words assuming the principle D for O^* (and keeping the rest of the logics as before) contradicts the possibility of normative conflict between different persons. We do, however, consider it plausible that weak obligation implies weak permission:

$$O^* A \supset P^+ A \tag{48}$$

Taking the contraposition of this, applying interdefinability principles for O^* and O^+, we find that it amounts to stating that strong obligation implies strong permission (i.e. $O^+ A \supset P^* A$), which is what we would expect. We will not go into the discussion of how to specify similar group notions, except to note that our strategy will be similar as to the one above.

In summary, given the bridge principles (39), (40), and (48), it seems that O^+ should have a modal logic of type KD, that O^* should have a modal logic of type K, and that to each person there should be associated an obligation operator ($_i O$) which have a modal logic of type KD. The following depicts the impersonal and personal notions currently under discussion in a scheme similar to Hansson's (c.f. section 5) (where the arrows indicate provable consequences):

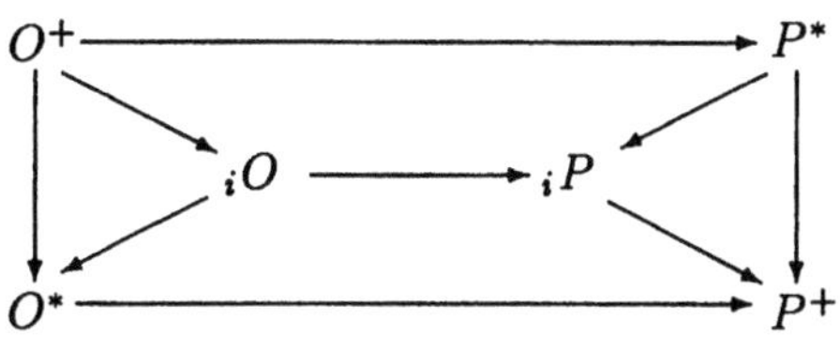

8 The logical price to pay

There is one slight problem with this, however. It is common when investigating certain aspects of norms by means of deontic logic that some unfortunate, or even paradoxical, aspects are left untouched. It is quite common, for instance, in investigations such as ours to accept for technical reasons 'norms' with logically determined contents, even though it is widely accepted that these are not real norms. In other words, the normal modal logical principle $O\,True$ is usually taken as a price one has to pay for having a simple and streamlined logical tool to work with. On our analysis, however, this problematic principle cannot be ignored.[9] If we let O^+True be a principle, it follows from $O^+A \supset P^*A$ that P^*True is a principle. By closure under equivalence and propositional logic, it follows that $P^*(\neg A \vee A)$. Assuming normality of P^*, we have by the principle $P^*.C$ that $(P^*\neg A) \vee (P^*A)$. By interdefinability of O^* and P^* and propositional logic, we have that the problematic D-principle for O^* has slipped back in: $O^*A \supset P^*A$. As we saw above, apart from being unintuitive (weak obligation implies strong permission), this rules out conflicts of obligations between different persons.

There are several ways in which this problem may be solved. The cause appears to be the validity of the already unintuitive principle O^+True. Rejecting this, we reject the normality (in a modal logical sense) of O^+. It seems that, if we do so, we should also reject the equally unintuitive (but not yet problem–causing) principles O^*True and $¡O\,True$. Several approaches reject similar principles for similar reasons (e.g., Jones and Pörn in [26]). That our reason for doing so is technical rather than intuitive seems to be no grounds for criticism, as the principles are usually kept for technical reasons.

Another solution would be to give up principle (48). This, however, leaves us with no adequate solution to the problem of asymmetry. Yet another solution would be to block the inference of P^*A from O^+A, when A is a tautology. In other words: to ensure that (48) only holds in cases where A is contingent. There may be other plausible solutions. We admit to not knowing which is the better approach. For the time being, however, we shall offer the following restricted bridge principle instead of (48), where $\diamond$ is the possibility operator of some suitable notion of necessity:

$$((\diamond A \wedge \diamond\neg A) \supset (O^+A \supset P^*A)) \tag{49}$$

We note that this blocks the problematic inference of P^*True. It can easily be seen that (49) is equivalent to: $((\diamond A \wedge \diamond\neg A) \supset (O^*A \supset P^+A))$.

[9] We owe this observation to José Carmo.

9 Ought–to–be versus ought–to–do

Our discussion was initially motivated by Hilpinen's consideration that a personal obligation for i that A can be represented as an impersonal obligation that it be the case that i sees to it that A (e.g., $O({}_iEA)$, cf. [27, page 42]). This proposal may, however, be seen as an attempt to kill two birds with one stone. First, there is the idea that personal obligations can be reduced to impersonal obligations. Second, there is the idea that what i ought to do can be represented as an ought–to–be statement concerning i's action.

Let us consider the idea that personal obligations can be reduced to impersonal obligations first. This idea may be held to be expressed by Hilpinen's reduction: ${}_iOA \equiv O_iEA$. We have argued that this reduction has to be rejected in view of employing only one kind of impersonal notion. It may be tempting to try to reconstruct Hilpinen's reduction by means of the enhanced set of impersonal notions as described above. In other words, we may consider offering the following reductions:

$$ {}_iOA \equiv O^*({}_iEA) \tag{50} $$
$$ {}_iPA \equiv P^+({}_iEA) \tag{51} $$

Even though this handles the problem of asymmetry, we note, however, that it still leaves us with the unwanted principle (10) discussed in section 3.1:

$$ ({}_iOA \wedge {}_jO\neg A) \supset ({}_kOA \wedge {}_kO\neg A) $$

Because of this, we will not yield to the temptation of offering yet another reduction of personal to impersonal notions. We thus reject the first idea.

Concerning the second idead, Castañeda writes:

> Deontic statements divide neatly into: (i) those that involve agents and actions and support imperatives, and (ii) those that involve states of affairs and are agentless and have by themselves nothing to do with imperatives. The former belong to what used to be called the *Ought–to–do* and the latter to the *Ought–to–be*. [28, page 452]

Hence Castañeda seems to accept two different meanings of ought–statements, on the one hand expressing obligations or duties (i.e. imperatives) and on the other hand expressing what is desirable, or preferable, or good (i.e. value judgements). This distinction is common to moral philosophy, though what consequences to draw from it is controversial (cf. [29], [15], [30]). Castañeda considers this a *neat* distinction where statements of preference are always impersonal and concerning states of affairs and statements of obligation are always personal and concerning actions. In our view this distinction is not so neat as we accept obligations to be of different types - personal as well as impersonal, concerning actions as well as concerning states of affairs. Even though we reject the idea that that what i ought to do can be represented as an ought–to–be statement concerning i's action, we do consider our previously discussed personal obligation operator suited for this purpose. To express Castañeda's *Ought–to–do* statements, and the corresponding statements of permission, we shall write:

$$ {}_iO({}_iEA) \tag{52} $$
$$ {}_iP({}_iEA) \tag{53} $$

We note that (52) and (53) do not have a useless iteration of indices. First, we hold that an impersonal notion of obligation is unsuited to convey ought–to–do expressions satisfactorily without the complement expressing actions in some sense. Second, expressions such as ${}_iOB$, where B does not express any action of i, seems to have their own separate meaning. Consider the case where the manager of a firm is under an obligation that the company's financial status is reported to the company board once a month. Let us assume that this manager has a particularly helpful assistant. Without the manager's consent this assistant sends the financial status to the board each month, thus seeing to it that the manager's obligation is fulfilled. In such a situation, we would formalise the manager's obligation as ${}_iOA$, where A expresses that the company's financial status is reported to the company board. As far as we are concerned, the managers obligations are personal, but may be fulfilled by someone else.

Expressions such as ${}_iO({}_iEA)$, however, are useful when analysing ought–to–do expressions. Consider the situation where a manager is under an obligation to report his personal opinion regarding the company's financial status. This is not something easily discharged by the helpful assistant. Here, the manager's obligation concerns a particular act of his on the reporting of his personal opinion. Here, we would formalise the manager's obligation as ${}_iO({}_iEA)$, where A expresses that the manager's personal opinion is reported. For this obligation to be fulfilled, the manager has to (personally) see to it that A is the case.

In summary, we hold (52) and (53) to be the best proposals for personal ought–to–do and permitted–to–do expressions within the current framework.

10 Conclusion

We would like to point out, once again, that having a clear conception of the relationship between personal and impersonal deontic notions is important for two application areas, with which we have been concerned: the analysis of rights in terms of deontic logic, and the characterisation of multiagent systems.

The contribution of this paper lies in the analysis of the relationship between impersonal and personal deontic notions, keeping as close to standard deontic logic as possible. We have given arguments against both the reduction of personal notions to impersonal notions, and impersonal notions to personal notions. Finally, we have offered our view on how to represent personal ought–to–do statements.

We have not intended to offer a full analysis of impersonal and personal notions in their own right. Our concern has been to point out some problems pertaining to their relationship, and to offer remedies for these.

Acknowledgments

We would like to thank Andrew J.I. Jones, José Carmo, Risto Hilpinen, David Makinson, Layman Allen, and the anonymous referees for offering valuable comments on earlier versions of this paper. Lee Bygrave and Jens Petter Berg has supplied us with jurisprudential intuitions. Thanks to Andrew Preston for language corrections.

References

[1] Jaakko Hintikka. Some main problems of deontic logic. In Risto Hilpinen, editor, *Deontic Logic*, pages 59–104. D. Reidel Publishing Company, Dordrecht - Holland, 1971.

[2] Risto Hilpinen. On the semantics of personal directives. In Carl H. Heidrich, editor, *Semantics and Communication*, pages 162–179. North-Holland Publishing Company, Amsterdam, 1974. An earlier version of this paper was published as [19].

[3] Henning Herrestad and Christen Krogh. Obligations directed from bearers to counterparties. In *Proceedings from ICAIL'95*, Washington, May 1995. ACM Press. Electronically available at http://www.uio.no/~christek/papers/-ICAIL95b.ps.

[4] Christen Krogh. Obligations in multiagent systems. In *Proceedings of the fifth Scandinavian conference on AI*, Trondheim, May 29 - 31 1995. ISO Press. Electronically available at http://www.uio.no/~christek/papers/SCAI95.ps.

[5] Christen Krogh. The rights of agents. In Michael Wooldridge, Klaus Fischer, Piotr Gmytrasiewicz, Nicholas R. Jennings, Jörg P. Müller, and Milind Tambe, editors, *Proceedings of IJCAI'95 WS on Agent theories, languages and architectures*, pages 140–151, Montreal, August 20 - 25 1995. Preproceedings. Proceedings to be published by Springer Verlag, 1996. Electronically available at http://www.uio.no/~christek/papers/IJCAI95.ps.

[6] E.J. Lemmon. *An Introduction to Modal Logic.* American Philosophical Quarterly, monograph series vol. 11, Basil Blackwell, Oxford, 1977. Written in 1966, in collaboration with Dana Scott. Edited by Krister Segerberg.

[7] Brian F. Chellas. *Modal Logic - an Introduction.* Cambridge University Press, Cambridge, 1980.

[8] Risto Hilpinen and Dagfinn Føllesdal. Introduction to deontic logic. In Risto Hilpinen, editor, *Deontic Logic*, pages 1–35. D. Reidel Publishing Company, Dordrecht - Holland, 1971.

[9] Stig Kanger. Law and logic. *Theoria*, 38:105–132, 1972.

[10] Ingmar Pörn. *Action Theory and Social Science.* D.Reidel Publishing Company, Dordrech - Holland, 1977.

[11] Dag Elgesem. *Action Theory and Modal Logic.* PhD thesis, Department of Philosophy, University of Oslo, 1993.

[12] Roderick M. Chisholm. The ethics of requirement. *American Philosophical Quarterly*, 1(2):147–153, 1964.

[13] Bas C. van Fraassen. Values and the heart's command. *The Journal of Philosophy*, 70:5–19, 1973.

[14] John F. Horty and Nuel Belnap. The deliberate stit: A study of action, omission, ability, and obligation. *Journal of Philosophical Logic*, 1995. Forthcoming.

[15] J.L.E Garcia. The tunsollen, the seinsollen, and the soseinsollen. *American Philosophical Quarterly*, 23(3):267–276, July 1986.

[16] Stig Kanger. New foundations for ethical theory. Technical report, Stockholm University, Stockholm, 1957. Modified version republished as [27].

[17] Alan Ross Anderson. Logic, norms and roles. *Ratio*, 4(36):36–49, 1962.

[18] Carl R. Kordig. Relativized deontic modalities. In Alan Ross Anderson, Ruth Barcan Marcus, and R.M. Martin, editors, *The Logical Enterprise*, pages 221–257. Yale University Press, New Haven, 1975.

[19] Risto Hilpinen. On the semantics of personal directives. *Ajatus*, 35:141–157, 1973.

[20] Ingmar Pörn. *The Logic of Power*. D. Reidel Publishing Company, Dordrecht - Holland, 1970.

[21] David McNaughton. *Moral Vision - An Introduction to Ethics*. Blackwell Publishers, Oxford, 1988.

[22] Bengt Hansson. Deontic logic and different levels of generality. *Theoria*, 36:241–248, 1970.

[23] Georg Henrik von Wright. Deontic logic. *Mind*, 60:1–15, 1951.

[24] Simo Knuuttila. Deontic logic in the forteenth century. In Risto Hilpinen, editor, *New Studies in Deontic Logic*, pages 225–248. D. Reidel Publishing Company, Dordrecht - Holland, 1981.

[25] Wayne Madsen. *Handbook of Personal Data Protection*. Macmillan Publishers, UK, 1992.

[26] Andrew J.I. Jones and Ingmar Pörn. Ideality, sub-ideality and deontic logic. *Synthese*, 65:275–290, 1985.

[27] Stig Kanger. New foundations for ethical theory. In Risto Hilpinen, editor, *Deontic Logic*, pages 36–58. D. Reidel Publishing Company, Dordrecht - Holland, 1971. (Published as a privately distributed pamphlet in 1957).

[28] Hector Neri Castañeda. On the semantics of the ought–to–do. *Synthese*, 21:449–468, 1970.

[29] Gilbert Harman. Human flourishing, ethics and liberty. *Philosophy and Public Affairs*, 12:307–322, 1983.

[30] Sven Ove Hansson. Norms and values. *Crítica*, 23(67):3–13, 1993.

Must I Do What I Ought?
(or Will the Least I Can Do Do?)

Paul McNamara
Philosophy, University of New Hampshire
Durham, USA

Abstract

Some key pre-theoretic semantic and pragmatic phenomena that support a negative answer to the main title question are identified and a conclusion of some significance is drawn: a pervasive bipartisan presupposition of twentieth century ethical theory and deontic logic is false. Next, an intuitive model-theoretic framework for "must" and "ought" is hypothesized. It is then shown how this hypothesis helps to explain and predict all the pre-theoretic phenomena previously observed. Next, I show that the framework hypothesized possesses additional expressive and explanatory power, thus adding further confirmation that it is on the right track.

1 Introduction

With the linguistic turn in philosophy this century, philosophers began to pay close attention to language, in particular, to words and phrases that were taken to express concepts of traditional philosophical concern. Although the importance of language may have been exaggerated at times, the basic point seemed undeniable: language contains vital clues about concepts of traditional philosophical concern and it is simply foolish not to garner the potential benefits for philosophical theory. But this venture is not without risks. A community of scholars might mis-take an expression to be continuous with some concept of philosophical concern that it is in fact not continuous with, and they might do so largely unreflectively, with all the risks of potential confusion that unexamined assumptions can typically engender. I would like to suggest that just such an assumption has pervaded ethical theory and deontic logic this century. And it is this assumption that I will argue is mistaken--by arguing for a negative answer to the main title question.[1] The latter will be done by reflecting

---

1 This paper is roughly a concise version of [1]. However, in addition to being more detailed, [1] also discusses utilitarianism and supererogation, and does not focus exclusively on <u>deontic</u> contexts for the modal auxiliaries, but rather on a general account of the modals. I originally suggested a negative answer to the main title question in [2]. Two other philosophers who have argued nay (in the

on various pre-theoretic pragmatic and semantic data. I will then turn to the constructive task of sketching and supporting a simple semantic and logical framework that appears to account for all the earlier data cited--and thus does not make the problematic assumption in question. I then reply to the main objection that I anticipate. I then identify some additional independent confirming evidence for the framework, and expand it accordingly. This can be seen as an attempt to make good on my claim in the concluding remarks of [8]. Finally, I consider a slightly weaker semantic and logical framework that allows for conflicting "ought"s, but not "must"s.

2 Some Data & Its Significance

Consider the following groups of expressions:

S *must* see to it that p.
S has to see to it that p.
S is obligated to see to it that p.
S is required to see to it that p.
It is S's duty to see to it that p.
It is imperative that S see to it that p.
It is incumbent on S to see to it that p.

S *ought* to see to it that p.
S should see to it that p.
It is morally advisable that S see to it that p.
It is morally preferable that S see to it that p.
It is morally best that S sees to it that p.
It is morally most appropriate that S see to it that p.
It is morally ideal that S see to it that p.

Ignoring subtleties, the grouping seems quite natural.[2] The members of the first group appear to be uniformly stronger than those of the second. Let's focus on "must" and "ought", and let's assume throughout that "S must (ought\can't) __" is short for "S *overridingly* must (ought\can't) __". Similarly, read "x can __" as "all things weighed, x can __". Doing so simplifies matters and the main point would not be substantially affected by recasting things to allow for the general case ([9],[10]). Now it looks like "must" properly implies "ought": what I must do, I ought to do, but not necessarily vice versa. And appearances aren't deceiving here.

---

context of general reflections on the modals) are Wertheimer [3] and White [4]. Jones and Porn [5] are the only deontic logicians I know of who expressly try to accommodate a negative answer to our main title question. Cf. also Robertson [6] and Chisholm [7] (in passing).

2 No pretense is made that there are not nuanced differences within each group. The grouping is just an intuition pump to get us started.

Imagine that we drive to work together and there are two routes we often take, respectively involving two exits on a certain highway. I, but not you, know that a certain bridge on the route involving the second exit is closed for repairs this week. As a result, the route involving the first exit is the only acceptable route today. At the last minute, as we come to Exit 1, I remember and say: "You *ought* to turn here". What I have said is true of course, but it is also inappropriate. For by saying "You *ought* to turn here", I leave open the possibility that the second route (the one you were intending to take) is still an acceptable alternative. I have thus misleadingly suggested that the first exit is merely preferable and hence optional. In contrast, had I said "You *must* turn here", there would have been no such suggestion of optionality. To say "You must turn here" is not to assert that turning here is the preferable alternative (though it implies this), but that it is the only acceptable alternative--period! Thus, by saying what I did, and not what I could have said, I have misled. And we can easily imagine that when you find out about the bridge, you will be a bit irked: "Paul, if you knew Exit 2 was out of the question today, why didn't you say 'You *must* turn here' at Exit 1? Had you, I would have veered off and we wouldn't be wasting our time now doubling back, destined to be late!"

This simple pragmatic phenomena, "ought"'s conversational implication of optionality, itself provides a very strong argument that "ought" does not semantically imply "must". For suppose it did. Well, to say that an action is *optional* is to say you can do it and you can skip doing it. But then since "must" clearly implies non-skip-ability, and hence non-optionality, "ought" would be semantically incompatible with something it nonetheless routinely conversationally implies. To be sure, this sort of thing does happen in extreme cases of sarcasm, such as when one says "Oh, now that's just GREAT!", precisely to conversationally imply that things are "anything but". However, there are none of the trademarks of such indirectness in the present case.

And what is it *to be late for work*, anyway? Is it to arrive later than you *ough*t? Surely this is too weak. To be late for work is to arrive later than you *must*, not merely later than you ought. More generally, what is a *deadline*? Is it merely a time by which something ought to be done? No, it is a time by which something must be done. Indeed, is it not often the case that as a deadline approaches, there is a period when you ought to be giving so much time and energy to meeting the deadline, then later, as the window of opportunity to fulfill your commitment diminishes, there is a period where you must do so? Similarly, to say that something is a *job requirement*, is to say that it is part of the job that the thing must be done, not merely that it ought to be done. To say that something is a *law* is to say that it is a legal directive that must be obeyed, not a legal counsel that ought to be followed. And so forth.

The above idioms seem to call for a strong modal. "Ought" seems too weak semantically to fit the bill. In contrast, "must" seems to give just the semantic strength called for. So this differential constitutional contribution also seems to support our semantic hypothesis: that "must" properly implies "ought".

And while we"re at it, what do we do with these words? If your boss tells you that you *must* do some task, she is typically *commanding* you to do it or *insisting* that you to do it. If she tells you that you *ought* to do it, she is typically *recommending* the action or *advising* you to do it. Ignore the one and you may not get a raise; ignore the other and you may find yourself on the soup line. If a queen addressing her subjects says that they ought to do such and such (e.g. stock food for the winter), she is advising or recommending a course of action. However, if she tells them they must do such and such (e.g. supply food to her warring army) she is commanding her subjects or *laying down the law*. Ignore the one and you may lose favor; ignore the other and you may lose your head. And what editor worth a hoot would fail to redmark a script where as they drag you out, you plea for your life by saying: "Your majesty, you *ought to* spare me! My children!"? Surely "must" provides the right drama here. And if any declarative routinely "implies" an imperative, it is surely the "must" declarative (cf. Hare [11]). If you don't believe me, the next time you want to make sure that Junior cleans his room, try: "Junior, you *ought to* clean your room before you go out", and let me know how you make out. And so forth.

Now where do "can" and "can't" fit in generally? To say you *can* turn here is to say that it is not the case that you must not turn here, to say you *can't* turn here is to say you mustn't turn here and to say you *must* turn here is to say you can't fail to turn here. Using obvious abbreviations:

a) CANp <-> ~MUST~p
b) CANTp <-> MUST~p
c) MUSTp <-> ~CAN~p

But if we substitute "ought" for "must" in these equivalences, all prima facia plausibility is lost. Although "it is not the case that you ought to not turn here" implies "you can turn here", "you can't turn here" implies "you ought to not turn here", and "you can't skip turning here" implies "you ought to turn here", the implications are all proper--their converses fail:

d) CANp <- ~OUGHT~p (holds)
e) CANp -> ~OUGHT~p (fails)
f) CANTp -> OUGHT~p (holds)
g) CANTp <- OUGHT~p (fails)
h) OUGHTp <- ~CAN~p (holds)
i) OUGHTp -> ~CAN~p (fails)

For example, to see that the first converse implication, e), fails, imagine that the bridge will merely slow us down a bit, so that although Exit 1 is preferable, Exit 2 will still get us to work on time. Then I can turn off at Exit 2 even though I ought to still turn at Exit 1 and not at Exit 2. Similarly for g) and i). There is nothing unusual about this. Often enough, you don't have to do what you ought to do, and so you can do what you ought not do, morally speaking.

In fact, the failure of each of these "ought"-"can" equivalences is already a logical consequence of "must"'s properly implying "ought", conjoined with the above "must"-"can" equivalences, propositional logic and the intersubstitutability of p and ~~p in the scope of "must" and "ought". Consider the conjunction of d) and e), the "ought" analog to the first "must"-"can" equivalence, a). Now in keeping with "must"'s properly implying "ought", assume OUGHTp, but ~MUSTp, for some p. Then given OUGHTp, OUGHT~~p. But now from e) we'd get ~CAN~p, and then from a), we'd get ~~MUST~~p, and thus MUSTp--contrary to the hypothesis. Similarly for the "ought" analogues to b) and c).

Now these implicational failures are crucial for ethical theory and deontic logic. For it is uncontroversial, or at least it should be, that "can" and "can't" express *permissibility* and *impermissibility*, respectively. I hesitate because some people think "can" and "can't" don't belong here, rather "may" and "may not" do instead. But this rests on a confusion. "May" is typically associated with the giving and getting of permission, not with permissibility. You ask Sister Marie "May I go the rest room?", by way of getting *her permission* to leave the room. If she says "Yes", she thereby makes it permissible to do so. On the other hand, if you are wondering whether it is permissible to park your car on a given spot, you might ask a passerby "Can I park here?" to find out if it is permissible to do so, not to get the *passerby's* permission. "May I park here?" is best saved for the parking attendant. So "can" and "can't" do routinely express *permissibility* and *impermissibility*, respectively. But then the moral is obvious. "Must", but not "ought", expresses whatever ethicists and deontic logicians have virtually uniformly taken "ought" to express: *moral or deontic necessity*. For the latter has routinely been taken to be whatever satisfies the familiar definitional equivalences involving permissibility and impermissibility. And this means that, contrary to a dominant bipartisan trend this century, we can't take "ought" as basic and then assume that what is permissible is whatever satisfies "~ought~", nor that what is impermissible is whatever satisfies "ought~"--and for reasons that have nothing to do with the possibility of moral dilemmas. For we now see that this move rests on a false presupposition about what "ought" expresses. Yet it is this presupposition that constitutes a major rationale for the fact that "ought" has probably been the most talked about moral expression in Anglo-American ethical theory. It has been taken to be the dual of permissibility and thus to be continuous with the traditional concern with *that* notion. Similarly, in deontic logic, one is hard-pressed to even find a text where deontic necessity is not paraphrased as "ought". Just consider the fact that contrary to duty imperatives are almost invariably expressed as iffy-oughts, and then monadic deontic necessity, permissibility and impermissibility are routinely defined in terms of iffy-oughts conditional on a tautology. Indeed, in one of the landmarks in our discipline, Stig Kanger's [12] deontic logic is characterized as "a logic with some means for expressing the notion of ought", thus suggesting that "ought" is definitive of our discipline. And as we will see momentarily, this conflation has tended to support the historically dominant paradigm in the semantics of deontic logic as well. So the twentieth century Anglo-American concentration on philosophically significant language has missed its intended mark here. By focusing on "ought", we have lost the continuity we thought we had with the traditional concern for moral permissibility and moral necessity. Derivatively, we have conflated impermissibility with what I ought not do and we have conflated permissibility with what it is not the case that I ought not do. I believe that this mistake has led to other confusions ([1], [13]), but I wish now to turn to a more constructive task.

3 The Semantic Framework

There is a simple abstract framework for representing all these modals. In a given context, there is a non-empty set of *acceptable* alternatives. Let's think of these alternatives as some sort of maximal objects, say worlds. (Thus the use of the highway example, where the actions in question are turns at exits, but the alternatives are routes.) Now although all the acceptable alternatives are just that--acceptable, they needn't be on a par morally speaking. Some may be ranked higher than others and there may be ties (and thus non-trivially, *levels*). Relative to any world, i, we have the following picture:

Figure 1

The vertical rectangle represents the ranked i-acceptable worlds; a rectangle rather than a line is used to represent the possibility of ties; a "*" indicates the presence of an i-acceptable alternative (ruling out conflicts). Formally we define the relevant structures as follows ("DWE" for "Doing Well Enough"):

$F = \langle W, A, \leq \rangle$ is a DWE-Frame:
1. W is non-empty;
2. $A \subseteq W^2$ and A is serial;
3. $\leq \subseteq W^3$ where:
   a) ($k \leq_i j$ or $j \leq_i k$) iff (Aij & Aik), for any i,j,k in W;
   b) if $j \leq_i k$ and $k \leq_i l$ then $j \leq_i l$, for any i,j,k,l in W.

(Note that aside from confining the relata of each $\leq_i$ to the i-acceptables, clause 3a) also implies both the reflexivity and connectivity of the $\leq_i$.) As is customary, we let a DWE-model, *M*, be a DWE-frame coupled with a function that assigns a proposition (set of worlds) to each variable.

With such structures, we can interpret the moral uses of the modals as follows. What I *must* do (per *M* and i) is anything that I do in all of the i-acceptable alternatives; what I *can* do is anything that I do in some i-acceptable alternative, what I *can't* do is anything that I do in no i-acceptable alternative; and what is *optional* is anything that I do in some i-acceptable alternative but fail to do in some other. Finally, what I *ought* to do is anything I do in all i-acceptable alternatives ranked at least as high as some such alternative. Pictorially (for the case where there are upper and lower bounds):

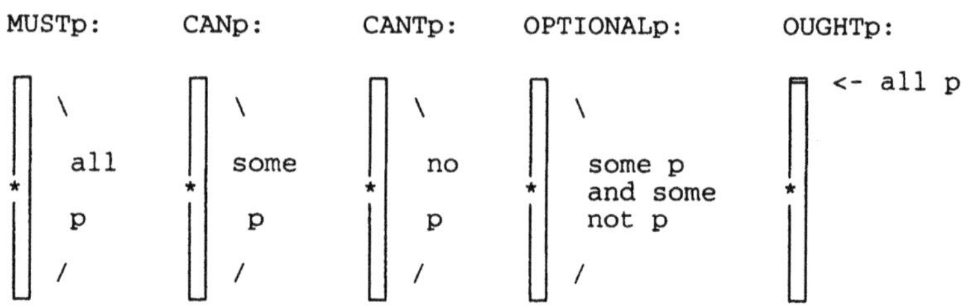

Figure 2

We might also want to enrich these structures and then add what is needed to get the result that the highest ranked acceptable alternatives are the same as the morally best alternatives ([14],[13]). Here I will just assume that this equivalence holds.

Of course, we need only take MUST and OUGHT as primitive, as the remaining three operators can be defined via "MUST", "~", and "&", as indicated earlier. Formally, where *M* is one of our models, we propose the following official truth-conditions:

$M]\!=_i$ MUSTp: (j)(if Aij then $M]\!=_j$ p).
$M]\!=_i$ OUGHTp: Ej(Aij & (k)(if j $\leq_i$ k then $M]\!=_k$ p)).
$M]\!=_i$ CANp: Ej(Aij & $M]\!=_j$ p).
$M]\!=_i$ CANTp: (j)(if Aij then $M]\!=_j$ ~p).
$M]\!=_i$ OPTIONALp: Ej(Aij & $M]\!=_j$ p) and Ej(Aij & $M]\!=_j$ ~p).

This scheme generates the following respective extensions of the familiar traditional deontic square (interpreted in the usual manner) and the traditional threefold partition of actions into those that are obligatory, forbidden or neither (optional):

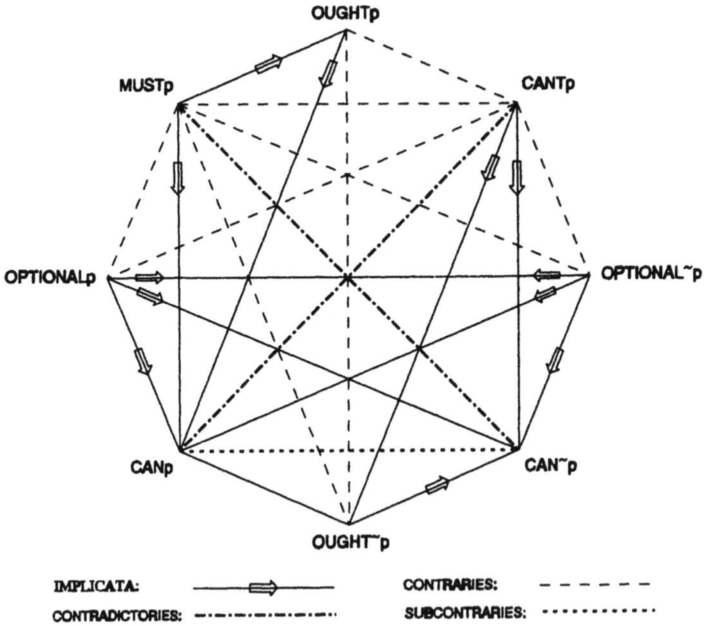

Figure 3: Deontic Octagon[3]

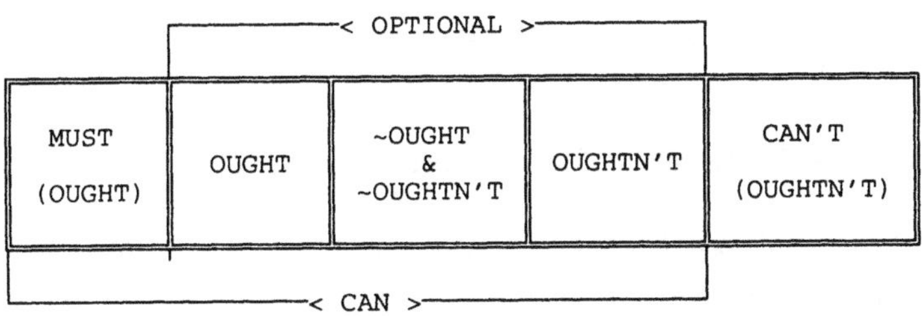

Figure 4: Five-Fold Partition

3 Given that the conditions at the six non-"ought" nodes can be expressed via negation and "must", we could generate a dodecagon by adding the analogous four missing nodes for "ought" (i.e. "~OUGHT~p", "~OUGHTp", "~OUGHTp & ~OUGHT~p" and "~OUGHT~p & ~OUGHT~~p). However, notice that there do not appear to be any single expressions reserved in English to express the "ought"-analogues of "can", and "optional" (which is why I left them out). This reveals yet another asymmetry between "ought" and "must". This Octagon is similar to deontic polygons independently arrived at by Joachim Hrushka and Jan C. Joerden [15]. See [13], Appendix D.

(The Five-Fold Partition is the double-lined area. The five cells designate mutually exclusive and jointly exhaustive classes. Operators tagged to single lines outside the partition highlight the location of non-exclusive classes within the partition.)

Although I will be turning to an extension of the above framework, my suspicion is that the following logic, call it "M-O", is determined by the set of DWE- models, (where "MUST" and "OUGHT" are the deontic operators added to a classical propositional language in the usual way, and "CAN" is defined as "~MUST~"):

A0. All tautologous DWE-wffs;
A1. a) MUST(p -> q) -> (MUSTp -> MUSTq)
b) OUGHT(p -> q) -> (OUGHTp -> OUGHTq)
A2. MUSTp -> OUGHTp
A3. OUGHTp -> CANp

R1:]- p and]- p -> q =>]- q
R2:]- p =>]- MUSTp.

Now where L is a language that results from adding a single unary operator, "*" to a classical propositional language, it is well known (Hilpinen [16], Aqvist [17]) that an SDL logic for L is axiomatizable as follows:

SA0. All tautologous L-wffs;
SA1. *(p -> q) -> (*p -> *q)
SA2. *p -> ~*~p

SR1:]- p and]- p -> q =>]- q
SR2:]- p =>]- *p.

Theorem: M-O contains an SDL fragment for "MUST", and one for "OUGHT".

Proof: (1) For "MUST" we need only note that "MUSTp -> ~MUST~p", the only missing SDL ingredient from M-O, follows trivially from A2 and A3 and the definition for "CAN". (2) In the case of "OUGHT", we need only "OUGHT" instances of both SA2 and SR2. The latter is trivially derivable from R2 and A2. The former requires a moments reflection. From R2, we get]- MUST(p -> (~p -> (p & ~p))). By A2, this yields]- OUGHT(p -> (~p -> (p & ~p))). But then from this and A1b), we get]- OUGHTp -> OUGHT(~p -> (p & ~p)). Now from A1b) itself, we have]- OUGHT(~p -> (p & ~p)) -> (OUGHT~p -> OUGHT(p & ~p)). So by hypothetical syllogism, we get]- OUGHTp -> (OUGHT~p -> OUGHT(p & ~p)), which by importation gives us]- (OUGHTp & OUGHT~p) -> OUGHT(p & ~p). Now from A3, and the definition of "CAN", we also have]- OUGHT(p & ~p) -> ~MUST~(p & ~p). But]- MUST~(p & ~p) follows directly from R2 (and A0), so we have]- ~OUGHT(p & ~p). But then from this and the second to last conditional, we get]- ~(OUGHTp & OUGHT~p), and so]- OUGHTp -> ~OUGHT~p.

Notice that our semantics diverges from the dominant classical paradigm in deontic logic. Here, deontic necessity is interpreted in terms of a set of *acceptable* worlds, a set whose members needn't be on a par morally speaking. In contrast, in the classical paradigm, deontic necessity is routinely interpreted as truth in all the *ideal* worlds. For example, in another landmark in deontic logic, Bengt Hansons's [18], when setting out the semantics for a reduction of deontic logic to alethic modal logic with an Anderson-Kanger constant, he tells us, regarding Q, the set of worlds interpreting the constant that ". . . the possible worlds in Q must be (morally) superior. There may be many worlds in Q, *but the differences between them cannot be morally relevant.* I will label the possible worlds in Q (morally) *ideal* worlds" (stress added). I think this is a mistake (cf. [19]), and I think it is no coincidence that on the ethical theory front this century, utilitarianism has been the dominant paradigm against which competing ethical theories have had to be measured. Surely there is little pre-theoretic support for the contention that morally ideal behavior is always morally mandatory. (Cf. the first observation in Makinson [20], pp.373-4.) Does the moral exemplar not at once often do what is morally ideal *and* what is morally optional? And isn't this precisely why we deem the moral exemplar exemplary? Morality recommends the exemplar's action, but it does not demand it. So we should expect morally relevant differences between the members of the set of worlds used to interpret deontic necessity and that set should not be identified with the ideal worlds. I think that the ideal worlds paradigm for deontic necessity has tended to support the conflation between deontic necessity and "ought", and more importantly, the conflation of deontic necessity with "ought" has tended to support the ideal worlds paradigm. Please don't misunderstand. "Ought" is an important moral notion, as is the notion of what is morally ideal. Without the higher flights of morality, although our life might not be nasty, brutish and short, it would surely be less rich, inspiring and promising than it sometimes is. But it is no mere quirk that deontic necessity has been of central and long standing traditional concern. It, after all, mandates morality's *demands*, not merely its recommendations. It tells us what is necessary, what we absolutely must do, whether or not we opt to do all that we might on morality's behalf. It draws the line below which we cannot go, and thus it is a moral notion of some urgency.

4. Interlude: Ockham's Erasure

It might be objected that "ought to do" is quite obviously ambiguous in moral contexts and that I am being naive not to recognize this. In particular, it might be protested that even though "ought to do" sometimes means something weaker than "must do", as I suggest, it nonetheless often means precisely what "must do" means, especially in moral contexts. For example, it might be suggested that it means just this ordinarily when one says "You ought to keep your promise".

I don't think it is obvious at all that "ought to do" is ambiguous in the way suggested and there are good theoretical reasons to deny the claim. We are often too quick in the face of objections to retreat with an "Oh, that's one sense of the term, but a different one than the one I'm explicating, so your objection doesn't apply." Often enough when this happens, we make a mistake that ordinary speakers make: we recognize that there are two salient sorts of situations in which a term applies and then wrongly conclude that the term is ambiguous, having one sense that applies only in one of the two sorts of situations. Consider for a moment the case of the grammatical conjunctive "or". "Or" is used in exclusive *cases* and in inclusive *cases*. This has lead many logicians, linguists and ordinary speakers to think there is an ambiguity. They speak of the "exclusive sense" of "or" and of the "inclusive sense" of "or". The exclusive sense is said to be stronger and to only apply to a subset of the situations where the inclusive sense applies. Notice that what drives the dual interpretation of "or" in this dispute is the exclusiveness of *the sentential complements* in supportive examples: "Either it's Tuesday or its Thursday", "Either she is dead or she is just missing", or even the infamous "Either you have been beating your wife again lately or you haven't. Just answer the question!". Such examples are used to support the alleged ambiguity. But this ambiguity contention for "or", however natural, is waning and univocality is winning among theoreticians. I think the same thing will be the case for "ought to do". There aren't two meanings of "ought to do" in moral contexts, one stronger and one weaker. There is just one meaning. What tends to drive the ambiguity hypothesis is that in uncomplicated circumstances, "You ought to keep your promise", "You ought not steal the car", "You ought to feed your children", "You ought not beat your children, etc. are true in cases where what you ought or ought not do, you also must or must not do. The reason is simple enough: the *things* the complements happen to express *are* things you must, or must not, do. But consider in contrast these: "You ought to give more to charity?", "You ought to do Bob the favor?", "You ought not do the favor only for Bob." "You ought not stay too long?". In uncomplicated moral contexts, these are naturally interpreted as *not* entailing their "must" analogues. So the fact that for some "ought to do" sentences it is natural to assume that the complements express things one typically must do, should not lead us to conclude that "ought" is there being used in "its strong sense" where it means what "must" means. No more than we should conclude from the fact that there are lots of situations where a sentence of the form *p or q* is true and it is plain that only one of the disjuncts can hold, that in such situations "or" is used in its "strong exclusive sense." Rather, in both case, we should follow *Ockham's Erasure* (Ziff's term, I believe):

Don't multiply *meanings* beyond necessity!

Certainly as a theoretical strategic maxim, this is a sensible one. Once we recognize that "ought" does have *a* sense that is weaker than "must", and one for which it is not the dual of "can", there seems little to be gained by insisting that it still has yet another stronger sense, one that is equivalent to "must" and is the dual of "can". (Much less that it is this meaning that is the *dominant* one, something that seems quite implausible once we recognize the various default pragmatic differences between "ought" and "must" noted above.) We can't always trust unreflective judgments of ambiguity, whether by an ordinary speaker or a professional philosopher, linguist or logician. Rather we have to look at a large cross-section of our perhaps conflicting intuitions and try to come up w\ the best overall theoretical account of the mess.

Although it is beyond the scope of this paper to delve into the modals generally, I believe that if we consider other contexts, we will find that in most of them, the same point I am making about "must" and "ought" in moral contexts holds as well. "Bob, you ought to see a doctor" is weaker than "Bob, you must see a doctor.", "Bob ought to use a napkin" is weaker than "Bob must use a napkin", "Bob ought to be home by now" is weaker than "Bob must be home by now", etc. And in each case their is a parallel conversational implication of optionality or openness, as well as parallel differences in default illocutionary force. We do not find any uniform ambiguity between a strong and weak meaning of "ought" across other contexts. What we do find is a striking uniformity of the relations of modals to one another across a wide variety of contexts. We should want to explain this fact, and I think that in devising a comprehensive theory for the modals, there will be strong theoretical pressure to deny the ambiguity alleged for "ought to do" in moral contexts.[4]

5 Explanation of the Data

This framework provides a simple explanation of the prior semantic and pragmatic observations, and thus perhaps serves as a simple model of what we should be looking for in an account of moral language. First, it predicts the semantic observations: "must"'s properly implying "ought, the validity of the equivalences involving "must", "can" and "can't"; the validity of the one-way implications involving "ought", "can" and "can't"; and the invalidity of the converse implications. For example, if all of my acceptable alternatives include p, then so do all of the best of these, but obviously not vice versa. Thus "must"'s proper implication of "ought".

Secondly, our proposal is not only compatible with "ought"'s conversational implication of "optionality", it explains it. For consider some relevant facts. We observed that "must" precludes optionality, and it does on our proposal, since if I must do something, then there is no acceptable alternative where I don't, as would be required if it were optional. In contrast, "ought" does not preclude optionality on our proposal. For the best of my acceptable alternatives can all involve something that fails to occur in some suboptimal but acceptable alternative. However, on our

---

4 Cf. [1] and Kratzer [21]. Although Kratzer's work on the semantics of modal auxiliaries is the best I am aware of, discussion of "ought", as opposed to "must" and "can", is conspicuously absent, and the truth-conditions she offers for 'must' (and derivatively for "can" and "can't") seem better suited to 'ought' (and derivatively to "not ought not" and "oughtn't").

proposal, "ought" is compatible with non-optionality, for "must" implies "ought". But now notice the crucial consequence. *The only* case where "ought" is taken to apply to something that is nonetheless non-optional is the case where "must", also applies to that thing. So the fact that "ought" rolls off my lips when the stronger and more informative "must" could have rolled off just as easily leads the hearer to infer that I take "ought", but not "must", to apply. And as we just saw, in all such case our semantic proposal dictates that the item in question is optional. Thus "ought"'s conversational implication of optionality. (Since "must" semantically precludes optionality on our proposal, it comes as no surprise that "must" has no such conversational implication.)

Thirdly, our proposal helps to explain the well-suitedness of "must", but not "ought", to the constitution of various modally loaded notions. Consider deadlines again. We saw that a deadline is not a time by which something ought to be done, but a time by which it must be done. Why? Well, to say that you *ought* to get something done by noon is merely to say that it is preferable or best to get it done by noon. It leaves open the possibility that it will be acceptable, though sub-optimal, to not get it done by noon. But to say you *must* get something done by noon is to say that it is simply unacceptable to not get it done by noon. It is to say that there are no acceptable alternatives, optimal or suboptimal, that don't involve getting the thing done by noon. Deadlines, by their very nature, restrict the acceptable execution of a task to times at or before the time of the deadline. Thus deadlines are not mere "ought"s, but "must"s. Similarly for the other cases.

Finally, our proposal makes the differences in illocutionary force unsurprising. If "must" means there are no acceptable alternatives that leave the item it is applied to out, then no wonder we use it to command or insist that the item be done. In doing so, we try to close off all alternatives that preclude the item by forcing the choice to be exclusively among alternatives that include it. And if "ought" means that the best of the acceptable alternatives include the item it is applied to, then no wonder we use "ought" to recommend and advise, and no wonder we don't typically use it to command or insist. For how can I non-covertly make an item mandatory for someone by merely saying it is preferable, when this is compatible with its being acceptable that the item is foregone?

6 Additional Expressive & Explanatory Power

As a bonus, the simple framework we have motivated here by considering modal auxiliaries in moral contexts also provides interesting analyses of other moral notions and thus provides some indication of what we should expect for a conceptual framework for common-sense morality. Here I provide a quick sketch, by way of confirming that our framework is on the right track.[5] We can say that p is *morally indifferent* if and only if p holds at some world in each level of the agent's acceptable alternatives, and similarly for ~p. We can thereby distinguish optionality from indifference, and it can be seen readily that *Urmson's Criterion* ([22],[6]), that indifference properly implies optionality, is satisfied.[6] Since p is *morally significant* if and only if it is not morally indifferent, p will be morally significant if and only if either p or ~p holds in all worlds at some level of the agent's acceptable alternatives. p is *involved in doing the minimum morality demands* if and only if there is some acceptable world where all acceptable worlds ranked as low are p-worlds. A proper scheme for common sense morality (and especially for going beyond the call) must represent this neglected notion. p will be *beyond the call* if and only if p is permissible but ~p is involved in doing the minimum, that is, if and only if there is some acceptable world where all acceptable worlds ranked as low are ~p-worlds while nonetheless p still occurs in some acceptable (and hence higher ranked) alternative. Thus as we should expect, p will be beyond the call only if it is optional, non-indifferent and precluded by doing the minimum. p will be *permissibly suboptimal* if and only if p is permissible but you ought to see to ~p, that is, if and only if there is some acceptable world where all acceptable worlds ranked as high are ~p-worlds while nonetheless p still occurs in some acceptable (and hence lower-ranked) alternative. Thus permissible sub-optimality turns out to be the mirror-image of going beyond the call (permissible "supra-minimality"), and what is involved in doing the minimum is the mirror-image of doing what one ought. Diagrammatically, and with obvious abbreviations (assume for the moment for that there are upper and lower bounds):

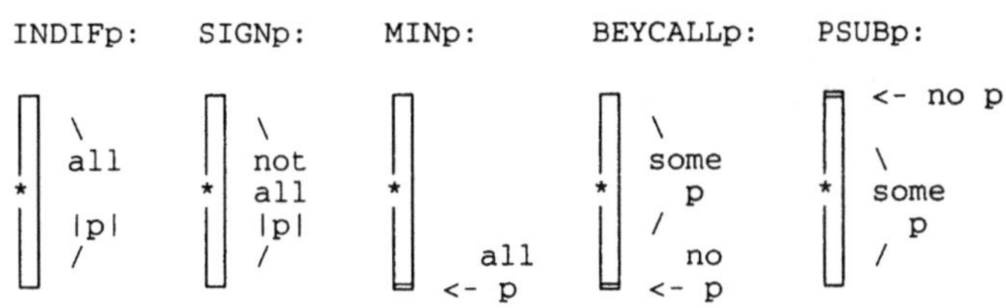

Figure 5

---

5 I discuss these extensions more fully in McNamara [6], which is essentially a companion to this paper.

6 In [23] I employ two orderings to argue that, contrary to the tradition, act utilitarianism is compatible with the possibility of *some* supererogation. Zimmermann [24] cites the possibility of likewise employing two orderings to argue for an obvious generalization: that act *consequentialism* is compatible with supererogation. He also argues that supererogation calls for two such orderings. But the argument that it calls for *two* such orderings--we've seen why it calls for one--loses some of its substantive interest given its reliance on the meta-theoretical claim that virtually *any* theory of right and wrong can be construed as an act optimizing theory, if we are but willing to wink at the feigned orderings "called for" by the construals (cf. [14]).

Here, we could take MUST, OUGHT, MIN and INDIF s the deontic primitives, with the remaining operators defined according to he following equivalences:

CANp <-> ~MUST~p
CANTp <-> MUST~p
OPTIONALp <-> (~MUSTp & ~MUST~p)
SIGNp <-> ~INDIFp
BEYCALLp <-> (CANp & MIN~p)
PSUBp <-> (CANp & OUGHT~p)

Formally, we extend our earlier language accordingly and provide official truth-conditions per DWE-models for the additional operators along the lines suggested in our glosses:

$M]=_i$ INDIFp: (j)[if Aij then Ek(k $=_i$ j & $M]=_k$ p) & Ek(k $=_i$ j & $M]=_k$ ~p)].
$M]=_i$ SIGNp: Ej[Aij & either (k)(if k $=_i$ j then $M]=_k$ p) or (k)(if k $=_i$ j then $M]=_k$ ~p)].
$M]=_i$ MINp: Ej(Aij & (k)(if k $\leq_i$ j then $M]=_k$ p)).
$M]=_i$ BEYCALLp: Ej(Aij & $M]=_j$ p) & Ej[Aij & (k)(if k $\leq_i$ j then $M]=_k$ ~p)].
$M]=_i$ PSUBp: Ej(Aij & $M]=_j$ p) & Ej[Aij & (k)(if j $\leq_i$ k then $M]=_k$ ~p)].

Now consider the following logic, "DWE" ([6]), where "*" ranges over the operators "MUST", "OUGHT" and "MIN":

A0. All tautologous DWE-wffs;
A1. *(p -> q) -> (*p -> *q)
A2. MUSTp -> (MINp & OUGHTp)
A3. (MINp v OUGHTp) -> CANp
A4. INDIFp -> INDIF~p
A5. INDIFp -> (~MINp & ~OUGHTp)
A6. MUST(p -> q) & MUST(q -> r) & INDIFp & INDIFr .->. INDIFq

R1:]- p and]- p -> q =>]- q
R2:]- p =>]- MUSTp.

Fact: DWE is determined by the class of all DWE-models.[7]

Theorem: DWE contains SDL fragments for "MUST", "OUGHT" and "MIN".
Proof: Trivial adaptation of the previous theorem for M-O.

Let me also identify a logic weaker than DWE, called "DWE^{m}" (for "DWE minus") with the same language as DWE:

---

7 The determination results cited in this section are from Mares and McNamara [25].

A0. All tautologous DWE$^-$-wffs;
A1. MUST(p -> q) -> (*p -> *q)
A2. MUSTp -> (MINp & OUGHTp)
A3. (MINp v OUGHTp) -> CANp
A4. INDIFp -> INDIF~p
A5. INDIFp -> (~MINp & ~OUGHTp)
A6. MUST(p -> q) & MUST(q -> r) & INDIFp & INDIFr .->. INDIFq

R1:]- p and]- p -> q =>]- q
R2:]- p =>]- MUSTp.

The only change is in A1, but it is significant. In DWE^m, although an aggregation theorem for MUST follows, no such theorem for OUGHT nor for MIN is forthcoming in DWE^m. In DWE^m, we can still derive a no-conflicts theorem for MUST, but not for OUGHT nor MIN. MUST still satisfies the axioms and rules of SDL in DWE^m, but neither OUGHT nor MIN do. We might consider reading the three *-operators in DWE^m as all-things-considered operators, rather than as overriding operators, as initially suggested. The logic can then be interpreted as allowing for conflicts between OUGHTs and between MINs so interpreted, but as disallowing such conflicts between MUSTs. Put more positively in the case of MUST, the logic can be interpreted as asserting that all-things-considered-MUSTs are always overriding. Gowan seems to be alluding to a view like this for "must" and "ought" in [26].

At the semantic level, what we do essentially is decompose clause 3a) from the definition of DWE-Frames and then withdraw just the connectivity of $\leq_i$, thus allowing that some pairs of i-acceptable worlds may not be mutually comparable. We do so while retaining the reflexivity of $\leq_i$, as well as the confinement of $\leq_i$ to the i-acceptables, which are both implied by clause 3a) in the definition of DWE-Frames. The result is:

F = < W, A, ≤ > is a DWE$^-$-Frame:
1. W is non-empty;
2. $A \subseteq W^2$ and A is serial;
3. $\leq \subseteq W^3$ where:
   a) $k \leq_i j$ only if (Aij & Aik);
   b) if Aij then $j \leq_i j$;
   c) If $j \leq_i k$ & $k \leq_i l$ then $j \leq_i l$;

We then define DWE^m-models in the usual way.

Fact: DWE^m is determined by the set of all DWE^m-models.

With DWE^m as our base logic, we can add back aggregation principles, (*p & *q) -> *(p & q), for both OUGHT and MIN to get a logic easily shown to be equipollent with DWE itself. One interesting fact is that not only is DWE complete with respect to the set of $\leq_i$-connected DWE^m-models (equivalently, the set of DWE models) as indicated earlier, but the logic is also complete (and sound) with respect to the set of DWE^m-models where the following upper and lower bound principles for $\leq_i$ are met:

LB: If Aij & Aik then El($l \leq_i j$ & $l \leq_i k$)
UB: If Aij & Aik then Em($j \leq_i m$ & $k \leq_i m$).

Fact: DWE is determined by the class of DWE^m-models where LB and UB hold.

Such models can contain pairs of i-acceptable worlds, j and k, that are not mutually comparable, but there must then be some world that is comparable with both j and k, since in particular, there must be one that is ranked at least as high as both, and one that is ranked as least as low as both. These constraints on DWE^m-models thus suffice to validate aggregation and rule out conflicts for OUGHT and MIN, but without requiring that the set of i-acceptable worlds are "commensurable".

Let me briefly point out that there are various issues of both traditional and recent concern in analytic ethical theory that our framework seems ripe to accommodate.[8] It provides an explanation of why utilitarianism is overdemanding and why it is appealing: utilitarianism conflates moral necessity with moral optimality by assuming that "ought" expresses the former, and thus it demands too much; but it plausibly endorses the well-spring for utilitarianism, that we ought to do the best we can. Whether the utilitarian interpretation of its well-spring is sound or not, is another matter. My only point here is that our reflections tend to confirm that "ought" is an optimizing notion *of some sort*.[9] It also explains why a conceptual framework for action beyond the call has been program resistant: parties on both sides of the debate about action beyond the call implicitly conflated what morality demands with what morality recommends by conflating "must" with "ought". Furthermore, in order to have anything like a robust account of action beyond the call, we must have a set of acceptable alternatives that are ranked, so that there are both minimal and better than minimal ways to satisfy morality's demands. Acting beyond the call requires "a line" below which we cannot go. Note also how the notion of an agent-centered prerogative might fit in. Although perhaps an agent ought to put in the best performance she can, she needn't. For if she has done all that she must, then she will have a prerogative to not do the remaining things that she ought. Similarly, permissible sub-optimizing requires the possibility of operating at a less than optimal but nonetheless acceptable level. Finally, as we saw above, moral optionality needn't be conflated with moral indifference, and thus action beyond the call is not ruled out derivatively. For optionality just requires that the contrasting negative and positive options are among some acceptable alternatives, but indifference would seem to

8 For example, see Kagan [27], Mellema [28], Scheffler [29] and [30], and Slote [31] and [32].

9 See McNamara [1] and [13] for further evidence.

require that they occur at each level of the acceptable alternatives, so that no level of value among the acceptable alternatives hinges on either option.

7 Conclusion

I suspect that any adequate scheme for common sense morality will be able to generate structures such as those of the simple framework sketched here. Although I have only been able to provide a sketch here, I think the sketch does support my contention at the end of [6] that reflection on these modals and quasi-modals, *as well as* the evidence given there drawn from intuitive cases of going beyond the call, each provide independent evidential paths to the sort of framework sketched herein, thus collectively boosting the evidence beyond the mere sum of either path taken in isolation.

I also believe that the answer to our parenthetical title-question is "Yes". For it sure looks like *the least I can do* is the minimum that morality demands. But this uniformly overlooked expression, deserves separate treatment. Ditto for "doing more than you had to".[10] Let me just note here that, as with "must" and "ought", the use of these modally loaded expressions of common-sense morality is pervasive. They constitute rather fundamental coinage for us. They probably reflect notions that have the deepest and least transient roots in our pre-theoretic conceptual scheme. They are reasonable candidates for universals. It is thus perhaps unwise for ethicists and deontic logicians to ignore them.[11]

References

1. McNamara, Paul. Must I do what I ought? The Society for Exact Philosophy, Austin, 1994. Typescript.
2. McNamara, Paul F. Doing well enough. University of Massachusetts--Amherst. Typescript, 1981.
3. Wertheimer, Roger. *The significance of sense*. Cornell University Press, Ithaca, 1972.
4. White, Alan R. *Modal thinking*. Oxford University Press, Oxford, 1975.

---

10 These idioms are discussed in detail in [1] and [13].

11 I had the honor of presenting many of the ideas of this paper at the Society for Exact Philosophy in Austin, May 1994. I benefitted from discussion with Nicholas Asher, Marvin Belzer, Dan Bonevac, Nicholas Denyer, Susan Dimock, Lou Goble, Ray Jennings, Michael Kremer, Rob Koons, Mark Lance, Ed Mares, Michael Morreau, Andy Schwartz and Peter Scott. Dan, who hosted the conference, deserves special thanks. Likewise for various participants of the Northern New England Philosophy Conference at Dartmouth, Oct. 1994, including Bernie Gert, Walter Sinnott-Armstrong and Ernest Sosa Jr. Bernie deserves special thanks. In addition to hosting that conference and providing helpful commentary, he provided astute practical guidance on this project.

5. Jones, Andrew J. I. and Ingmar Porn. 'Ought' and 'must'. *Synthese* 1986; 66:89-93.
6. Robertson, John. Moral dilemma and practical necessity. Typescript, 1995.
7. Chisholm, Roderick M. Supererogation and offense: a conceptual scheme for ethics. *Ratio* 1963; 5:1-14.
8. McNamara, Paul. Doing well enough: toward a logic for common sense morality. *Studia logica*. Forthcoming.
9. McNamara, Paul. Chisholm on supererogation and deontic logic. In: Lewis E. Hahn (ed.). *The library of living philosophers: Roderick Chisholm*. Open Court, Illinois. Forthcoming.
10. McNamara, Paul. Toward a framework for deontic conditionals. In: Nute, Donald. *Defeasible deontic logic: essays in nonmonotonic normative reasoning*.D. Reidel, Dordrecht. In Process.
11. Hare, R.M. *Moral thinking: its levels, method and point*. Oxford University Press, Oxford, 1981.
12. Kanger, Stig. New foundations for ethical theory. In: [16], 1970, pp 36-58.
13. McNamara, Paul F. The deontic quadecagon. PhD thesis, University of Massachusetts, Amherst, 1990.
14. McNamara, Paul. Making room for going beyond the call. Typescript, 1994.
15. Joachim Hrushka and Jan C. Joerden. Supererogation: vom deontologischen sechseck zum deontologischen zehneck. *Archiv fu rechts und sozialphilosophie* 1987; 73:93-123.
16. Risto Hilpinen (ed.). *Deontic logic: introductory and systematic readings*. D. Reidel, Dordrecht, 1970.
17. Aqvist, Lennart. Deontic Logic. In: Gabbay D. and F. Guenthner (eds.). *Handbook of Philosophical Logic*, Vol. II. D. Reidel Publishing Co., Dordrecht, 1984, pp 605-714.
18. Hanson, Bengt. An analysis of some deontic logics. *Nous* 1969; 3:373-398. Also in: [16], pp 105-147.
19. McNamara, Paul F. Andersonian DWE-ish deontic logics. Typescript, 1995.
20. Makinson, David. Five faces of minimality. *Studia Logica* 1993; 152:339-379.
21. Kratzer, Angelika. What 'must' and 'can' must and can mean. *Linguistics and philosophy* 1977; 1:337-355.
22. Urmson, J.O. Saints and Heroes. In: Melden, A. I. (ed.) *Essays in moral philosophy*. University of Washington Press, Washington, 1958, pp 198-216.
23. McNamara, Paul F. Supererogation, utilitarianism and Urmson's criterion: a compatibility thesis. Philosophy Colloquium at the University of North Carolina, Greensboro, 1988. Typescript.
24. Zimmermann, Michael. Supererogation and doing the best one can. *American philosophical quarterly* 1993; 30:373-380.
25. Mares, Edwin and McNamara. Supererogation in deontic logic: metatheory for dwe and some close neighbors. Typescript, 1995.
26. Gowan, Christopher W. Introduction. *Moral dilemmas*. Oxford University Press, Oxford, 1987, pp 3-33.
27. Kagan, Shelly. *The limits of morality*. Oxford University Press, Oxford, 1989.

28. Mellema, Gregory. 1991. *Beyond the call of duty*. Albany: SUNY Press.
29. Scheffler, Samuel (ed.). *Consequentialism and its critics* (Revised Edition). Oxford University Press, Oxford, 1988.
30. Scheffler, Samuel. *The rejection of consequentialism* (Revised Edition). Oxford University Press, Oxford, 1994.
31. Slote, Michael. *Common-sense morality and consequentialism*. Routledge & Kegan Paul, London, 1985.
32. Slote, Michael. *Beyond optimizing*. Harvard University Press, Cambridge, 1989.

Defeasible Reasoning with Legal Rules

Lambèr Royakkers*
Center for Law and Informatization, Tilburg University
Tilburg, The Netherlands

Frank Dignum
Dept. of Mathematics and Computer Science
Eindhoven Universtity of Technology
Eindhoven, The Netherlands

Abstract

The last few years several defeasible deontic reasoning formalisms are developed as a way to solve the problem of deontic inconsistency. However, these formalisms are unable to deal with some very common forms of deontic reasoning, since e.g. their expressiveness is restricted. In this paper we will establish a priority hierarchy of legal rules to solve the problem of deontic conflicts and we will give a mechanism to reason about nonmonotonicity of legal rules over the priority hierarchy. The theory presented here, based on default logic and a modification and extension of the argumentation framework of Prakken, properly deals with some shortcomings of other defeasible deontic reasoning approaches.

1 Introduction

Logical analysis of reasoning with inconsistent rules is a very relevant area for AI-and-Law research, since rules used in the legal domain are often conflicting. 'Prioritized' rules have received attention in research on formalizing nonmonotonic reasoning, particularly as a way of modelling the choice criterium in dealing with exceptions (cf. [1,2,3,4]).

To deal with the inconsistencies various sorts of consistency-based approaches have been developed, such as the nonmonotonic logic of McDermott and Doyle ([5]) and the default logic of Reiter ([6]). But these approaches fail to reason about conflicting norms, since they are all based on non-modal logics. As a way to solve the problems of deontic conflicts, forms of defeasible reasoning (cf. [7]) are adopted, which provide a mechanism to establish preference

*This article was supported by a grant from the Legal Research Foundation, which is part of the Netherlands Foundation for Scientific Research.

hierarchies of norms and to select a more applicable norm among conflicting ones in a specific situation (cf. [8,9]). The existing formalisations of defeasible deontic reasoning approaches ([10,11,12]) are unable to deal with several highly common forms of deontic logic (cf. [3]).

A first problem is *the lack of notion of permission* of certain approaches (cf. [10,12]). In these approaches $O(p/q)$: 'p is obligated in case of q' is treated as a normal default and can be read as a Reiter-default $q : p/p$: a non-deontic Reiter-default. Inherent in this treatment is the absence of a reasonable Reiter-default for the *negated obligation* (permission).

Another problem is *the defeasibility of only a single opposing statement* in some approaches (cf. [10,12,13]). In these approaches only couples of statements are considered to check inconsistencies. For instance, take the three statements $O(a)$, $O(\neg a \vee b)$ and $\neg O(b)$. No single statement is in conflict with the other single statements. However, the group of statements $O(a)$ and $O(\neg a \vee b)$ implies $O(b)$ in standard deontic logic, which is in conflict with the statement $\neg O(b)$.

The third problem is that most approaches (cf. [10,11,12]) can only deal with defeasible conditionals that are deontic. But deontic defaults are not the only defaults in legal reasoning. Consider the deontic default $a : O(b)/O(b)$. With this default it is very often the case that a is derived by another default rule, e.g. $c : a/a$, which is called a 'classification rule' or an 'interpretation rule'. In the legal domain it is accepted that these rules are also defeasible (cf. [15]). Prakken shows this by extending Hart's standard example on a park regulation that forbids vehicles to enter the park:

> "not only this rule itself may turn out to be defeasible, for example, if the vehicle is an ambulance, but also rules on when something counts as a vehicle may be defeasible: imagine that a court says that objects on wheels that are meant for normal transport are vehicles: then roller skates used by people on their way to the office might be recognised as an exception."

In this paper we will develop a theory of defeasible deontic reasoning which adequately deals with the above-mentioned problems. The theory is an extension and modification of the argumentation framework in default logic developed by Prakken ([3,13]). Further, our theory is an extension of the theory of Dung (citeDu), which only considers argumentation frameworks with one kind of conflicts between arguments.

The structure of this paper is as follows. In section 2 we give the representation of legal rules, which we subdivide in rules and conditional norms. In section 3, we will discuss the argumentation framework for rules. The argumentation framework for norms, depending on rules selected from the argumentation framework for rules, will be discussed in section 4. We will concentrate here on defeasibility and violation. We end with some conclusions.

2 Legal Rules: Rules and Conditional Norms

The fundamental logical structure of legal knowledge gives rise to the non-monotonicity of legal reasoning ([6,16]): the consequences derivable from a set of legal and factual premises can be invalidated by further information. This means that rules can be 'defeated' by other or new rules and facts. The principle idea of this paper, which goes back to Rescher ([17]), is to allow the rules to be ordered and to use this ordering in such a way that conflicts can be solved in a logical argumentation framework using nonmonotonic logic. Such an ordering can often be discerned when considering the rules in a legal code. The 'Lex Superior' principle, for instance, is based on the general hierarchy of a legal system; the rules are divided along the lines of the hierarchical structure of the normative system. Rules with a lower rank of priority have to respect the consequences that follow from a higher ranked rule. For example, a constitutional rule is hierarchically superior to a rule enacted by the ordinary legislature. The latter, in its turn, is superior to a rule enacted by a local government. More criteria or rules of preferences are used to resolve normative inconsistencies, such as 'Lex Specialis' and 'Lex Posterior'.[1] Thus, the law is not merely a set of rules, but a hierarchical system.[2] The aforementioned hierarchy principles are thus used to prioritize normative rules.[3] To describe the ordering between the formulas, we use the following notation. Let x and y be legal rules, then '$x \preceq y$' means that y is preferred over x; '$x \sim y$' is an abbreviation for '$x \preceq y$ and $y \preceq x$'; and '$x \prec y$' is an abbreviation for '$x \preceq y$ and $y \not\preceq x$'. The ordering relation $\preceq$ is reflexive and transitive.

Legal rules, or at least most of them, subordinate a legal effect to a legal condition. With legal effect we mean every qualification generated by a legal norm: the ascription of deontic or normative modalities, status, professional titles, other legal qualities of persons and things. With legal condition we mean every condition to which a legal effect is subordinated. The legal rules are represented as conditional statements of the type

$$a_1 \wedge a_2 \wedge \ldots \wedge a_n \Rightarrow \theta,$$

where θ is the legal effect and $a_1, a_2, \ldots, a_n$ are the elements of the antecedent: the conjunction of literals[4], representing the legal condition. If θ is a norm: an obligation ($O(\phi)$) or a permission ($P(\phi)$), with ϕ a formula of the propositional logic, then the conditional statement is called a *conditional norm*. If θ is a literal, then the statement is called a *rule*.

The statement $A \Rightarrow B$ has to be interpreted as a *normal default* under the theory of Reiter ([6,19]) $A : B/B$: "If A and it can be consistently assumed B, then we can infer B". This means that $\Rightarrow$ is *not* interpreted as the material

[1] These rules of preferences are mostly not codified.

[2] This is emphasised by several legal philosophers (see [15] and [18]).

[3] This gives a priority logic in the tradition of Brewka ([1], see also [3]).

[4] A literal is any atomic propositional formula and any negation of an atomic propositional formula.

implication, but as an inference rule, that can be defeated. From A and $A \Rightarrow B$, we can infer B unless $\neg B$ can be proven. This representation corresponds to the formalisations usually proposed by legal theory and legal logic ([20]).

In our theory we distinguish between rules and norms for the following reasons:

- Rules cannot be violated;
- The defeasibility of rules is different from the defeasibility of norms (cf. definition 3.5 and definition 4.8), which is the most important difference.

The most important thing about the difference between rules and norms is not *which* differences there are, but simply *that* there are differences. That is why we discuss our theory on different levels: first on the level of rules (section 3) and secondly on the level of norms based on a given set of rules (section 4).

The set of rules will be denoted by W and the set of conditional norms by Δ. Furthermore we have a factual sentence F representing the factual situation, which consists of background knowledge and contingent facts. The background knowledge consists of necessary conditions, for example a human is mortal. A set of conditional norms, a set of rules and a factual sentence will be called a deontic context.

Definition 2.1 *A deontic context $T = (\Delta, W, F, \prec)$ consists of a set Δ of conditional norms, a set W of rules, a factual propositional sentence F: the conjunction of background knowledge F_b and contingent facts F_c, and an ordering $\prec$ over rules and conditional norms.*

3 Rules

Facts (formalised by the sentence F) can contain material implications. However, rules are represented by normal defaults written as a conditional statement of the type $a_1 \wedge a_2 \wedge \ldots \wedge a_n \Rightarrow \theta$, with θ the legal effect formalised by a literal. The theory of defeasible reasoning for rules here is based on four notions:

- the notion of an *argument* (definition 3.4);
- the notion of *defeating* (definition 3.5);
- the notion of a *defeasibility chain* (definition 3.7);
- the notion of *justified, defensible and overruled arguments* (definition 3.9).

At the end of this section we will define *maximal coherent argument sets* of rules that we will use for the notion of the applicability of norms and the violation of obligations in section 4.

Before we discuss the notion of an argument, we give three definitions which we will use in the sequel.

Definition 3.1 *Let V be a set of rules and r a literal, then V explains r ($V \cup \{F\} \models r$) iff*

$$\{F\} \vdash r \text{ or } \exists_{a_1 \wedge a_2 \wedge \ldots \wedge a_n \Rightarrow r \in V} \{V \cup \{F\} \models a_i \mid i \in \{1, 2, \ldots, n\}\}.$$

Definition 3.2 *Let V be a set of rules, then the consequences of $V \cup \{F\}$ ($Cons(V \cup \{F\})$) is defined as*

$$Cons(V \cup \{F\}) := \{r \mid r \text{ is a literal and } V \cup \{F\} \models r\}.$$

Definition 3.3 *Let V be a set of rules. Then $V \cup \{F\}$ is coherent iff*

$$\neg\exists_{r \text{ is a literal}} (r \in Cons(V \cup \{F\}) \wedge \neg r \in Cons(V \cup \{F\})).$$

Definition 3.4 *Let $M \subseteq W$, ϕ a literal and $M \cup \{F\}$ coherent. Then M explains ϕ minimally iff*

- $\{F\} \cup M \models \phi$ *and*
- $\neg\exists_{\phi_1 \in M} (\{F\} \cup M \setminus \{\phi_1\}) \models \phi$.

We call M a minimally explaining set or an argument. The set of all arguments will be denoted as $\mathcal{M}_r$. The ϕ-relevant set of W, denoted by $[\phi]\mathcal{M}_r$ is the set of all arguments in $\mathcal{M}_r$, that explain ϕ minimally.
M_1 is a subargument of M iff $M_1 \subset M$ and M_1 is an argument. If there is an argument for ϕ, thus $[\phi]\mathcal{M}_r \neq \emptyset$, then ϕ is called an outcome.

Definition 3.5 *Let $M_1 \in [\phi]\mathcal{M}_r$ and $M_2 \in [\phi']\mathcal{M}_r$. Then M_1 is defeated by M_2 ($M_1 \prec^* M_2$) iff*

$$\exists_{\phi_1 \Rightarrow \phi_2 \in M_1} \exists_{\phi_3 \Rightarrow \phi_4 \in M_2} \{\phi_3 \Rightarrow \phi_4 \succeq \phi_1 \Rightarrow \phi_2\} \wedge \{\phi_2\} \cup \{\phi_4\} \cup \{F\} \text{ is inconsistent.}$$

Thus, an argument M_2 defeats an argument M_1 iff M_1 and M_2 have contradictory conclusions ϕ_2 and ϕ_4 with respect to the factual sentence F and the rule $\phi_3 \Rightarrow \phi_4 \in M_2$ (responsible for the conflict) does not have a lower priority then the rule $\phi_1 \Rightarrow \phi_2 \in M_1$. Note that $\{\phi_2\} \cup \{F\}$ and $\{\phi_4\} \cup \{F\}$ are consistent, which directly follows from definition 3.4.

$\prec^*$ is not transitive and not asymmetric. It is possible that $M_1 \prec^* M_2$ and $M_2 \prec^* M_1$ both hold.

Example 3.6
(1) $a \Rightarrow b$
(2) $c \Rightarrow \neg a$
(3) $d \Rightarrow a$
(4) $b \Rightarrow \neg c$
(5) $e \Rightarrow c$
F: $f \wedge (f \rightarrow d) \wedge e$
with $(5) \prec (4) \prec (3) \prec (2) \prec (1)$.

Let $M_1 = \{e \Rightarrow c, c \Rightarrow \neg a\}$ and $M_2 = \{d \Rightarrow a, a \Rightarrow b, b \Rightarrow \neg c\}$. Then $M_1 \prec^* M_2$, since $e \Rightarrow c \prec b \Rightarrow \neg c$ and $M_2 \prec^* M_1$, since $d \Rightarrow a \prec c \Rightarrow \neg a$.

Definition 3.7 *A defeasibility chain is a sequence of arguments in $\mathcal{M}_r$:*

$$M_1 \prec^* M_2 \prec^* \ldots \prec^* M_n$$

with the following conditions:

- $\forall_{k,l=1,2,\ldots,n \wedge k \neq l} M_k \neq M_l$;
- $\neg\exists_{M_{n+1} \in \mathcal{M}_r}(M_{n+1} \neq M_1, \ldots, M_n) \rightarrow (M_n \prec^* M_{n+1})$.

We define $Ch(\mathcal{M}_r)$ as the set of all defeasibility chains of arguments in $\mathcal{M}_r$.

The first condition ensures that cycles in defeasibility chains are avoided. Suppose that $M_1 \prec^* M_2$ and $M_2 \prec^* M_3$, with $M_1 \subset M_3$. We thus would end up with the endless chain $M_1 \prec^* M_2 \prec^* M_3 \prec^* M_2 \prec^* M_3 \ldots$. The second condition provides that a chain stops if there is no 'higher' argument than the last argument in the chain.

Take the example above, then

$$\begin{aligned} Ch(\mathcal{M}_r) = \{ & \{d \Rightarrow a, a \Rightarrow b\} \prec^* \{e \Rightarrow c, c \Rightarrow \neg a\} \prec^* \{d \Rightarrow a, a \Rightarrow b, b \Rightarrow \neg c\}, \\ & \{d \Rightarrow a, a \Rightarrow b, b \Rightarrow \neg c\} \prec^* \{e \Rightarrow c, c \Rightarrow \neg a\}, \\ & \{e \Rightarrow c\} \prec^* \{d \Rightarrow a, a \Rightarrow b, b \Rightarrow \neg c\}\}, \\ & \{d \Rightarrow a\} \prec^* \{e \Rightarrow c, c \Rightarrow \neg a\} \prec^* \{d \Rightarrow a, a \Rightarrow b, b \Rightarrow \neg c\}, \\ & \{e \Rightarrow c, c \Rightarrow \neg a\} \prec^* \{d \Rightarrow a, a \Rightarrow b, b \Rightarrow \neg c\}. \end{aligned}$$

Definition 3.8 *$Ch(M)$ is the set of all defeasibility chains in $Ch(\mathcal{M}_r)$ starting with M.*

The defeasibility chains in $Ch(\mathcal{M}_r)$ take as input the set of all possible arguments and their mutual relations of defeat. They produce as output a distinction between arguments in three classes:[5]

1. *justified* arguments;
2. *overruled* arguments;
3. *defensible* arguments.

A justified argument is a 'winning' argument. Such an argument can be defeated by another argument, but that argument will be overruled. An overruled argument is a 'losing' argument. A defensible argument is an argument that is neither justified nor overruled. In other words, an 'undeciding' argument.

Definition 3.9 *Let $M \in \mathcal{M}_r$. Then*

[5]The terms justified, overruled and defensible argument have been introduced by Prakken and Sartor ([21]).

- *M is a justified argument iff for all chains $M \prec^* M_1 \prec^* \ldots \prec^* M_n \in Ch(\mathcal{M}_r)$ it holds*

$$n \text{ is even } \wedge \neg\exists_{M' \in \mathcal{M}_r} M_n \prec^* M'.$$

- *M is an overruled argument iff there is a chain $M \prec^* M_1 \prec^* \ldots \prec^* M_n \in Ch(\mathcal{M}_r)$*

$$n \text{ is odd } \wedge \neg\exists_{M' \in \mathcal{M}_r} M_n \prec^* M'.$$

- *M is a defensible argument iff M is neither a justified argument nor an overruled argument.*

Notice that M_n in the chain of the definition 3.9 is a justified argument, since $Ch(M_n) = \{M_n\}$, which is equivalent with $\neg\exists_{M' \in \mathcal{M}_r} M_n \prec^* M'$.

Let $M \prec^* M_1 \prec^* \ldots \prec^* M_n$ be a chain in $Ch(\mathcal{M}_r)$, then we call the arguments M_i with i is odd *odd arguments* and the arguments M and M_i with i is even *even arguments*.

In a defeasibility chain $M \prec^* M_1 \prec^* \ldots \prec^* M_n$ with n is even (odd), we stipulate that the odd (even) arguments are the *attacked* arguments and the even (odd) arguments the *not attacked* arguments. If the chain ending with M_n, then M_{n-1} is an attacked argument, because it is defeated by a not attacked argument. M_{n-2} is not attacked, because it is defeated by an attacked argument (M_{n-1}) and so on. For example, M is defeated and overruled if M_1 is a non-attacked argument and this follows if n is odd.

Proposition 3.10

1. *Let M be a justified argument. Then all odd arguments in the chains of $Ch(M)$ are overruled arguments.*

2. *Let M be an overruled argument. Then there is a chain in $Ch(M)$ with all even arguments overruled or defensible.*

Proof.

1. Let M be a justified argument. Then for all chains $M \prec^* M_1 \prec^* \ldots \prec^* M_n$ in $Ch(M)$ it holds that n is even and $\neg\exists_{M' \in \mathcal{M}_r} M_n \prec^* M'$. Let M_k with k is odd an argument of a chain $M \prec^* M_1 \prec^* \ldots \prec^* M_n$ in $Ch(M)$. Then this chain without the first k arguments, thus $M_k \prec^* M_{k+1} \prec^* \ldots \prec^* M_n$, is a chain in $Ch(M_k)$ which satisfies the conditions of an overruled argument M_k, since the chain contains an even number of arguments and $Ch(M_n) = M_n$. Thus all odd arguments are overruled arguments.

2. Let M be an overruled argument. Then there is a chain $M \prec^* M_1 \prec^* \ldots \prec^* M_n$ with n is odd and $\neg\exists_{M' \in \mathcal{M}_r} M_n \prec^* M'$. Suppose an even argument M_k is justified, then the chain $M \prec^* M_1 \prec^* \ldots \prec^* M_n$ without the first k arguments, i.e. $M_k \prec^* M_{k+1} \prec^* \ldots M_n$ is a chain in $Ch(M_k)$ and does not satisfy the conditions for a justified argument M_k, since the chain contains an even number of arguments. Thus the even arguments is such chains in $Ch(M)$ are overruled or defensible.

It is not the case that all even arguments in the chains of $Ch(M)$ with M a justified argument have to be justified. See the following example:

Example 3.11
(1) $a \Rightarrow \neg b$
(2) $c \Rightarrow a$
(3) $b \Rightarrow \neg d$
(4) $e \Rightarrow b$
(5) $\neg g \Rightarrow \neg f$
(6) $d \Rightarrow \neg h$
(7) $f \Rightarrow d$
(8) $i \Rightarrow f$
(9) $h \Rightarrow \neg j$
(10) $k \Rightarrow h$
(11) $j \Rightarrow \neg g$
(12) $l \Rightarrow j$
F: $c \wedge e \wedge i \wedge k \wedge l$
with $(12) \prec (11) \prec \ldots \prec (1)$.

Let $M_1 = \{a \Rightarrow \neg b, c \Rightarrow a\}$;
$M_2 = \{b \Rightarrow \neg d, e \Rightarrow b\}$;
$M_3 = \{\neg g \Rightarrow \neg f, j \Rightarrow \neg g, l \Rightarrow j\}$;
$M_4 = \{d \Rightarrow \neg h, f \Rightarrow d, i \rightarrow f\}$;
$M_5 = \{h \Rightarrow \neg j, k \Rightarrow h\}$.
$Ch(M_3) = \{M_3 \prec^* M_5 \prec^* M_4 \prec^* M_2 \prec^* M_1\}$ and $Ch(M_1) = \{M_1\}$, thus M_3 is a justified argument. However, M_4 is not a justified argument. The set of defeasibility chains starting with M_4 is $Ch(M_4) = \{M_4 \prec^* M_2 \prec^* M_1, M_4 \prec^* M_3 \prec M_5\}$. Since M_5 is an overruled argument (the chain $M_5 \prec^* M_4 \prec^* M_2 \prec^* M_1$ satisfies the conditions for the overruled argument M_5), the chain $M_4 \prec^* M_3 \prec^* M_5$ does not satisfy the conditions for a justified argument M_4. Thus, M_4 is not a justified argument. Notice that M_4 is a defensible argument, since neither of the two chains in $Ch(M_4)$ satisfies the conditions for an overruled argument M_4.

Corollary 3.12

1. *A justified argument can only be defeated by an overruled argument;*
2. *If there is no justified argument, then there is no overruled argument;*
3. *There is no justified argument iff all arguments are defensible;*
4. *There is a justified argument iff there is a defeasibility chain with one argument.*

Proof.

1. Let M be a justified argument and defeated by M_1. We have to prove that M_1 is an overruled argument. For all chains $M \prec^* M_1 \ldots \prec^* M_n$

in $Ch(M)$ it holds that n is even and $Ch(M_n) = \{M_n\}$. For all these chains without the first argument M, i.e. $M_1 \prec^* \ldots \prec^* M_n$, it holds that they are elements of $Ch(\mathcal{M}_r)$ and satisfy the conditions of an overruled argument. Thus M_1 is an overruled argument.

2. Suppose there is no justified argument. Then there is no chain $M \prec^* M_1 \prec^* \ldots \prec^* M_n$ in $Ch(\mathcal{M}_r)$ with $Ch(M_n) = \{M_n\}$. Hence, there is no overruled argument.
The converse does not hold. For example, let $W = \{a \Rightarrow b\}$ and a a fact. Then the only argument is $\{a \Rightarrow b\}$ and this argument is justified.

3. Suppose there is no justified argument. Then there is no overruled argument, thus all arguments are defensible.
Evidently, if all arguments are defensible, then there is no justified arguments.

4. Suppose M is a justified argument. Then for all chains $M \prec^* M_1 \prec^* \ldots \prec^* M_n$ in $Ch(M)$ it holds that $Ch(M_n) = \{M_n\}$. Thus there is a defeasibility chain with one argument.
If there is a chain with one argument, say M, then M is a justified argument. Hence there is a justified argument.

The converse of 3.12.1 does not hold: An overruled argument need not necessary be defeated by a justified argument.

Example 3.13
(1) $a \Rightarrow b$
(2) $c \Rightarrow d$
(3) $\neg b \Rightarrow \neg e$
(4) $f \Rightarrow \neg b$
(5) $e \Rightarrow \neg d$
(6) $g \Rightarrow e$
(7) $h \Rightarrow d$
F: $a \wedge c \wedge f \wedge g \wedge h$
with $(7) \prec (6) \prec \ldots \prec (1)$.

Let $M_1 = \{a \Rightarrow b\}$;
$M_2 = \{c \Rightarrow d\}$;
$M_3 = \{\neg b \Rightarrow \neg e, f \Rightarrow \neg b\}$;
$M_4 = \{g \Rightarrow e, e \Rightarrow \neg d\}$;
$M_5 = \{h \Rightarrow d\}$.
Then $Ch(M_5) = \{M_5 \prec^* M_4 \prec^* M_3 \prec^* M_1, M_5 \prec^* M_4 \prec^* M_2\}$. M_5 is an overruled argument, since there is a chain $M_5 \prec^* M_4 \prec^* M_3 \prec^* M_1$ with $Ch(M_1) = \{M_1\}$. Further, M_5 is only be defeated by argument M_4 which is an overruled argument, since $M_4 \prec^* M_2 \in Ch(M_4)$ and $Ch(M_2) = \{M_2\}$. Thus, an overruled argument can be defeated by an overruled argument.

The following three problems form the main problems in the literature of defeasible arguments. We will show that these problems can be adequately handled by the theory as it follows from definition 3.6.

Example 3.14 *The intermediate conclusion*
(1) $a \Rightarrow b$
(2) $c \Rightarrow \neg b$
(3) $d \Rightarrow a$
F: $c \wedge d$
with $(3) \prec (2) \prec (1)$.

Here a conflict arises between the rules (1) and (2). By definition 3.5 the choice is made between the norms which are certainly in conflict with each other. The rule (3) with *the intermediate conclusion* a, necessary to derive the outcome b, is irrelevant to the conflict.

The minimally explaining sets (arguments) are $M_1 = \{d \Rightarrow a, a \Rightarrow b\}$, $M_2 = \{c \Rightarrow \neg b\}$ and $M_3 = \{d \Rightarrow a\}$. The sets of defeasibility chains are $Ch(M_1) = \{M_1\}$, $Ch(M_2) = \{M_2 \prec^* M_1\}$ and $Ch(M_3) = \{M_3\}$. M_1 and M_3 are justified, since they are not defeated by an argument. M_2 is overruled, because it is defeated by a justified argument M_1. Thus, a and b are the outcomes.

Example 3.15 *Iterated conflicts*
(1) $c \Rightarrow a$
(2) $d \Rightarrow \neg a$
(3) $\neg a \Rightarrow \neg b$
(4) $a \Rightarrow b$
F: $c \wedge d$
with $(4) \prec (3) \prec (2) \prec (1)$.

Here a conflict arises between the rules (1) and (2) and between the rules (3) and (4). The rules (1) and (2) have intermediate conclusions a and $\neg a$ for the final conclusions $\neg b$ and b. This type of problem is called *iterated conflicts*: conflicts on both intermediate and final conclusions.

The minimally explaining sets are $M_1 = \{c \Rightarrow a\}$, $M_2 = \{d \Rightarrow \neg a\}$, $M_3 = \{d \Rightarrow \neg a, \neg a \Rightarrow \neg b\}$ and $M_4 = \{c \Rightarrow a, a \Rightarrow b\}$. The sets of defeasibility chains are $Ch(M_1) = \{M_1\}$, $Ch(M_2) = \{M_2 \prec^* M_1\}$, $Ch(M_3) = \{M_3 \prec^* M_1\}$ and $Ch(M_4) = \{M_4 \prec^* M_3 \prec^* M_1\}$. M_1 and M_4 are justified arguments. M_2 and M_3 are overruled arguments, because they are defeated by the justified argument M_1: a and b are the outcomes. Note that M_4 is defeated by M_3, but not overruled, since M_3 is an overruled argument (defeated by M_1).

Example 3.16 *Circular conflicts*
(1) $a \Rightarrow b$
(2) $c \Rightarrow \neg b$
(3) $b \Rightarrow \neg a$
(4) $d \Rightarrow a$
(5) $d \Rightarrow b$
F: $c \wedge d$
with $(5) \prec (4) \prec (3) \prec (2) \prec (1)$.

Here we have conflicts between the rules (1) and (2), the rules (3) and (4) and the rules (2) and (5). The applicability of rule (1) depends on rule (4), with the intermediate conclusion a. Rule (4) is in conflict with a higher rule (3) (iterated conflict). And the applicability of (3) depends on (5) and this rule is in conflict with (2). Rule (2) is in conflict with rule (1) and the applicability of (1) depends on rule (4) and so on. This will never stop, reason why we call this *the problem of circular conflicts.*

The minimally explaining sets are $M_1 = \{d \Rightarrow a, a \Rightarrow b\}$, $M_2 = \{c \Rightarrow \neg b\}$, $M_3 = \{d \Rightarrow b, b \Rightarrow \neg a\}$, $M_4 = \{d \Rightarrow a\}$ and $M_5 = \{d \Rightarrow b\}$. The sets of defeasibility chains are $Ch(M_1) = \{M_1 \prec^* M_3 \prec^* M_2\}$, $Ch(M_2) = \{M_2 \prec^* M_1 \prec^* M_3\}$, $Ch(M_3) = \{M_3 \prec^* M_2 \prec^* M_1\}$, $Ch(M_4) = \{M_4 \prec^* M_3 \prec^* M_2 \prec^* M_1\}$ and $Ch(M_5) = \{M_5 \prec^* M_2 \prec^* M_1 \prec^* M_3\}$. All the arguments are defensible, since there is no justified argument. Thus, b, $\neg b$, $\neg a$ and a are the outcomes.

Definition 3.17 *Let $\mathcal{M}_r^*$ be the set of all defensible and justified arguments in $\mathcal{M}_r$. Then w is a maximal coherent argument set iff w is the set of all rules of the arguments in $\mathcal{M}_r^*$ with*

- *$Cons(w \cup \{F\})$ is consistent and*
- *$\neg\exists_{M \in \mathcal{M}_r^* \setminus w}(Cons(w \cup \{F\} \cup M)$ is consistent).*

$\mathcal{W}$ is defined as the set of all maximal coherent argument sets $w \subseteq \mathcal{M}_r$. $Out(w)$ is defined as the set of all outcomes of the arguments $M \subseteq w$.

Consider example 3.16. Then $\mathcal{W} = \{w_1, w_2, w_3\}$ with $w_1 = \{d \Rightarrow a, a \Rightarrow b, d \Rightarrow b\}$, $w_2 = \{c \Rightarrow \neg b, d \Rightarrow a\}$ and $w_3 = \{b \Rightarrow \neg a, d \Rightarrow b\}\}$. It follows that $Out(w_1) = \{a, b\}$, $Out(w_2) = \{a, \neg b\}$ and $Out(w_3) = \{\neg a, b\}$. There is no $w \in \mathcal{W}$ with $Out(w) = \{\neg a, \neg b\}$, since M_3 is the argument for $\neg a$ and $M_3 \cup \{F\} \models b$. Thus if $\neg a \in Out(w)$, then b must also be an element in $Out(w)$, since otherwise we cannot derive $\neg a$: The literal b is necessary to derive $\neg a$. In other words, if an argument M is a subset of a maximal coherent argument set w, then all outcomes of the subarguments of M are elements of the set $Out(w)$.

Corollary 3.18 *Let $w \in \mathcal{W}$. Then for all justified arguments M in $\mathcal{M}_r$ it holds that $M \subseteq w$.*

The following definition is needed for the definition of arguments of norms in the next section.

Definition 3.19 *Let W be the set of rules and $w \in \mathcal{W}$, then $W(w)$ is defined as the maximal coherent set of rules in W with respect to w and F. Let $\phi \Rightarrow \phi_1 \in W$, then*

$$\phi \Rightarrow \phi_1 \in W(w) \text{ iff } \{\phi \Rightarrow \phi_1\} \cup w \cup \{F\} \text{ is coherent.}$$

Note that all rules in w are in $W(w)$.

Example 3.20
(1) $c \Rightarrow a$
(2) $d \Rightarrow \neg a$
(3) $d \Rightarrow b$
(4) $\neg c \Rightarrow \neg b$
F: $c \wedge d$
with (4) $\prec$ (3) $\prec$ (2) $\prec$ (1).

Then $w = \{c \Rightarrow a, d \Rightarrow b\}$ and $W(w) = \{c \Rightarrow a, \neg c \Rightarrow \neg b, d \Rightarrow b\}$. The rule $d \Rightarrow \neg a$ is not an element of $W(w)$, since the rule is incoherent with $w \cup \{F\}$. Further, note that $\neg c \Rightarrow \neg b$ is an element of $W(w)$, although not of w. We will see in the following section that such a rule can be applicable for a derivation of a certain norm.

4 Norms

The defeasible deontic reasoning is based on five notions: The notion of *applicable norms* based on a set $w \in \mathcal{W}$ (definition 4.1) and the same four notions of the defeasible reasoning for rules (definitions 4.2, 4.7, 4.9 and 4.11).

We also will define *maximal coherent argument sets* for norms that we will use for the definition of violated obligations in subsection 4.2. At the end of this subsection we give some examples of defeasible deontic reasoning with violated norms.

Definition 4.1 *Let $w \in \mathcal{W}$ and $a_1 \wedge a_2 \wedge \ldots \wedge a_n \Rightarrow \psi$ be a conditional norm in Δ. Then*

$$\psi \in \Delta(w, F) \text{ iff } \forall_{i \in \{1,2,\ldots,n\}} w \cup \{F\} \models a_i.$$

ψ is called an applicable norm in Δ with respect to w and F. Thus, $\Delta(w, F)$ is the set of all applicable norms in Δ with respect to w and F.

Example 4.2

W: *(1)* $a \Rightarrow b$	Δ: *(1')* $b \Rightarrow O(\neg a)$	*F:* $a \wedge e \wedge h$
(2) $b \Rightarrow c$	*(2')* $d \wedge h \Rightarrow O(i \vee l)$	
(3) $e \Rightarrow d$	*(3')* $g \Rightarrow O(b)$	
(4) $f \Rightarrow g$		
(5) $h \Rightarrow \neg c$		

with (5) $\prec$ (4) $\prec$ (3) $\prec$ (2) $\prec$ (1) *and* (3') $\prec$ (2') $\prec$ (1').

The arguments are:
$M_1 = \{a \Rightarrow b\}$;
$M_2 = \{a \Rightarrow b, b \Rightarrow c\}$;
$M_3 = \{e \Rightarrow d\}$;
$M_4 = \{h \Rightarrow \neg c\}$.
It is easy to see that M_1, M_2 and M_3 are justified arguments, since they are

not defeated by an argument. M_4 is defeated by M_2, thus M_4 is an overruled argument. There is one maximal coherent argument set w: $M_1 \cup M_2 \cup M_3$. Thus $w = \{a \Rightarrow b, b \Rightarrow c, e \Rightarrow d\}$. Now we can give the applicable norms:

1. $O(\neg a) \in \Delta(w, F)$, since $w \cup \{F\} \models b$;
2. $O(i \vee l) \in \Delta(w, F)$, since $w \cup \{F\} \models d$ and $w \cup \{F\} \models h$;
3. $O(b) \notin \Delta(w, F)$, since $w \cup \{F\} \not\models g$.

Thus $\Delta(w, F) = \{O(\neg a), O(h \vee l)\}$.

For $\Delta(w, F)$ we will use the standard deontic logic (SDL) satisfying the two axioms:

(K) $O(p \rightarrow q) \rightarrow (O(p) \rightarrow O(q))$;

(D) $\neg(O(p) \wedge O(\neg p))$.

We use the following two inference rules:[6]

(RK1)

$$\frac{O(p_1), O(p_2), \ldots, O(p_n), p_1 \wedge p_2 \wedge \ldots \wedge p_n \rightarrow q}{O(q)}$$

(RK2)

$$\frac{O(p_1), O(p_2), \ldots, O(p_n), p_1 \wedge p_2 \wedge \ldots \wedge p_n \Rightarrow q}{O(q)}$$

Analogue with the definition of arguments for rules, we define arguments for norms.

Definition 4.3 *Let $N \subseteq \Delta(w, F)$, ψ a norm and $N \cup \{F_b\} \cup W(w)$ consistent. Then N explains ψ minimally iff*

- $\{F_b\} \cup W(w) \cup N \vdash \psi$ *and*
- $\neg \exists_{\psi_1 \in N} \{F_b\} \cup W(w) \cup N \setminus \{\psi_1\} \vdash \psi$.

We call N a minimally explaining set or an argument. The set of all arguments will be denoted as $\mathcal{N}(w, F)$. The ψ-relevant set of $\Delta(w, F)$, denoted by $[\psi]\mathcal{N}(w, F)$ is the set of all arguments, that explain ψ minimally.
Let $\mathcal{N} \subseteq \mathcal{N}(w, F)$. If there is an argument for ψ, thus $[\psi]\mathcal{N} \neq \emptyset$, then ψ is called an outcome.

We do not use F, but F_b in this definition, because otherwise we would derive ridiculous conclusions. Consider the following example.

[6] For a short discussion about these rules, see section 5.

Example 4.4
(1) $O(a)$: It is obligated to go to school
(2) $O(b)$: It is obligated to behave
$F = F_c$: $a \wedge \neg b$: you go to school and you do not behave
with (2) $\prec$ (1).

Now we can derive $O(\neg b)$ from $O(a)$ and F_c, because $\{F_c\} \vdash a \rightarrow \neg b$ and by (RK1) $O(a) \rightarrow O(\neg b)$. Thus if we use F instead of F_b, then $\{O(a)\}$ would be an argument for the outcome $O(\neg b)$ and this is not a desirable result.

Definition 4.5 *Let $\mathcal{N} \subseteq \mathcal{N}(w, F)$ and $N_1, N_2 \in \mathcal{N}$. Then N_1 and N_2 are in conflict iff*

$$\{F_b\} \cup W(w) \cup N_1 \vdash \psi \textit{ and } \{F_b\} \cup W(w) \cup N_2 \vdash \neg\psi.$$

Example 4.6
(1) $O(a \vee b)$
(2) $O(\neg a)$
(3) $O(c \vee \neg b)$
(4) $O(\neg c)$
with (4) $\prec$ (3) $\prec$ (2) $\prec$ (1).

Let $N_1 = \{O(a \vee b), O(c \vee \neg b), O(\neg c)\}$ and $N_2 = \{O(\neg a)\}$ be arguments in $\mathcal{N}$. Then N_1 is an argument for $O(a)$ and N_2 an argument for $\neg O(a)$, since $O(\neg a) \rightarrow \neg O(a)$ is deduced by the axiom (D). Thus N_1 and N_2 are in conflict.

Example 4.7
(1) $O(a)$
(2) $O(\neg b)$
(2) $\prec$ (1) *and* $W(w) = \{a \Rightarrow b\}$.

Let $N_3 = \{O(a)\}$ and $N_4 = \{O(\neg b)\}$ be arguments in $\mathcal{N}$. Then N_3 is an argument for $O(b)$, since $\{a \Rightarrow b\} \cup N_3 \vdash O(b)$ and N_4 is an argument for $\neg O(b)$. Thus N_3 and N_4 are in conflict.

4.1 Defeasibility

Because the arguments in $\mathcal{N}(w, F)$ can be in conflict, we resolve these deontic conflicts by adopting defeasible reasoning. Defeasibility of norms in $\Delta(w, F)$ determines the validity of these norms. The validity depends on the rules in w. For instance, let $\Delta(w, F) = \{O(a), O(b)\}$, $O(b) \prec O(a)$ and $w = \{a \Rightarrow \neg b\}$. Then $\{O(a)\}$ is an argument for $\neg O(b)$, since $w \cup \{O(a)\} \vdash \neg O(b)$. Now we say that $\{O(a)\}$ defeats $\{O(b)\}$ and that $O(b)$ is not valid. Suppose now that $w = \emptyset$, then $O(a)$ and $O(b)$ are both valid, because $\{O(a)\}$ and $\{O(b)\}$ are not in conflict.

Definition 4.8 *Let N_1 and N_2 be arguments. Then N_1 is defeated by N_2 ($N_1 \prec^* N_2$) iff N_1 and N_2 are in conflict and*

$$\exists_{\psi_1 \in N_1} \forall_{\psi_2 \in N_2} \psi_1 \preceq \psi_2.$$

The main difference between this definition and definition 3.5 is that we here look at the statements of the two arguments with the lowest priority and in definition 3.5 we look at the statements of the two arguments with conflicting conclusions. Note that it is possible that $N_1 \prec^* N_2$ and $N_2 \prec^* N_1$ both hold, but only if the statements with the lowest priority of the two arguments have the same priority. $\prec^*$ is not transitive and not asymmetric.

Example 4.9 *Consider example 4.6. Then N_1 is defeated by N_2, since $O(\neg c)$ has a lower priority than all the norms in N_2: $O(\neg c) \preceq O(\neg a)$.*
Consider now example 4.7. N_4 is defeated by N_3, since $O(\neg b) \preceq O(a)$. If $(1) \sim (2)$, then $N_3 \prec^* N_4$ and $N_4 \prec^* N_3$.

The following three definitions for arguments in $\mathcal{N}$ concerning defeasibility chains, $Ch(N)$ and justified, defensible and overruled arguments are exactly the same as the definitions for arguments in $\mathcal{N}_r$ (cf. definitions 3.7, 3.8 and 3.9).

Definition 4.10 *Let $\mathcal{N} \subseteq \mathcal{N}(w, F)$. A defeasibility chain is a sequence of arguments in $\mathcal{N}$: $N_1 \prec^* N_2 \prec^* \ldots \prec^* N_n$ with the following conditions:*

- $\forall_{k,l=1,2,\ldots,n \wedge k \neq l} N_k \neq N_l$;
- $\neg\exists_{N_{n+1} \in \mathcal{N}} (N_{n+1} \neq N_1, \ldots, N_n) \rightarrow (N_n \prec^* N_{n+1})$.

We define $Ch(\mathcal{N})$ as the set of all defeasibility chains in $\mathcal{N}$.

Definition 4.11 *Let $\mathcal{N} \subseteq \mathcal{N}(w, F)$. Then $Ch(N)$ is the set of all defeasibility chains in $Ch(\mathcal{N})$ starting with N.*

Definition 4.12 *Let $\mathcal{N} \subseteq \mathcal{N}(w, F)$ and $N, N_1, \ldots N_n \in \mathcal{N}$. Then*

- *N is a justified argument iff for all chains $N \prec^* N_1 \prec^* \ldots \prec^* N_n$ it holds*
$$n \text{ is even} \wedge \neg\exists_{N' \in \mathcal{N}} N_n \prec^* N'.$$
- *N is an overruled argument iff there is a chain $N \prec^* N_1 \prec^* \ldots \prec^* N_n$ with*
$$n \text{ is odd} \wedge \neg\exists_{N' \in \mathcal{N}} N_n \prec^* N'.$$
- *N is a defensible argument iff N is neither a justified argument nor an overruled argument.*

Notice that $\neg\exists_{N' \in \mathcal{N}} N_n \prec^* N'$ is equivalent with $Ch(N_n) = \{N_n\}$.

Example 4.13 *Consider example 4.6. Let $\mathcal{N} = \{N_1, N_2, \ldots, N_6\}$, with*
$N_1 = \{O(a \vee b), O(c \vee \neg b), O(\neg c)\} \in [O(a)]\mathcal{N}$
$N_2 = \{O(\neg a)\} \in [\neg O(a)]\mathcal{N}$
$N_3 = \{O(a \vee b), O(\neg a)\} \in [O(b)]\mathcal{N}$
$N_4 = \{O(c \vee \neg b), O(\neg c)\} \in [\neg O(b)]\mathcal{N}$
$N_5 = \{O(a \vee b), O(c \vee \neg b), O(\neg a)\} \in [O(c)]\mathcal{N}$
$N_6 = \{O(\neg c)\} \in [\neg O(c)]\mathcal{N}$

Argument N_1 is defeated by the justified argument N_2, since $O(\neg c) \prec O(\neg a)$ and $Ch(N_2) = \{N_2\}$, thus N_1 is an overruled argument. N_3 and N_5 are justified arguments, since $Ch(N_3) = \{N_3\}$ and $Ch(N_5) = \{N_5\}$. N_4 and N_6 are overruled arguments, since $N_4 \prec^* N_3 \in Ch(N_4)$ and $N_6 \prec^* N_5 \in Ch(N_6)$. If (4) $\sim$ (3) $\prec$ (2) $\prec$ (1), then N_5 and N_6 are defensible arguments, since $Ch(N_5) = \{N_5 \prec^* N_6\}$ and $Ch(N_6) = \{N_6 \prec^* N_5\}$.

Example 4.14 *Consider example 4.7. N_4 is defeated by the justified argument N_3 and $Ch(N_3) = \{N_3\}$, thus N_4 is an overruled argument.*

Definition 4.15 *Let $\mathcal{N} \subseteq \mathcal{N}(w, F)$ and $\mathcal{N}^*$ the set of all defensible and justified arguments in $\mathcal{N}$. Then o is a maximal consistent argument set iff o is a union of a part of the arguments in $\mathcal{N}^*$ with*

- *$o \cup \{F_b\} \cup W(w)$ is consistent and*
- *$\neg\exists_{N \in \mathcal{N}^* \setminus o}(o \cup \{F_b\} \cup W(w) \cup N$ is consistent).*

$\mathcal{O}(\mathcal{N})$ is defined as the set of all maximal consistent argument sets o. $Out(o)$ is defined as the set of all outcomes of the arguments $N \subseteq o$.

Corollary 4.16 *Let $o \in \mathcal{O}(\mathcal{N})$. Then for all justified arguments N in $\mathcal{N}$ it holds that $N \subseteq o$.*

Example 4.17 *Consider example 4.13 with the justified arguments N_2, N_3 and N_5. There is one maximal consistent argument set o: $o = N_2 \cup N_3 \cup N_5 = N_5 = \{O(a \vee b), O(c \vee \neg b), O(\neg a)\}$.*

If (4) $\sim$ (3) $\prec$ (2) $\prec$ (1), then N_5 and N_6 are defensible (see example 4.13), then there are two maximal consistent argument sets: $o_1 = N_2 \cup N_3 \cup N_5 = N_5$ and $o_2 = N_2 \cup N_3 \cup N_6 = \{O(a \vee b), O(\neg a), O(\neg c)\}$, thus $\mathcal{O}(\mathcal{N}) = \{o_1, o_2\}$.

4.2 Violation

An obligation is violated iff the obligation is not fulfilled. In SDL we can represent the violated obligation by $O(p) \wedge \neg p$. Analogue we define the violated obligation in our theory.

Definition 4.18 *Let $\mathcal{N} \subseteq \mathcal{N}(w, F)$, $o \in \mathcal{O}(\mathcal{N})$ and $o \cup W(w) \cup \{F_b\} \vdash O(\phi)$. The norm $O(\phi)$ is violated in o iff $w \cup \{F\} \models \neg\phi$. The set of violated norms in o will be denoted as $V(o)$.*

Example 4.19 *Let $o = \{O(a), O(b)\}$, $W(w) = \{a \Rightarrow c\}$, $\{F_b\} = w = \emptyset$ and F_c: $\neg b \wedge \neg c$. Then $O(c) \in V(o)$, since $o \cup W(w) \vdash O(c)$ and $\{F\} \vdash \neg c$ and $O(b) \in V(o)$, since $o \cup W(w) \vdash O(b)$ and $\{F\} \vdash \neg b$.*

Definition 4.20 *Let $\mathcal{N} \subseteq \mathcal{N}(w, F)$, $o \in \mathcal{O}(\mathcal{N})$ and $V(o)$ be the set of violated norms. Then $N \subseteq o$ is a violated argument iff*

$$\exists_{\psi \in V(o)} N \cup W(w) \cup \{F_b\} \vdash \psi.$$

$\mathcal{N}(V(o))$ is the set of all the violated arguments with respect to $V(o)$.

Example 4.21 *Consider example 4.6:*
(1) $O(a \vee b)$
(2) $O(\neg a)$
(3) $O(c \vee \neg b)$
(4) $O(\neg c)$
with $(4) \prec (3) \prec (2) \prec (1)$, $W(w) = \{F_b\} = \emptyset$ *and* F_c*:* $\neg c$*. Suppose further (see example 4.13) that* $\mathcal{N} = \{N_1, N_2, \ldots, N_6\}$*, with*
$N_1 = \{O(a \vee b), O(c \vee \neg b), O(\neg c)\} \in [O(a)]\mathcal{N}$
$N_2 = \{O(\neg a)\} \in [\neg O(a)]\mathcal{N}$
$N_3 = \{O(a \vee b), O(\neg a)\} \in [O(b)]\mathcal{N}$
$N_4 = \{O(c \vee \neg b), O(\neg c)\} \in [\neg O(b)]\mathcal{N}$
$N_5 = \{O(a \vee b), O(c \vee \neg b), O(\neg a)\} \in [O(c)]\mathcal{N}$
$N_6 = \{O(\neg c)\} \in [\neg O(c)]\mathcal{N}$.

From example 4.13 and 4.17 it follows that $o = N_2 \cup N_3 \cup N_5$, thus $\mathcal{O}(\mathcal{N}) = \{N_5\}$. $O(c) \in V(o)$, since $N_5 \vdash O(c)$ and $\{F\} \vdash \neg c$. Then N_5 is a violated argument. N_2 and N_3 are not violated arguments, since $N_2 \not\vdash O(c)$ and $N_3 \not\vdash O(c)$.

Definition 4.22 *Let* $\mathcal{N} \subseteq \mathcal{N}(w, F)$ *and* $\psi_1, \ldots, \psi_n \in \Delta(w, F)$*. Then*

$$\mathcal{N} \setminus^* \{\psi_1, \ldots, \psi_n\} = \cup\{N | N \in \mathcal{N} \wedge \forall_{i=1,\ldots,n} \psi_i \notin N\}.$$

Thus $\mathcal{N} \setminus^* \{\psi_1, \ldots, \psi_n\}$ is the set of arguments in $\mathcal{N}$ without the arguments containing a norm of the set $\{\psi_1, \ldots \psi_n\}$.

Example 4.23 *Consider a programme committee which requires that submissions (papers) for the conference need to be sent by mail. However if your paper is not sent by mail then your paper must be sent by fax. And if no faxmachine is available one should ultimately try to send by email.The facts are that the paper is sent by fax and if you send your paper by fax, then you do not send the paper by email. The formalisation is as follows:*

W: (1) $email \Rightarrow \neg mail$	Δ: (1') $O(mail)$	F_c: fax
(2) $fax \Rightarrow \neg mail$	(2') $\neg mail \Rightarrow O(fax)$	F_b : $fax \rightarrow \neg email$
	(3') $\neg fax \Rightarrow O(email)$	

with $(2) \prec (1)$ *and* $(3') \prec (2') \prec (1')$.

Let us consider this example. If the paper was sent by mail, then there is no violation; this is the best situation. However, the norm $O(mail)$ is violated, because the paper is not sent by mail. Then the second best situation is that your paper is sent by fax. We can formalise this process as follows:
$Out(w) = \{\neg mail\}$, $W(w) = \{fax \Rightarrow \neg mail, email \Rightarrow \neg mail\}$
$\Delta(w, F) = \{O(mail), O(fax)\}$
$\mathcal{N}(w, F) = \{N_1, N_2\}$, with $N_1 = \{O(mail)\}$ and $N_2 = \{O(fax)\}$.
Notice that $O(email)$ is not an element of $\Delta(w, F)$, since it is not an applicable norm: $w \cup \{F\} \not\models \neg fax$.

N_1 is a justified argument, since $Ch(N_1) = \{N_1\}$ and N_2 is an overruled argument, since it is defeated by N_1: N_1 and N_2 are in conflict, since $\{F_b\} \cup W(w) \cup N_1 \vdash O(mail)$ and $\{F_b\} \cup W(w) \cup N_2 \vdash \neg O(mail)$ and $O(fax) \prec O(mail)$. Thus, $o = \{N_1\}$.

The norm $O(mail)$ has a higher priority then the norm $O(fax)$. However the norm $O(mail)$ is violated:

$$V(o) = \{O(mail)\} \text{ and } (\mathcal{N}(w, F))(V(o)) = \{N_1\}.$$

Now we consider the norms again without the violated norm, because they can become valid because of the violation of the norm $O(mail)$. Let $\mathcal{N} = \mathcal{N}(w, F) \setminus^* \{O(mail)\}$. Then $\mathcal{N} = \{N_2\}$. The argument N_2 is a justified argument. Let $o' \in \mathcal{O}(\mathcal{N})$, then it follows that $o' = \{N_2\}$. N_2 (with as an outcome $O(fax)$) is a justified and not violated argument with respect to the violation of the norm $O(mail)$. Thus sending your paper by fax is the second best situation.

5 Conclusions

In this paper we have presented a theory of the defeasible deontic reasoning dealing with some very common problems of other approaches:

- Defeasibility between *groups of conditional norms*;
- The combining of defeasibility of rules and norms;
- The lack of notion of permission.

However, we do not claim that our theory is completely free from the above-mentioned problems. Adopting defeasible reasoning for non-deontic constraints (rules) is not trivial, such as the study of the validity of deducing of $O(b)$ from $O(a)$ and $a \Rightarrow b$. This corresponds with the choice of the rule (RK2):

$$\frac{O(p_1), O(p_2), \ldots, O(p_n), p_1 \wedge p_2 \wedge \ldots \wedge p_n \Rightarrow q}{O(q)}.$$

However, it is also possible to add the default rule schema (RK)

$$(p_1 \wedge p_2 \wedge \ldots \wedge p_n \Rightarrow q) \rightarrow (O(p_1) \wedge O(p_2) \wedge \ldots O(p_n) \Rightarrow O(q)),$$

which also allows defaults between norms.

The consequence of the use of the (RK2)-rule is that norms are dependent on rules and that rules are not dependent on norms. This means that rules have a higher priority then norms. Thus an argument of rules cannot be defeated by an argument of norms. If we had not made this separation between rules and conditional norms (i.e. we opted for (RK) in stead of (RK2)) then we would get for example the following situation:

$W \cup \Delta$: (1) $O(a)$ F: $\emptyset$
(2) $O(b)$
(3) $a \Rightarrow \neg b$
with (3) $\prec$ (2) $\prec$ (1).

From $O(a)$ and $a \Rightarrow \neg b$ we can deduce $O(\neg b)$, thus $\{O(a), a \Rightarrow \neg b\}$ would be an argument for $O(\neg b)$. Thus the set $\{O(a), (O(b), a \Rightarrow \neg b\}$ is incoherent. We cannot deduce $\neg(a \Rightarrow \neg b)$ from $O(a)$ and $O(b)$ with the (RK2)-rule. However, with (RK) we could derive $\neg(a \Rightarrow \neg b)$ from $O(a)$ and $O(b)$:
$O(a) \wedge O(b) \rightarrow O(a) \wedge \neg O(\neg b)$;
$O(a) \wedge \neg O(\neg b) \rightarrow \neg(\neg O(a) \vee O(\neg b))$;
$\neg(\neg O(a) \vee O(\neg b)) \rightarrow \neg(O(a) \rightarrow O(b))$;
$\neg(O(a) \rightarrow O(b)) \rightarrow \neg(a \Rightarrow \neg b)$.
Thus $\{O(a), O(b)\}$ is an argument for $\neg(a \Rightarrow \neg b)$.[7]

The advantage of changing the (RK2)-rule by a stronger rule is that rules and conditional norms do not have to be separated in the arguments and that rules do not have a higher priority than conditional norms by definition, which is a consequence from the theory presented in this paper. At first sight this seems a good concept to solve the problem of the separation of rules and norms. However, this concept gives rise to some serious problems, since the definition of defeating arguments for rules is different from the definition for norms. In the definition of defeating arguments for rules we look only at single statements with conflicting final conclusions in contrast with the definition for norms where we not only look at final conclusions but at groups of statements deriving conflicting conclusions. Furthermore, the applicability of a conditional norm depends on the facts and rules, thus this means that we have to separate rules and conditional norms. We leave this issue of the separation of arguments for rules and conditional norms in deontic reasoning as a future research topic.

Most deontic defeasible reasoning approaches are based on specificity considering the amount of relevant information or supporting evidence. Our approach is based on the more general idea of priority (authority). Other defeasibility criteria can easily be converted to some form of defeasibility on the basis of priority.

References

[1] Brewka G. *Preferred subtheories: An extended logical framework for default reasoning.* In: Proceedings IJCAI-1991, pp 1043-1048.

[2] Poole D. *A logical framework for default reasoning.* Artificial intelligence 1988; 36:27-47.

[3] Prakken H. *Logical tools for modelling legal argument.* PhD thesis, Amsterdam, 1993.

[7]As a consequence we allow negations of default rules. It is not trivial to decide the meaning of these formulas.

[4] Shoham Y. *Reasoning about change*, MIT Press, Cambridge, 1988.

[5] McDermott D and Doyle J *Non-monotonic logic I.* Artificial intelligence 1980; 13:41-72.

[6] Reiter R. *A Logic for default reasoning.* Artificial Intelligence 1980; 13:81-132.

[7] Pollock J. *Defeasible reasoning.* Cognitive Science 1987; 11(4):481-518.

[8] Alchourrón C and Makinson D. *Hierarchies of regulations and their logic.* In: Hilpinen R (ed) New studies in deontic logic. D. Reidel Publishing Company, 1981, pp 125-148.

[9] Royakkers L and Dignum F. *Deontic inconsistencies and authorities.* In: Breuker J (ed) Normative reasoning, 1994.

[10] Horty J. *Moral dilemmas and nonmonotonic logic.* Journal of philosophical logic 1994; 23:35-65.

[11] Ryu Y. *Conditional deontic logic augmented with defeasible reasoning*, submitted to Data & knowledge engineering.

[12] Tan Y.-H and Van der Torre L. *Multi preference semantics for a defeasible deontic logic.* In: Prakken H (ed) Legal knowledge based systems. The relation with legal theory. Koninklijke Vermande, Lelystad, 1994, pp 115-126.

[13] Prakken H. *Two approaches to defeasible reasoning.* In: Jones A and Sergot M (eds) Proceedings of second international workshop on deontic logic in computer science. Tano, Norway, 1994, pp 281-295.

[14] Dung P. *On the acceptability of arguments and its fundamental role in nonmonotonic reasoning and logic programming.* In: Proceedings IJCAI-93, pp. 852-857.

[15] Hart H. *The concept of law.* Clarendon Press, Oxford, 1961.

[16] Delgrande J. *An approach to default reasoning based on a first-order conditional logic.* Artificial intelligence 1988; 36:63-90.

[17] Rescher N. *Hypothetical reasoning.* North-Holland, Amsterdam, 1964.

[18] Kelsen H. *Reine rechtslehre.* Zweite auflage, Wien, 1975.

[19] Reiter R. *Nonmonotonic reasoning.* Annual review of computer science 1987; 2:147-186.

[20] Sartor G. *Defeasibility in legal reasoning.* Rechtstheorie 1993; 24:281-316.

[21] Prakken H and Sartor G. *On the relation between legal language and legal argument: assumptions, applicability and dynamic priorities.* In: Proceedings of the fifth international conference on artificial intelligence and law. University of Maryland, MD USA, 1995, pp 1-11.

Indirect Action, Influence and Responsibility

Filipe Santos[1] and José Carmo[2]
[1] Department of Management Science, ISCTE,
Av. das Forças Armadas, 1600 Lisboa Codex, Portugal
fas@inesc.pt
[2] Department of Mathematics, IST,
Av. Rovisco Pais, 1096 Lisboa Codex, Portugal
jcc@inesc.pt

Abstract

In this paper a new action concept is proposed, within the same tradition of the logical characterization of act descriptions as employed by Kanger, Pörn and Lindhal. The concept sums up a successful goal oriented social activity into a unique assertion referring to the agent who has initiated it. It is argued that this notion allows us to increase the expressiveness of the characterization of agents´ activity within organizations.

For describing the agents´ interaction the *normative systems perspective* is adopted, viewing such interactions as governed by norms and expressing them in terms of what agents are permitted and obliged to do. In this context, the new concept allows us to cope with normative descriptions that play an important role in organizations: *responsibilities.*

The minimal properties of this *"indirect" agency operator* are discussed, and an *"influence" operator* is also introduced in order to cope with interactions between the different agents. The interest of these concepts within the analysis and specification of organizations is illustrated.

1 Introduction

We are particularly interested in the formal analysis and specification of organizations. We adopt the *normative system perspective* proposed in [1], seeing organizations as instances of normative systems, *i.e.* a set of interacting agents whose behaviour can be seen as norm-governed, and thus expressing interaction in terms of what agents are permitted and obliged to do, and also other complex normative relations between them.

In their pioneering work, Kanger, Pörn and Lindahl have combined deontic logic and action logic as basic building blocks to describe social interaction and complex normative concepts (see e.g [2, 3, 4, 5, 6]). Their logics have considerable expressive power, and they are able to articulate several distinctions at an appropriate level of abstraction, mainly in virtue of the modal logic of action they employed, which ignores temporal aspects. They introduce a relativised modal operator E_i, where expressions of the form E_iA are read "the agent i brings it about that A" or "agent i sees to it that A is the case".

The main core of this paper is to propose a new *"indirect" agency concept*, within a logical approach similar to the one employed by Kanger, Pörn and Lindhal, in order to increase the expressiveness of the characterization of agents´ activity within organizations. It is argued that this "indirect" agency concept is important to deal with types of normative interactions that commonly occur in organizations, namely those represented by *responsibilities*. As we shall indicate, the agency operator proposed by Chellas [7] has some properties similar to the ones here suggested for our new operator; however, we think that the distinction between direct and indirect agency is necessary, and that it has not been clearly described before.

A brief description of the rest of the paper follows. In Section 2 we make a very brief presentation of the main modal logics of action/agency that have been proposed in the literature[1], discuss their main differences and their expressiveness in the characterization of social interaction. We end Section 2 by arguing that there is room for a distinction between a "direct" and an "indirect" agency concept. In Section 3 we show that the "indirect" agency concept is needed in order to cope with the notion of responsibility; we propose a new modal operator for it and discuss its basic principles and its relationships with the "direct" agency operator. In Section 4 we define its semantics and in Section 5 we discuss further ideas to be considered in a complete characterization of the new agency concept, namely ideas related with *influence*. Conclusions and directions for future research appear in Section 6, including a very brief discussion of the specification of organizations (in regard to the aspects herein analyzed).

2 The modal logic of action

The action concept herein introduced follows the same tradition of the logical characterization of act descriptions as employed by Kanger, Pörn and Lindhal (see e.g. [2, 3, 6, 8]). They introduce a relativised modal operator E_i, where expressions of the form E_iA are read "the agent i brings it about that A" or "agent i sees to it that A is the case". The logical development of this approach is due to Pörn [3],

---

[1] It is not intended here to make an overview of all the different modal logics of action/agency that have been proposed in the literature, but only of what we could call the representatives of the main modal approaches to these logics. On the other hand, we restrict ourselves to the modal logics of action where no explicit reference to the specific undertaken actions is made, which we also call logics of agency (following other authors). Besides these approaches, mainly coming from the philosophical area, another proposal for logics of actions has appeared in Computer Science, namely the dynamic logic approach (see e.g. [12, 13]), where the modal operators are indexed by terms that explicitly refer to the actions. We think that these two approaches to the logics of actions both have virtues, and a discussion is needed of their differences and advantages, as well as of how to combine them, particularly in the context of specifications and refinements; this is however completely outside the scope of this paper.

with more refined versions presented in [2, 4, 9]. An important feature of these logics is that actions are taken to be relationships between agents and states of affairs that they bring about, and are considered in abstraction from temporal aspects.

In the accounts of their logics, Kanger and Pörn [2, 4] introduce two propositional normal modalities, D_i of type KT and D'_i of type KD (according to Chellas classification [10]), where sentences of the form D_iA and D'_iA are respectively read "it is necessary for something i does that p" and "but for i's action it would have been the case that p". The semantics of these modalities is defined using standard models through two accessibility relations R_i^D (reflexive) and $R_i^{D'}$ (serial and irreflexive) with the following readings:

$(\alpha,\beta)\in R_i^D$	iff	everything which i brings about in α is the case in β
$(\alpha,\beta)\in R_i^{D'}$	iff	not everything that i brings about in α is the case in β

As usual, truth at a world α in a model $\mathcal{M}$ is defined for the operators D_i and D'_i as follows:

$\mathcal{M}, \alpha \models D_iA$	iff	$\mathcal{M}, \beta \models A$, for all β such that $(\alpha,\beta)\in R_i^D$
$\mathcal{M}, \alpha \models D'_iA$	iff	$\mathcal{M}, \beta \models A$, for all β such that $(\alpha,\beta)\in R_i^{D'}$

Kanger and Pörn define E_i by combining the modalities above, getting a non-normal one as follows (where $C'_iA =_{def} \neg D'_i\neg A$ is read "but for i's action it might have been the case that A"):

(def.Kanger [2])	$E_iA =_{def} D_iA \wedge D'_i\neg A$
(def.Pörn)	$E_iA =_{def} D_iA \wedge C'_i\neg A$

Their resulting logics yield as valid the schemata C, K, T and No (i.e., $\neg E_i$True). The counterfactual component introduced by the operator D'_i / C'_i allows them to surpass some problems of the action logic previously proposed in [3][3], where it can be shown that the agents "bring about" logical truths and other counter-intuitive results due to the closure under logical consequence of the operator E_i. It should be noted that in [9] it is argued that the counterfactual component employed by Pörn is too strong because it collapses two different notions of *avoidability* relevant to the characterization of agency: "what is brought about by an agent must not be necessarily true" and "agent's activity is instrumental in the production of what he brought about".

In [9], Dag Elgesem develops a semantical framework where he includes the notion of agent's control in the analysis of the action notion, following in this

---

[2] In [4] it is pointed out that Andrew Jones proved that Kanger's definition is too strong, since it follows from the definition that the conjunction $E_i(p\rightarrow q) \wedge E_ip$ is inconsistent, and it is argued that there are many actions which can be adequately described by means of such conjunctions.

[3] Note that in [3] the operator D_i alone is taken to represent the "brings it about" notion (giving a normal modal logic).

respect the analysis of the teleological notion of *objective goal directness* due to Sommerhoff (see [11]), i.e. the phenomenon of an organism's activity being directly correlated with some goal. The semantics of Elgesem's action modality is based on minimal models (see [10]), using functions $f_i : W \times 2^W \rightarrow 2^W$ where $f_i(\alpha,X)$ means the set of worlds where agent i realizes his ability to bring about the goal denoted by X w.r.t. (with respect to) an actual world α. Truth at a world α in a model $\mathcal{M}$ for expressions of the form E_iA is defined by (where $\|A\|^{\mathcal{M}}$, in the sequel written only as $\|A\|$, represents the truth-set of A in the model $\mathcal{M}$):

(def.Elgesem) $\quad \mathcal{M}, \alpha \models E_iA \quad \text{iff} \quad \alpha \in f_i(\alpha, \|A\|^{\mathcal{M}})$

Elgesem constrains the function f_i in order to re-construct Sommerhoff's analysis in a possible world semantics[4], getting as a result a propositional classical modal system of type ECT with also the schema (No). His model allows him to explore further notions and their relationships with the action concept, distinguishing e.g. the interpretations of "can" as ability and as opportunity. For instance, the former is defined by:

(def. ability) $\quad \mathcal{M}, \alpha \models \text{Ability}_i\, A \quad \text{iff} \quad f_i(\alpha, \|A\|) \neq \emptyset$

yielding as valid the schema $E_iA \rightarrow \text{Ability}_iA$, reflecting the idea that agency is the manifestation of a capacity to act.

It must be stressed that these approaches offer, in some respect, additional expressive power when compared with other action logics based on an action ontology (see e.g. [12, 13] and footnote 1). In fact, these theories allow us to describe the possible atomic individual agent act-positions w.r.t. each formula A: E_iA, $E_i\neg A$ and $\neg E_iA \wedge \neg E_i\neg A$ ("i remains passive w.r.t. A"), as a starting point to the analysis of social interaction (see in regard to this Lindahl's *theory of normative positions* developed in [6] and also [14] for a systematic generation of different normative positions within Lindahl's theory). Additional concepts related with this action operator can be found in [4].

There is another important approach to the modal logic of agency: the "stit theory" of Nuel Belnap and Michael Perloff (see [15, 16, 17, 18])[5]. The notion of "agent i sees to it that A" is represented in their logic by expressions of the form $STIT_iA$ and semantically follows the idea that "the present fact that A is guaranteed by a prior choice of i". Models for their "STIT" operator are based on tree like structures[6] representing branching time with the same past and open to the future, using the primitive notions of instants, moments, histories and choices for an agent. Histories are constructed from instants to moments (states of affairs), where *h(t)* is

---

4 The constraints employed by Elgesem follow three criteria proposed by Sommerhoff for *objective goal directness*: success, avoidability and non-accidence.

5 Franz von Kutschera has proposed a similar approach prior to the work of Belnap and Perloff (see [22]). See [23] for a detailed discussion of the differences between Kutschera's account and Belnap and Perloff's account.

6 Giving a probably more intuitive model than the one provided by the minimal models, although with some drawbacks, as we mention next.

the moment in history *h* at instant *t* [7]. Histories having the same present and past at instant *t* are connected by the equivalence relation $\equiv_t$ among histories: $h \equiv_t h'$ iff $h(t')=h'(t')$ at every instant $t' \leq t$. The following constraint imposes future branching only: if $h(t)=h'(t)$ then $h \equiv_t h'$.

The central element in Belnap and Perloff's analysis is the notion of agent's choice. A choice for an agent *i* at a moment *h(t)* (the "choice point") is a non empty set of histories *h'* passing through that moment (i.e., $h' \equiv_t h$). All i's choices at moment *h(t)* constitutes a partition on the set of all histories *h'* passing through that moment, i.e., a partition yielded by the equivalence relation between histories $C^i{}_t$, with the reading: $C^i{}_t(h,h')$ iff "histories h and h' are together in a choice for i at t". Belnap and Perloff impose the following constraints on agent's choices: "no choice between undivided histories" (if $t1<t$, then $h \equiv_t h'$ only if $C^i{}_{t1}(h,h')$) and "something happens" (the intersection of any combination of choices made by distinct agents at exactly the same moment must contain at least one history).

Truth at an instant t in a history h for expressions of the form $STIT_iA$ is defined by

(def. Belnap) $h, t \models STIT_iA$ iff there is a time $t1 < t$ such that
(1) $h', t \models A$ for every history h' such that $C^i{}_{t1}(h,h')$
and (2) $h',t \not\models A$ for some history h' such that $h \equiv_{t1} h'$

getting as a result a logic with schemata C, K, T, 4 (again according to Chellas classification), REFREF (i.e., $STIT_iA \leftrightarrow STIT_i \neg STIT_i \neg STIT_iA$) and No (i.e., $\neg STIT_i True$).

Belnap and Perloff's definition resembles Pörn's definition with respect to their positive and negative components. According to Elgesem the same criticisms against the negative component of Pörn's definition applies to Belnap and Perloff's approach. Moreover, Elgesem also presents arguments against the schemas (REFREF) and (4).

Another important criticism of Belnap's account is described in [19]. Chellas points out that the principle "something happens" and the negative component of Belnap and Perloff's definition are responsible for getting the schema $\neg STIT_iSTIT_jA$ (for $i \neq j$). This result excludes (at least) some notions of direct control within "stit theory". Moreover, the validity of the schema $STIT_iSTIT_jA \rightarrow STIT_iA$ (for $i \neq j$) becomes trivial in this theory.

Within a similar semantic framework, Chellas presented an earlier account to the modal logic of agency (see [7]). The time structures employed by Chellas in the characterization of his "sees to it" operator Δ_i are not restricted by the kinds of constraints adopted by Belnap and Perloff. Semantically, Δ_iA follows the idea that "A is true at all instigative alternatives for the agent i". Chellas represents this idea by using the reflexive relation between histories $R^i{}_t$, with the reading $R^i{}_t(h,h')$ iff

---

7 We follow here the notation used by Brian Chellas in [19] where he criticizes Belnap's account within the semantic framework of Chellas's previous account to agency (advanced in [7]). The main reason to take this option is due to the fact that we also want to mention Chellas's account and its differences w.r.t. Belnap's theory.

"for agent i, at time t, h' is an instigative alternative to h", just constrained by the "historical relevance" condition: $R^i_t(h,h')$ only if $h=_t h'$ (where $h=_t h'$ iff $h(t1)=h'(t1)$, at every instant $t1<t$). Truth at time t in a history h for expressions of the form $\Delta_i p$ is defined by

(def.Chellas) $h, t \models \Delta_i A$ iff $h', t \models A$, for every history h' such that $R^i_t(h,h')$

giving a normal modal logic of type KT . Moreover, contrary to what happens in the previously described logics of actions, in Chellas's logic the schema $\Delta_i\Delta_j A \rightarrow \Delta_i A$ (schema Q for Chellas) is valid. Chellas defends this schema as an important principle to be considered in the characterization of the "sees to it" notion (see [7, 19]), and calls it the "*Qui facit per alium*" principle. Although we agree with his arguments, we think that there is another important distinction to be drawn between the different agency logics related with this principle, as we shall explain below.

To conclude this very brief comparison, although the formal properties of the action operator vary among the authors mentioned above, their logical systems all have in common:

(T)	$\vdash E_i A \rightarrow A$
(C)	$\vdash (E_i A \wedge E_i B) \rightarrow E_i(A \wedge B)$
(RE)	If $\vdash A \leftrightarrow B$ then $\vdash E_i A \leftrightarrow E_i B$

The (T) schema captures the intuition that if agent i brings it about A then A is indeed the case (E_i is a "success" operator), and co-temporality is presupposed when imposing (C). ((RE) is just closure under logical equivalence.) Thus we may consider that these principles are the core of any action logic and we will also adopt them for the new action operator introduced below.

On the other hand, the previous logics differ in regard to some properties of E_i. For instance, considering E_i in isolation (i.e., considering only one agent "i"), we see that (besides [3]) only Chellas's logic verifies the (RM) rule (i.e., closure under consequence) and (of course) does not have schema (No). Although we are also going to accept the schema (No) for our new action operator, we admit that there may exist contexts where (No) can be abandoned without problems and where (RM) may be acceptable (or at least some weaker versions of it: e.g. when the negation of the consequent is possible). In fact, if conceptually it makes no sense that an agent "brings about" a necessary truth, in some contexts, like e.g. in organization specifications, we may take a more pragmatic view with respect to the "sees to it" notion and get a much simpler logic that suffices for the job at hand. However, we do not have space to discuss this in more detail in this paper.

Besides the previous differences concerning E_i in isolation, there is a main difference between the above logics with regard to the interaction between the agency of the different agents. In particular, only Chellas has the schema: $E_i E_j A \rightarrow E_i A$ (Belnap also has it but merely because the antecedent is always false for $i \neq j$). This suggest that there is a main difference between Chellas's action operator and the ones in [2, 4, 9]: according to our point of view, accepting $E_i E_j A \rightarrow E_i A$ means that we are accepting that an agent may "bring about" indirectly (by using another agent)[8]. On the other hand, accepting $E_iE_jA \rightarrow \neg E_iA$ (for $i \neq j$)[9] is a way of insisting that E_iA means that agent i must bring it about that A is the case by itself (directly). Thus, it seems that we may have two different action operators: a "direct" action concept and an "indirect" one (the former will be referred to in the sequel as E_i and the latter as G_i). Moreover, we think that both are useful and needed (at least in some contexts). In what follows we want to defend this point of view, and to study in more detail the indirect action operator. Of course, since the logics in [2, 4, 9] do have neither of the two previous schemas, we may say that they are neutral with respect to this important aspect of agency.

3 Indirect action and responsibility

In Organization Theory (see e.g. [20]), the concept of responsibility[10] plays an important role concerning organizational design with respect to the work differentiation needed to attain the objectives of the organization. Work differentiation is obtained along a horizontal and a vertical dimension, respectively associated with the division of work and management positions. In this context, responsibilities are considered essential ingredients to match agents with groups of tasks, allowing the creation of positions within organizations. Roughly speaking, responsibilities are seen to be duties that members of an organization must fulfil, in accordance with procedures, policies and strategies. On the one hand, agents in operational positions of the organization fulfil their duties by acting themselves according to their capabilities, and on the other hand, agents in management positions fulfil their duties by ensuring that certain results are obtained via other agents, by using existent organizational procedures or creating new ones, by attributing responsibilities, etc. Of course, the same agent may be in an "operational position" w.r.t. some task and in a "management position" w.r.t. another task (or may even be in both positions w.r.t. one and the same task).

The concrete actions of agents in management positions depend on the underlying structure of the organization (e.g. power structure and task decomposition), and even depend on specific circumstances (e.g. availability of specific agents and resources). That is why responsibilities allocated to this type of agent often refer to specific results that must be achieved within the organization, without entering into details about the concrete actions that must be performed. The point which we want to stress here is that, for the sake of the characterization of the agents´ responsibilities and interactions within organizations, we need something more than a "direct" action concept, if we want to be able to express such responsibilities in a simple and abstract way.

---

[8] It should be noted that in [3] this schema is also accepted and it is argued that it follows from it that "E_iA does not permit the interpretation that i alone is the agent of the performance of A".

[9] In the sequel we denote this schema by (EE¬E).

[10] It should be mentioned that the word "responsibility" may have very different meanings. Here we are thinking in terms of "responsibility for some task".

Let us consider the following very simple organization pictured in diagram 1, representing an hierarchy with only two agents: *a* in a management position and *b* in an operational position (w.r.t. some "task" *p*). We assume that agent *a* has "some power" over agent *b* (w.r.t. *p*) but that he does not have the capability to (directly) bring it about that *p* ; and we assume that agent *b* does have the capability to bring it about that *p*[11].

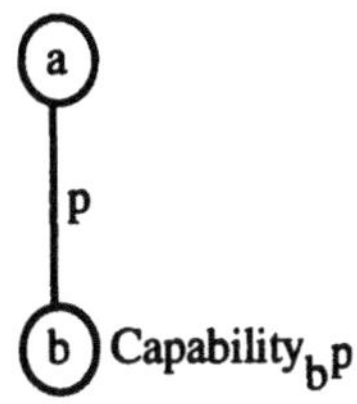

diagram 1

Suppose now that we want to specify that agent *a* has the responsibility to obtain *p*. Using the "direct" action operator we cannot represent a's responsibility as OE_ap because it is impossible for *a* to bring it about that *p* is the case by himself: the only way *a* has to fulfil his responsibility is to exercise his "power" over *b* in order that *b* brings it about that *p*. Thus, using the "direct" action operator, in combination with some obligation operator O, in order to represent a's responsibility to obtain *p* we need to know exactly the form of his "power" and explicitly make it obligatory for *a* to exercise that power. Supposing, for instance, the simple case where the power takes the form of a direct control, i.e., an ability of *a* to bring it about that *b* brings it about that *p*, then we could represent a's responsibility to obtain *p* as OE_aE_bp.

Of course, it could be argued that instead of considering a "direct" action operator we could consider an action operator like the ones in [2, 4, 9]: since they do not necessarily require the agency to be direct, we could represent a's responsibility to obtain *p* as OE_ap. However, in such cases we could not derive that *a* has fulfilled his obligation by acting as expected (i.e. by E_aE_bp), since we do not have the schema $E_aE_bA \rightarrow E_aA$. Thus again we would be inclined to represent a's responsibility to obtain *p* as OE_aE_bp.

Consider now the organization pictured in diagram 2, representing an hierarchy with four agents: *a* and *b* are in management positions and agents *c* and *d* in operational ones. To simplify we assume that agent *a* has "power" over agent *b* w.r.t. any action this agent is able to perform (and for this reason we do not put in the diagram any label in the link between the two agents) and that agent *b* has a similar power over agents *c* and *d*, but that agent *a* has no direct power over agents *c*

---

[11] Herein the concept of capability follows the same idea as that of Elgesem's concept of ability mentioned above. We just want to establish the difference between formal abilities within organizations (our capabilities) and proper abilities of the agents. On the other hand, in order to simplify as far as possible the presentation, we do not include in the formal language any operator to express this notion of "capability" (and refer to it only in the description of the organization examples). This allows a simpler model, and suffices for our present purposes.

or *d* (as follows intuitively from the fact that there is no line linking *a* to *c* or to *d* in the diagram). Moreover, we assume that both *c* and *d* have the capability to bring it about that *p* (but neither *a* nor *b* have this capability).

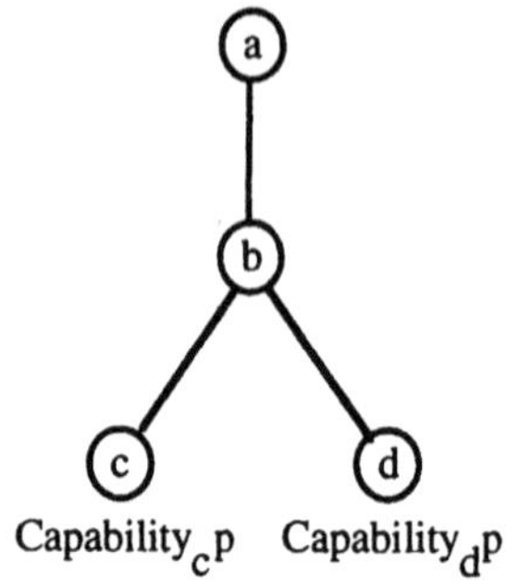

diagram 2

Suppose again that we want to specify that agent *a* has the responsibility to obtain *p*. In this case the only way *a* has to fulfil his responsibility is by exercising his power over *b* in order that he exercises his power over agents *c* or *d*, both having the capability to bring it about that *p*. Thus, assuming again that the "power" in question takes the form of a "direct control", by similar arguments we would represent a's responsibility to obtain *p* by $OE_aE_b(E_cp \vee E_dp)$. Let's now consider that *a* knows merely that *b* has "authority" over agents in operational positions which are able to secure *p*. In this case we could even consider $OE_aE_b(\exists_x)E_xp$ to be an acceptable characterization of a's responsibility. Add to the example more organizational details, e.g. more steps in the hierarchy and task decomposition, or assume (as we want to do) that the "power" involved may take other forms besides a direct control (for instance, the ability to create normative positions in other agents, by attributing obligations/responsibilities to them) and we will encounter difficulties in the attempt to give a complete characterization of responsibilities using the "direct" action operator.

The previous examples illustrate that we need something more than a "direct" action concept, if we want to be able to express such responsibilities in a simple and abstract way. Our suggestion is to introduce an "indirect" action concept and deal with obligations and responsibilities in a unified way. We propose the word *ensure* (or *guarantee*) for this purpose (retaining the expression "bring it about" for the "direct" action concept) and we suggest the representation of expressions of the form "*agent i is responsible for A*" by expressions of the form "agent i is obliged to ensure that A" or "it is obligatory that agent i ensures that A".

Moreover, both action concepts are needed within the context of organizations, since we also need to represent situations were agents are obliged to bring about p without using other agents, and even situations where agents are obliged to ensure p but not allowed to do it by themselves.

Although we are as yet far from defining a precise meaning for this *ensures* notion, and although there may be other interpretations compatible with it, we have particularly in mind the following reading for the expression "agent i ensures that A": "agent i has successfully exercised his "power / influence" in order to obtain A", or more simply, "with the goal of achieving A, agent i has initiated a process that has succeeded in securing that goal". Note that one way in which agent i can successfully exercise his "power / influence" to obtain A is by bringing about A (i.e., bringing about is a particular case of ensuring). Note also that the type of "power / influence" involved may depend on the underlying structure of the organization (we will return to this point later).

Let us use a monadic operator G_i for representing this agency concept, where expressions of the form G_iA are read "the agent i ensures that A". By using this operator, the phrase "agent i is responsible for A" can now be represented by expressions of the form OG_iA, for some appropriate deontic obligation operator O[12].

With respect to the formal properties of G_i, we propose the use of a propositional classical system with the following schemata:

(T)	$G_iA \rightarrow A$
(C)	$(G_iA \wedge G_iB) \rightarrow G_i(A \wedge B)$
(No)	$\neg G_i\mathrm{True}$

The schemata (T), (C) and (No) follow the same intuitions as applied for Elgesem´s interpretation of the operator E_i. According to our discussion at the end of the previous section, the main difference between our logical system and that of Elgesem concerns the interaction of different agents, and is reflected in the adoption of Chellas's Q schema (which will be referred to here as GGG):

(GGG) $\quad G_iG_jA \rightarrow G_iA$

This schema has the following intuitive reading: "whenever agent i ensures that agent j ensures that A, agent i also ensures that A" and marks the key difference between a "direct" and an "indirect" action concept. It follows from (GGG) that G_iA cannot be interpret as "i alone ensures that A". Moreover, since $G_iG_jA \rightarrow (G_iA \wedge G_jA)$ (by (T) and (GGG)), each agent involved in G_iG_jp ensures that p.

Note that $(G_iG_kA \wedge G_kG_jA) \rightarrow G_iG_jA$ should not be a logical principle in the logic of the "indirect" action concept. Thus $G_iA \wedge G_jA \wedge \neg G_iG_jA \wedge \neg G_jG_iA$ may hold even in situations where both agents i and j are involved in the same organizational activity oriented towards obtaining A, in which case neither of the agents has explicitly ensured the participation of the other in order to obtain A. This point reveals an important aspect of the meaning of G_i: its argument just refers to the goal supported by agent i.

---

[12] In this paper we do not make any effort to discuss which kind of obligation operator suits our purposes. For this reason we do not define its semantics, leaving that open in the following models. This does not mean that we are not fully aware of the long dispute about deontic logic and its paradoxes, but only that it falls outside the scope of the present paper (see e.g. [24] for a first introduction to these matters, and the proceedings of the previous DEON workshops [25, 26] for deontic logics and their applications in these areas).

With respect to the relationships between the different agency operators (the direct and the indirect), we should accept (at least) the following reasonable principles:

(EG)	$E_iA \to G_iA$
(GEGG)	$G_iE_jA \to G_iG_jA$
(EEEG)	$E_iE_jA \to E_iG_jA$
(EGE)	$E_iA \to G_iE_iA$

The first principle is fundamental and states that "whenever agent i brings it about that A, agent i also ensures that A" (bringing about is a particular case of ensuring). The other two state respectively that "whenever i ensures that j brings it about A, i also ensures that j ensures A" and "whenever i brings it about that j brings it about A, i also brings it about that j ensures A"[13]. The last one states that "whenever agent i brings it about that A, i also ensures that i brings it about A".

Note that from these schemas (together with (GGG)) we can derive that $E_iE_jA \to G_iA$, as is desirable. Note also that from (EGE) and (T) we can derive $E_iA \leftrightarrow G_iE_iA$, i.e., we find no relevant distinction to be drawn between E_iA and G_iE_iA. This equivalence allows us to deal with responsibility in a uniform way, since we can represent "agent i is obliged to bring it about that A" as a particular form of responsibility: OG_iE_iA.

On the other hand, although $(G_iA \wedge \neg E_iA) \to (\exists_j)G_iG_jA$ (for $j \neq i$) looks natural given the intuitive ideas related with "indirect" and "direct" agency concept, it also should not be accepted as a principle in this logic, since we intend to allow situations where agent i ensures that A by bringing about some of it sub-tasks and by successfully exercising his "power / influence" in order to indirectly obtain the other remaining sub-tasks of A.

The operator presented above allows us to increase the expressiveness of the characterization of agents´ interactions within organizations. On the one hand, using the G operator in the scope of an obligation operator we can deal with the concept of *responsibility*. Within this approach, responsibilities are just obligations applied to actions of a more general character. One main advantage of adopting this view is that we can deal with the notion of *responsibility fulfilment* in the same way as we deal with the notion of *obligation fulfilment*, i.e., $OG_iA \wedge G_iA$. Note that an expression of the form $OG_iA \wedge E_iA$ represents just a particular form of responsibility fulfilment, because schema (EG) allows us to infer the schema $(OG_iA \wedge E_iA) \to (OG_iA \wedge G_iA)$.

On the other hand, the G operator offers us additional types of act positions to be considered in the generation of normative positions within an organization. For instance, by considering the operators E_i and G_i, we get five one-agent act-positions, instead of the three described in [6, 14] for the operator E_i:

(A_1) E_iA
(A_2) $E_i\neg A$

---

[13] If we have the (RM) rule for the action operators, or some weaker version of it of the kind described in the previous section, than (GEGG) and (EEEG) would follow from (EG).

(A3) $\neg G_iA \wedge \neg G_i\neg A$
(A4) $G_iA \wedge \neg E_iA$
(A5) $G_i\neg A \wedge \neg E_i\neg A$

Using these positions as a starting point for the generation of normative positions (w.r.t. an obligation operator O) offers us additional distinctions to be considered in the analysis of social interaction. We have not explored this possibility as yet, but we think that it deserves further research. Just as an example of the potentialities of the combination of the operators E_i and G_i, we may distinguish further concepts of interpersonal control w.r.t. A (see in regard to this [3, 4]), e.g. E_iE_jA and E_iG_jA for "direct" control, and G_iE_jA and G_iG_jA for "indirect" control.

4 Minimal semantics for the new operator

For the characterization of the operator G, we adopt a model $\mathcal{M} = \langle W, g_1, ..., g_n, P\rangle$, where: W is a set of possible worlds; P is a function applying each propositional atom p_i in the set of elements of W where p_i is true; and each g_i (i = 1...n) is a function with signature $2^W \rightarrow 2^W$,where $g_i(X)$ means the set of worlds where agent i *ensures* (the state of affairs described by) X[14]. Truth at a world in this model is defined in the standard way for expressions without modal operators, and for expressions of the form G_iA by (where $\| \ \|$ is the usual extension of P that applies each sentence A in its truth-set):

(def.g) $\mathcal{M}, \alpha \models G_iA$ iff $\alpha \in g_i(\|A\|)$

To validate the previous schemata for G_i (i.e. (T), (C), (No) and (GGG)), the following constraints on the model $\mathcal{M}$ are imposed[15]: for all X and Y (subsets of W),

(t_g) $g_i(X) \subseteq X$
(c_g) $g_i(X) \cap g_i(Y) \subseteq g_i(X \cap Y)$
(no_g) $g_i(W) = \varnothing$
(ggg) $g_i(g_j(X)) \subseteq g_i(X)$

---

14 The semantic characterization of G_i uses the suggestion proposed in [10] exercise 7.10, page 211. The main advantage of this semantic structure w.r.t. the minimal models is that we easily obtain the truth set of the expressions built from the modal operators: for instance, $\|G_iA\| = g_i(\|A\|)$. This allows an easier manipulation of the model restrictions needed to validate the schemata that deal with different modalities.

15 Note that in the context of the minimal models, the constraint (ggg) might be written as (where N^i_α represents the set of propositions ensured by agent i in the world α):

(ggg') for all $\alpha \in W$ and $X \subseteq W$, if $\{\beta \in W: X \in N^j_\beta\} \in N^i_\alpha$ then $X \in N^i_\alpha$

In order to consider the proposed relationships between the operators G and E, we extend the previous model with functions f_i ($i = 1...n$) with signature $2^W \rightarrow 2^W$, where $f_i(X)$ means the set of worlds where agent i *brings about* X; define

(def.e) $\mathcal{M}, \alpha \models E_iA$ iff $\alpha \in f_i(\|A\|)$[16],

and impose the constraints (to validate the schemata (T), (C), (No) and (EE¬E) for E_i):

(t_e)	$f_i(X) \subseteq X$
(c_e)	$f_i(X) \cap f_i(Y) \subseteq f_i(X \cap Y)$
(no_e)	$f_i(W) = \emptyset$
($ee\neg e$)	$f_i(f_j(X)) \subseteq -f_i(X)$ (for $i \neq j$)

Finally, the semantic counterparts of (EG), (GEGG), (EEEG) and (EGE) are obtained by imposing:

---

16 If we compare our semantics with the one in [9], we note that we are using functions with signature $2^W \rightarrow 2^W$ instead of functions with signature $W \times 2^W \rightarrow 2^W$ as does Elgesem, and some justification seems to be needed for this. First, if we look at the intuitive description of $f_i(\alpha, X)$ given in e.g. [9] page 54, we observe that the parameter α is absent from it: $f_i(\alpha, X)$ is, intuitively, the set of worlds where agent i realizes his ability to bring about the goal X. One possible simple motivation for introducing the actual world α as a parameter of f_i can be to state that the ability to bring about some goal depends on the actual world; but then it seems that the previous reading of $f_i(\alpha, X)$ would imply that if $\beta \in f_i(\alpha, X)$ then in β agent i has the ability to bring about goal X (or, more precisely, that $\beta \in f_i(\beta, X)$), which is not the case. Thus, in order to justify the introduction of α as a parameter of f_i we need to give a more complicated interpretation to $f_i(\alpha, X)$; something like: $f_i(\alpha, X)$ is the set of worlds where agent i would bring about the goal X if he had behaved there as he has behaved in the actual world. As a matter of fact, this is not the interpretation that seems to be employed in [9] (see e.g. pages 63, 68 and 69), which is a different and more complicated than this one. Our point is only this: although the introduction of α as a parameter of f_i makes the semantics more powerful, allowing us to express other possible relevant concepts (as Elgesem does), in trying to get a basic logic for these "indirect" and "direct" agency operators we can (and possibly should) keep our semantics as straightforward as possible, in order to get the most simple and intuitive readings for g_i and f_i compatible with the desired properties for the action operators; this is the reason why we have omitted any reference to the actual world from the functions g_i and f_i. (Note that we can still define some capability/ability operator in our logic that validates the schema $E_iA \rightarrow \text{Capability}_iA$: we only need to define $\mathcal{M}, \alpha \models \text{Capability}_i A$ iff $f_i(\|A\|) \neq \emptyset$; although in this way we get a "capability" notion that does not depend on the actual world, it suffices for our present purposes and it may be appropriate for the models we have in mind for an organization, where each world corresponds to the exercise of some abilities and of some influences: see the concluding remarks).

(eg)	$f_i(X) \subseteq g_i(X)$
(gegg)	$g_i(f_j(X)) \subseteq g_i(g_j(X))$
(eeeg)	$f_i(f_j(X)) \subseteq f_i(g_j(X))$
(ege)	$f_i(X) \subseteq g_i(f_i(X))$

In the next section we will present further notions relevant to a complete characterization of this new "indirect" action concept.

5 Influence

In section 3, we suggested the reading "agent i ensures that A" (represented by G_iA) as equivalent to the reading "agent i has successfully exercised his "power / influence" in order to obtain A". The logical system proposed for the G operator provides a way to describe successful influences. However, the basic properties considered so far are not enough to relate G_iA with the specific "direct" acts performed in the organization by the agent i and by the other agents. Note that the specific "direct" acts related with the exercise of influence just succeed in cases where the influenced agents ensure what they are influenced to ensure (consider, e.g., the attribution of responsibilities).

In order to illustrate this point, let us consider the following example of an organization with three agents *a*, *b* and *c* , where the agents have just the capabilities indicated in diagram 3:

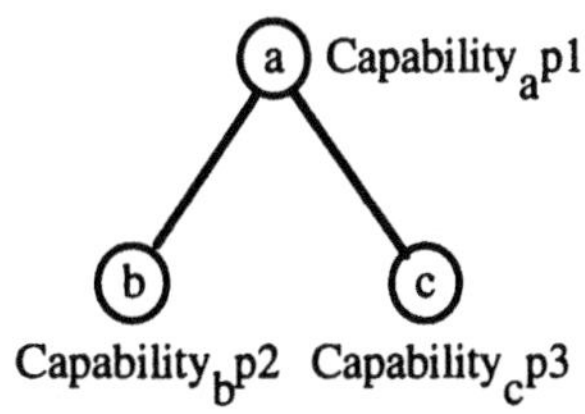

diagram 3

Within this example, there is just one possibility for achieving $p1 \wedge p2 \wedge p3$: each agent must act according to his capabilities. Thus, in order that *a* can ensure $p1 \wedge p2 \wedge p3$, he must himself do p1 and exercise a successful influence on agents *b* and *c* to ensure p2 and p3, respectively. In fact, the logical system provides the following deduction:

$$\{E_ap1, E_aG_bp2, E_aG_cp3\} \vdash G_a(p1 \wedge p2 \wedge p3)$$

However, there is no way to derive $G_a(p1 \wedge p2 \wedge p3)$ from the description of the specific "direct" acts performed in the organization. In fact, we cannot deduce $G_a(p1 \wedge p2 \wedge p3)$ from $\{E_ap1, E_bp2, E_cp3\}$, since this set of hypotheses does not mention the specific "direct" acts related with the exercise of a's influence over *b* and *c* in order to get p2 and p3.

In order to surpass this problem we must have a way of expressing the specific "direct" acts of influence (not necessarilly successful) and of relating the successful influences with these acts and the agency of the other agents. To this end, we start by proposing an additional monadic operator $_iI_j$ for dealing with the notion of *influence*, where expressions of the form $_iI_jA$ are read "i influences j to ensure A".

Note that we do not want to see this influence as a success operator (i. e., we don't want to have the schema (T): $_iI_jA \rightarrow A$). Otherwise we would still be faced with the same problem of not being able to represent the specific "direct" acts of influence which usually may fail to succeed.

For the sake of the applications we have in mind, related with organizations, we adopt the following relationship between the logic of influence and the logic of agency:

$$\text{(Ginf)} \qquad ({_iI_jA} \wedge G_jA) \rightarrow G_iA$$

with the following intuitive reading: "if agent i has influenced agent j to ensure A and agent j has ensured A, then agent i has ensured A".

Note that (Ginf) should not be seen as a general logical bridge between the logic of influence and the logic of agency, as may be seen by the following argument: "it can happen that agent i had influenced agent j to ensure A and agent j has ensured A but not because of i's influence, instead because of a third agent k's influence. Should we conclude in this situation that i has ensured that A is the case ?"[17]. However, within organizations, (Ginf) is usually accepted in the sense that "whenever agent i has influenced agent j (using his official power of influence) to ensure A and agent j has ensured A, then for the organization this *counts as* agent i´s ensuring A"[18]. Or, to put it another way, if agent i has the responsibility to ensure

---

17 We are grateful to Andrew Jones for pointing out this problem and also for further suggestions on how to solve it.

18 Although this would need further research, we could try to formalise this notion using a relativised conditional operator $\Rightarrow_o$ similar to the one presented in [21], where $A \Rightarrow_o B$ reads as follows "for the organization o, A counts as B". In that paper, the authors also introduced a unary operator D_o, where D_oA would be read as follows "it is a constraint of (operative in) the organization o that A ", and they defined it as a normal modality of type KD (but obviously not satisfying the (T) schema). The authors not only impose the obvious schema $(A \Rightarrow_o B) \rightarrow D_o(A \rightarrow B)$, but also adopt the schema $(A \Rightarrow_o B) \rightarrow (A \rightarrow D_oB)$.

Using this operator we could not only state that for any organization o: $({_iI_jA} \wedge G_jA) \Rightarrow_o G_iA$; we could also state that for some particular organizations o we have that: $({_iI_jA} \wedge A) \Rightarrow_o G_iA$ (or even that $OG_iA \rightarrow (A \Rightarrow_o G_iA)$); some specialists in Management Science say that the latter formula would correctly represent the actual practice in some organizations.

However, since in this paper we want to concentrate only on one organization, we do not need to complicate matters by considering "counts as" operators, and can simply stipulate that we adopt (Ginf) as an axiom schema of our logic.

A and he influences j to ensure A and j does that job, for the organization i has fulfilled his obligation[19].

Note that using the schema (EGE) and (Ginf) we can deduce the schema:

(EGinf) $({}_iI_jE_jA \wedge E_jA) \rightarrow G_iA$

Before discussing the properties that should be satisfied by this influence operator, it should be noted that the notion of influence proposed here can be refined according to the specific applications one has in mind, and that the schema (Ginf) just represents the general relationship we want to adopt between such acts of influence and the concept of ensuring. In fact, although an obvious way of influencing another agent is to exercise a direct control on him (whenever this is possible):

(dir.control) $E_iG_jA \rightarrow {}_iI_jA$

a much more common institutionalized form of exercising influence within organizations is the *attribution of obligations and responsibilities*. In such applications, this specific form of influence can be introduced in the logic by stipulating the schema below as a proper axiom

(atrib.resp) $E_iOG_jA \rightarrow {}_iI_jA$

that allows us to deduce also the schema

(Ginf.atrib.resp) $(E_iOG_jA \wedge G_jA) \rightarrow G_iA$ [20]

Coming back to the previous example, the logical system can now yield

---

19 Another more complicated problem is to decide what happens if an agent i has the responsibility to secure A and he has influenced another agent j to ensure A, but j failed to do it: has i fulfilled his responsibility in this case? Should we distinguish non-fulfilment of responsibility from failure to achieve goal A in cases of this kind? These are complicated issues that need much more attention, and possibly the answer will differ from organization to organization, and should be included in the description of the organization's policy. (We are grateful to one of the referees for pointing out this problem.)

Note also that the description of the organization's policy will in general include other requirements relevant to the topic of this paper. To give just one example: we could think that a possible requirement for the agents in management positions might be that whenever one of these agents exercises an influence on j to ensure A, this forbids the other agents from exerting a "contrary" influence on the same agent (which could be more or less expressed by a formula of the form ${}_iI_jA \rightarrow (\forall_k)O\neg_kI_j\neg A)$. However, this line of research is outside the scope of the present paper.

20 Using the equivalence $E_jA \leftrightarrow G_jE_jA$ and the (RE) rule, we can deduce from the previous schemas $E_iE_jA \rightarrow {}_iI_jE_jA$, $E_iOE_jA \rightarrow {}_iI_jE_jA$ and $(E_iOE_jA \wedge E_jA) \rightarrow G_iA$.

$$\{E_a p1, {}_aI_b p2, {}_aI_c p3, E_b p2, E_c p3\} \vdash G_a(p1 \wedge p2 \wedge p3)$$

as can be seen as follows (where we freely use the rules of propositional logic):

1.	$E_a p1$	hypothesis
2.	$G_a p1$	1 plus (EG)
3.	$G_b p2$	similar to the previous steps
4.	${}_aI_b p2$	hypothesis
5.	$G_a p2$	3, 4 and (Ginf)
6.	$G_a p3$	similar to the previous step
7.	$G_a(p1 \wedge p2 \wedge p3)$	2, 5, 6 and (C)

Although the formal characterization of the operator ${}_iI_j$ deserves further research, we can advance some basic properties to be adopted in a characterization of this notion of influence and its relationships with the other concepts. For instance, we think that, besides (Ginf), (RE) and (control), we should adopt the following schemata:

(C) $({}_iI_jA \wedge {}_iI_jB) \rightarrow {}_iI_j(A \wedge B)$

(Ginf∨) $({}_iI_j(A \vee B) \wedge G_jA) \rightarrow G_i(A \vee B)$

The schema (Ginf∨) is guided by the intuition that an influence of the form ${}_iI_j(A \vee B)$ represents an influence to choose one of the alternatives (it is up to j to choose between ensuring A and ensuring B). Note also that if we accept the rule below (as we shall do), then (Ginf∨) can be derived.

(Ginf→) if $\vdash A \rightarrow B$ then $\vdash ({}_iI_jB \wedge G_jA) \rightarrow G_iB$

By arguments similar to the ones given w.r.t. the action operators, we also impose the schema (No): $\neg {}_iI_j$True (it makes no sense to influence an agent to ensure a logical truth, although, as for the action operators, this may not be an important issue from a practical point of view). Finally, we also impose (¬If): $\neg {}_iI_j$False (which, by (C), implies that I verifies the (D) schema: ${}_iI_jA \rightarrow \neg {}_iI_j \neg A$). Although we do not completely exclude the possibility that for some applications ${}_iI_j$False may make some sense, we think that in general this is not the case and certainly it is not the case for the applications we have in mind where, intuitively, when an agent *i* influences *j* to ensure A, *i* believes that *j* will ensure A (or, considering a normative influence, ideally *j* ensures A). However, the schema ${}_iI_jA \rightarrow \neg {}_kI_j \neg A$ should not be considered as an axiom.

On the other hand, as we have already argued we do not want to have schema (T) for I. Moreover, we should not have the rule (RM) and for reasons stronger that the simple internal logical justification that otherwise we would derive ${}_iI_j$True from any influence of i over j. In fact, (RM) would allow us to derive the schema (M): ${}_iI_j(A \wedge B) \rightarrow {}_iI_jA$, and the acceptance of (M) would provide, for instance, the deduction (using (Ginf) and (C)):

$$\{{}_aI_b(p \wedge \neg q), {}_aI_c(\neg p \wedge q), E_bp, E_cq\} \vdash G_a(p \wedge q)$$

This deduction may be acceptable in cases where agents may achieve their goals by making contradictory influences. We may consider e.g. situations where *a* influences *b* to ensure $p \wedge \neg q$ because *a* believes that in this way *b* will ensure *p*, or even consider situations where *a* influences *b* to ensure A because *a* believes that this is the way to make *b* ensure $\neg A$. However, in the applications we have in mind not only when an agent *a* influences *b* to ensure A, *a* believes that *b* will ensure A, as these influences are seen as a mean to achieve some goal (e.g. $G_a A$) according to some rational strategy[21]. In these applications, we think that it is unreasonable to accept the previous deduction.

More generally, we think that (M) should not be a logical principle of a basic logic of influence, since it excludes the possibility of considering applications where there is an intention beyond these acts of influence and where by exerting an influence of the form ${}_iI_j(A \wedge B)$, *i* may have an intention that *j* ensures $A \wedge B$, but not A alone (that could even be an undesirable result for *i*). That is, both of the following conjunctions should be consistent: ${}_iI_j(A \wedge B) \wedge \neg {}_iI_jA$ and ${}_iI_j(A \wedge B) \wedge {}_iI_jA$. Moreover, it should be possible to consider logics of influence where the statements ${}_iI_j(A \wedge B)$ and ${}_iI_j(A \wedge B) \wedge {}_iI_j((A \wedge B) \vee (\neg A \wedge \neg B))$ are not logically equivalent, contrary to what happens if we consider the (RM) rule.

The example above also shows that we do not want the schema $({}_iI_j(A \wedge B) \wedge G_jA) \rightarrow G_iA$ (that can be derived from (M) and (Ginf)) as a basic principle of our logic. Also we do not want ${}_iI_jA \rightarrow {}_iI_j(A \vee B)$ (otherwise we would get (M) since $A \leftrightarrow (A \wedge B) \vee (A \wedge \neg B)$); nor do we want $({}_iI_j(A \vee B) \wedge G_jA) \rightarrow G_iA$ (otherwise we would get $\{{}_aI_bp, E_bp, E_bq\} \vdash G_a(p \wedge q)$, using a similar decomposition of p in $(p \wedge q) \vee (p \wedge \neg q)$). Finally, and analogously, we do not want to have ${}_iI_j(A \vee B) \rightarrow ({}_iI_jA \vee {}_iI_jB)$, since with it we would get $\{{}_aI_bp, \neg {}_aI_b(p \wedge \neg q)\} \vdash G_b(p \wedge q) \rightarrow G_a(p \wedge q)$.

In order to validate the desired schemata we need to extend our model with functions ${}_ih_j$ (i,j = 1...n) with signature $2^W \rightarrow 2^W$, where ${}_ih_j(X)$ means the set of worlds where agent i *has influenced* j *to ensure* X; define

$$\text{(def.I)} \quad \mathcal{M}, \alpha \models {}_iI_jA \quad \text{iff} \quad \alpha \in {}_ih_j(\|A\|),$$

and impose the following constraints:

---

[21] To give just one example of the kind of rational strategies we have in mind: suppose a situation where we have four agents: *a* is the only agent in a management position, having power over all the others (w.r.t. *p* and *q*), but has no capability to bring about either *p* or *q*; *b* has capability to bring about *p*, but not *q*; *c* has capability to bring about both *p* and *q*; and *d* has capability to bring about *q*, but not *p*. Suppose now that *a* wants to ensure $(p \wedge q)$. Then, two possible rational strategies for *a* to achieve that goal are: to influence *b* to bring about *p* and to influence *d* to bring about *q*, or to influence *c* to bring about both *p* and *q*, and in fact our logical system provides the following deductions: $\{{}_aI_bp, {}_aI_dq, E_bp, E_dq\} \vdash G_a(p \wedge q)$ and $\{{}_aI_c(p \wedge q), E_b(p \wedge q)\} \vdash G_a(p \wedge q)$. On the other hand, exerting the influences ${}_aI_b(p \wedge q)$ and ${}_aI_d(p \wedge q)$ would not be a good rational strategy, since none of the agents can satisfy the influences exerted (and there is no reason, a priori, to suppose that a partial fulfilment of an influence is better than nothing); and within our logical system we cannot deduce $G_a(p \wedge q)$ from $\{{}_aI_b(p \wedge q), {}_aI_d(p \wedge q), E_bp, E_dq\}$.

(cI)	${}_ih_j(X) \cap {}_ih_j(Y) \subseteq {}_ih_j(X \cap Y)$
(noI)	${}_ih_j(W) = \emptyset$
(¬if)	${}_ih_j(\emptyset) = \emptyset$
(ginf)	${}_ih_j(X) \cap g_j(X) \subseteq g_j(X)$
(ginf→)	if $X \subseteq Y$, then ${}_ih_j(Y) \cap g_j(X) \subseteq g_i(Y)$
(control)	$f_i(g_j(X)) \subseteq {}_ih_j(X)$

Finally, another problem with respect to the characterization of G_i has to do with a proper characterization of what we call *task decomposition*, i.e., the characterization of sub-tasks of a given task. This aspect is also important, since within organizations the performance of a task is frequently obtained by distributing sub-tasks to different agents. Note that, for instance, in the example in diagram 3 we dealt with an explicit decomposition of the task to be performed by agent *a* into *p1*, *p2* and *p3* (and then agent *a* has performed the sub-task *p1* himself and has distributed sub-tasks *p2* and *p3* to agents *b* and *c* respectively). A proper characterization of *task decomposition* would allow us to express responsibilities without referring explicitly to their possible decompositions.

Let us consider the following example of an organization with four agents *a*,*b*, *c* and *d* pictured in diagram 4, where: i) agent *a* is responsible for *q*; ii) *q* can be decomposed into *p1* and *r*; iii) *r* can decomposed into *p2* and *p3*; iv) influences are exerted by attribution of responsibilities.

What additional hypothesis Γ, axioms or proof rules are needed in order to provide the following deduction, concerning the fulfilment of *a*'s responsibility?

$$\Gamma \cup \{E_ap1, E_aOG_br, E_bOG_cp2, E_bOG_dp3, E_cp2, E_dp3\} \vdash G_aq$$

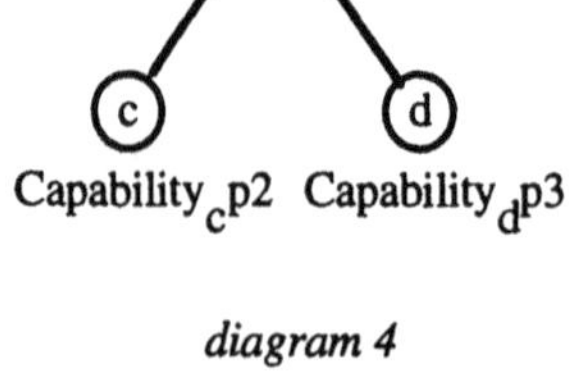

diagram 4

This deduction depends on the characterization of the task decomposition in this logical context. An economic way to do that is to take advantage of the (RE) rule and use the equivalence connective to represent the task decompositions described in ii) and iii), considering that the logic is extended with the following axioms: $q \leftrightarrow (p1 \wedge r)$ and $r \leftrightarrow (p2 \wedge p3)$. But, this is far from a realistic representation of task decompositions, because we may also consider alternative ways to decompose a task. Assuming that we have the implications $(p1 \wedge r) \rightarrow q$ and $(p2 \wedge p3) \rightarrow r$ instead of the previous equivalencies does not solve the problem, in virtue of the absence of (RM).

One practical way to surpass this problem is to accept the (RM) rule. Another way is the use of a relativised conditional operator $\Rightarrow_i$ to represent task decompositions, where expressions of the form $A \Rightarrow_i B$ are read "for agent i, ensuring A is a way of ensuring B"[22], or "for agent i, A counts as B", adapting the idea presented in [21] to this context. Note that this alternative has the advantage of making it possible to consider that the same task can be decomposed in different ways by different agents, and that a decomposition that is acceptable for some agent may not be acceptable for another agent, since the way in which a task is decomposed may depend e.g. on the agent's position within an organization; (depending on his position, an agent may have for instance a more abstract or a more concrete view of some task). This is also the reason why we have included an agent index on the operator. Using the $\Rightarrow_i$ operator, an expression of the form $((A \wedge B) \Rightarrow_i C)$ would represent a possible decomposition of C (for agent i), and an expression of the form $(A \Rightarrow_i C) \wedge (B \Rightarrow_i C)$ would represent two alternative ways to get C (for agent i). Thus, coming back to the previous example, we could then define Γ as $\{(p1 \wedge r) \Rightarrow_a q, (p2 \wedge p3) \Rightarrow_b r\}$.

The study of this conditional operator $\Rightarrow_i$ and of its relationships with the action operators is however outside the scope of this paper (not only because it needs further research, but also because of space limitations).

6 Concluding remarks

We have argued that there is an important distinction to be drawn between a "direct" and an "indirect" action concept, a distinction not explicitly addressed by previous authors; we have proposed a new action operator to represent this "indirect" agency and we have discussed its basic properties. The new operator allows us to increase expressiveness in the characterization of agents´ activities and interaction within organizations, by allowing an easy and abstract way of expressing the fundamental concept of responsibility (for some task). Although the study of the action concept here proposed needs further research, we have already introduced some fundamental related notions, such as influence, and have discussed their basic properties.

We think that these concepts play an important role in some aspects of the analysis and specification of an organization. For instance they can be used to answer such questions as: "can agent *a* be responsible by task *p*?" (interpreted as meaning "can agent *a* fulfil such a responsibility ?", i.e., "can the agent *a* (or has the agent *a* the means to) ensure *p* ?"). Another activity correlated with the *attribution of responsibilities* is the *delegation of "power" (influence)*. Thus, other kinds of interesting and useful questions are: "which power (influences) should *a* delegate to *b* in order that *a* can attribute (or assign) the responsibility of some task to *b* ?" and "has *a* the permission for such delegation of power (influence) ?", etc.

With respect to the specification of organizations, we are here thinking about an organizational model as a description of the agents and their capabilities, and the

---

22 Thus accepting the schema $(A \Rightarrow_i B) \rightarrow (G_i A \rightarrow G_i B)$. Moreover, using this schema and (Ginf) we could derive $(A \Rightarrow_j B) \rightarrow (({}_i I_j B \wedge G_j A) \rightarrow G_i B)$, which could replace the actual rule (Ginf→).

channels of influence (and also the task decompositions). The notion of "influence channel" is intended to capture some notion of "official permission" or "authorization" of influence (or of attribution of responsibilities) and will play an important role in the organizational model. Each channel represents a directional link between two agents, represented e.g. by $i >_A j$, with the reading "there is an influence channel from i to j w.r.t. A", and it is related with the notion of influence as the notion of "capability" is related with "bringing it about", in the sense that "if agent i influences agent j to ensure A, then there is an influence channel from i to j w.r.t. A". In this "semantic" context an organization can be informally seen as a graph of influence channels, and we can obtain a more intuitive (or operational) view of our indirect agency concept as follows: an agent *a* "can" ensure some task *p* if there is some "path" (or a set of "compatible" paths) in such a graph that "leads" to *p* starting in *a*, and an agent *a* ensures *p* if the "activated path(s)" leads to *p* from *a*.

More formally (and leaving out the problems of task decomposition and of permission to delegate power) we could represent an organization as a structure (Ag, >, C) where: Ag is a non-empty finite set of agents; > applies each pair of agents into a (possibly empty) set of formulas (not containing True, nor False); and C applies each agent into a set of propositional atoms. (We could write $i >_A j$ for $A \in >(i,j)$ and $Capability_i A$ for $A \in C(i)$.) Then, we could use the syntactic approach to derive our possible "worlds". Briefly, each world would be a kind of Hintikka set obtained from the organizational structure, considering that some of the capabilities of the agents have been realized (achieving the corresponding atomic tasks) and that some of the influence channels have been activated (describing the influences that have been exercised), and deriving from these E and I-formulas which formulas have been ensured by the different agents, using the schemata proposed here. Then we could say that in such an organization agent i "can" ensure A if there exists one such "world" where we can derive (or that contains) $G_i A$.

Of course, this is only a sketch of a possible organizational model and of a more intuitive (operational) semantics for the G operator within it. Its formal description must be done with care in order that it works properly. Although we have some ideas on how to do this, space limitations do not allow us to enter in details in this paper.

Another topic where much more must be done concerns the problem of task decomposition, and the study of the relativised conditional operator $\Rightarrow_i$. Nevertheless, we think that this paper provides some new and relevant insights regarding the formal theory of agency and its application within organizations.

Acknowledgments

This research has been carried out within the Portuguese research project n° STRDA/C/TIT/86/92-JURAD, financially supported by the Program STRIDE, FEDER and JNICT, and within the ESPRIT Basic Research Working Group 8319 MODELAGE ("A Common Formal Model of Cooperating Intelligent Agents"). The authors wish to thank the other participants of these projects, and in particularly Andrew Jones, for many useful discussions on the topics of this paper. The authors also want to thank Pedro Ramos and Luis Botelho for some helpful comments on this work. Finally, the authors wish to thank the anonymous referees for their suggestions and criticisms that helped improve the readability and effectiveness of the paper.

References

1. Normatics: the characterization of computer systems and complex organizations as normative systems, ESPRIT III BRA proposal, 1991
2. Kanger S. Law and logic. Theoria 1972; 38
3. Pörn I. The logic of power. Blackwell, Oxford, 1970
4. Pörn I. Action theory and social science: some formal models. Synthese Library 120, D. Reidel, Dordrecht, 1977
5. Pörn I. On the nature of social order. In: Fenstad J et al. (eds) Logic methodology, and philosophy of science VIII. Elsviere, 1989, pp 553-567
6. Lindahl L. Position and change - a study in law and logic. Synthese Library 112, D. Reidel, Dordrecht, 1977
7. Chellas B. The logical form of imperatives. Perry Lane Press, 1969
8. Kanger S. New foundations for ethical theory. Stockholm, 1957
9. Elgesem D. Action theory and modal logic. PhD thesis, Dept. of Philosophy, University of Oslo, 1993
10. Chellas B. Modal logic - an introduction. Cambridge University Press, 1980
11. Sommerhoff G. The abstract characteristic of living systems. In: Emery F (ed) Systems thinking. Penguin, Harmonsworth, 1969
12. Harel D. First-order dynamic logic. LNCS 68, Springer, 1979
13. Harel D. Dynamic logic. In: Gabbay D, Guenthner F (eds) Handbook of philosophical logic II. D Reidel, Dordrecht, 1984, pp 497-604
14. Jones A, Sergot M. On the characterisation of law and computer systems: the normative systems perspective. In: Meyer J-J Ch, Wieringa R (eds) Deontic logic in computer science: normative system specification. John Wiley & Sons, 1993, pp 275-307
15. Belnap N. Backwards and forwards in the modal logic of agency. Philosophy and Phenomenological Research 1989; 2: 777-807
16. Belnap N. Before refraining: concepts for agency. Erkenntniss 1990; 34:137-169
17. Belnap N, Perloff M. Seeing to it that: a canonical form for agentives. Theoria 1989, 54:175-199
18. Belnap N, Perloff M. In the realm of agents. Annals of Mathematics and Artificial Intelligence 1993; 9:25-48
19. Chellas B. Time and modality in the logic of agency. Studia Logica 1992; 51:485-517
20. Hodge B, Anthony W. Organization theory. Allyn and Bacon Inc, 1988
21. Jones A, Sergot M. Institutionalized power: a formal characterization. To be published in MEDLAR II special issue of Journal of IGPL, 1995
22. Kutschera F. Bewirken. Erkenntnis 1986; 24:253-381
23. Horty J , Belnap N. The deliberative stit: a study of action, omission, ability, and obligation. Unpublished manuscript, Dept of Philosophy, Univ. of Maryland, 1993
24. R. Hilpinen (ed). Deontic logic: introductory and systematic readings. D Reidel, 1971
25. Meyer J-J Ch, Wieringa R (eds). Proc. first international workshop on deontic logic in computer science (DEON'91). Amsterdam, 1991
26. Jones A, Sergot M (eds). Proc. second international workshop on deontic logic in computer science (DEON'94). Complex 1/94 NRCCL, Oslo, 1994

How to Combine Ordering and Minimizing in a Deontic Logic based on Preferences*

Yao-Hua Tan
EURIDIS
Erasmus University Rotterdam
P.O. Box 1738
3000 DR Rotterdam
The Netherlands
ytan@euridis.fbk.eur.nl

Leendert W.N. van der Torre
EURIDIS, Tinbergen Inst. and Dept. of CS
Erasmus University Rotterdam
P.O. Box 1738
3000 DR Rotterdam
The Netherlands
ltorre@euridis.fbk.eur.nl

Abstract

In this paper we propose a semantics for dyadic deontic logic with an explicit preference ordering between worlds, representing different degrees of ideality. We argue that this ideality ordering can be used in two ways to evaluate formulas, which we call *ordering* and *minimizing*. Ordering uses all preference relations between relevant worlds, whereas minimizing uses the most preferred worlds only. We show that ordering corresponds to strengthening of the antecedent, and minimizing to weakening of the consequent. Moreover, we show that in some cases ordering and minimizing have to be combined to obtain certain desirable conclusions, and that this can only be done in a so-called two-phase deontic logic. In the first phase, the preference ordering is constructed, and in the second phase the ordering is used for minimization. If these two phases are not distinguished, then counterintuitive conclusions follow.

1 Introduction

Preference-based deontic logics are deontic logics of which the semantics contains a preference ordering (usually on worlds of a Kripke style possible world model). This preference ordering reflects different degrees of 'ideality': a world is preferred over another world if it is, in some sense, better than the other world. For example, in some logics (e.g. [1, 2]) a value is associated with each world; in such cases, the ordering is connected (for all w_1 and w_2 we have $w_1 \leq w_2$ or $w_2 \leq w_1$). However, in general the preference ordering can be any partial pre-ordering. Hence, only reflexivity and transitivity are assumed. In such preference orderings there can be incomparable worlds. Incomparable worlds can be used to formalize moral dilemmas like $O(p) \wedge O(\neg p)$ in a consistent way, see [3].

In this paper we argue that a preference ordering can be used in two different ways to evaluate formulas. One way, which we call *Minimizing*, is to use the ordering to select the minimal elements that satisfy a formula. This minimizing approach is commonly taken in preferential semantics for non-monotonic logics, see for example [4, 5, 6, 7]. The other way, which we call *Ordering*, is to use the whole ordering to evaluate a

*This research was partially supported by the ESPRIT III Basic Research Project No.6156 DRUMS II and the ESPRIT III Basic Research Working Group No.8319 MODELAGE.

formula. One could explain the intuition behind the distinction between ordering and minimizing with the following metaphor. Ordering is what a person does when he envisions the message of the law, issued by the legislator, by determining the preference relations between the possible deontic states. In this envisionment process bad states are as important as good states. Minimizing is what the person does when he also tries to realize the best states. These two things are completely separated. A person might very well know how he should act, without acting accordingly.

The first deontic logic based on a preference ordering was introduced by Bengt Hansson [8]. It is a dyadic logic and it belongs to the first category. A dyadic obligation $O(\alpha \mid \beta)$ can be read as 'if β (the antecedent) is the case, then α (the consequent) ought to be the case'. Dyadic logics were developed to represent *Contrary-To-Duty* (CTD) obligations and to solve notorious CTD paradoxes like the Chisholm and Forrester paradox. For example, the Forrester paradox [9] 'you should not kill ($\neg k$), but if you kill you should kill gently ($k \wedge g$)', can be represented in a dyadic deontic logic by the two dyadic obligations $O(\neg k \mid \top)$ and $O(k \wedge g \mid k)$. The second obligation is a CTD obligation of the first one and, as B. Hansson put it, it instructs you 'how to make the best of your sad circumstances'. In B. Hansson's logic, a dyadic obligation $O(\alpha \mid \beta)$ is true iff α is true in the *minimal* (the preferred) β world. Hence, the first dyadic obligation says that k worlds are sub-ideal, and the CTD obligation says that optimal sub-ideal worlds are $k \wedge g$ worlds. B. Hansson's logic has been criticized because it lacks strengthening of the antecedent. For example, Alchourrón argues in [10] that lack of strengthening of the antecedent is acceptable for logics of defeasible reasoning (though it gives rise to the so-called irrelevance problem) or logics of defeasible obligations (traditionally called prima facie obligations, see [11]), but not for undefeasible obligations.

Recently, Sven Ove Hansson [12] introduced a preference ordering in a (monadic) deontic logic for a completely different reason. His preference-based deontic logic (PDL) avoids the Ross paradox by making $O(p) \rightarrow O(p \vee q)$ invalid. Hence, PDL is weaker than so-called 'standard' deontic logic (SDL), a normal monadic modal system of type KD according to the Chellas classification [13]. S.O. Hansson's PDL belongs to the second category of preference-based deontic logics, because the truth of $O(\alpha)$ depends on the whole ordering. In this paper we generalize the ordering approach to the dyadic case and we show that ordering validates strengthening of the antecedent, whereas minimizing validates weakening of the consequent. The following example shows that combining strengthening of the antecedent and weakening of the consequent is problematic for *any* deontic logic that can represent a-temporal Contrary-To-Duty obligations. It is not a problem for deontic logics that make a temporal distinction between antecedent and consequent (see e.g. [6, 10]). However, such logics cannot represent several paradoxes, for example the Forrester paradox [9], see [14, 15].

Example 1 *(SA+WC problem) Consider the obligations that (1) you should either serve in the army or do alternative service, and (2) you should not serve in the army. It is counterintuitive to derive that you should do alternative service, given that you already serve in the army. Still this counterintuitive conclusion follows in several dyadic deontic logics. Assume a dyadic deontic logic that validates at least substitution of logical equivalents and the following inference patterns* Strengthening of the Antecedent *(*SA*),* Weakening of the Consequent *(*WC*) and* Conjunction *(*AND*).*

$$\mathrm{SA}: \frac{O(\alpha \mid \beta_1)}{O(\alpha \mid \beta_1 \wedge \beta_2)}$$

$$\text{WC}: \frac{O(\alpha_1 \mid \beta)}{O(\alpha_1 \vee \alpha_2 \mid \beta)}$$

$$\text{AND}: \frac{O(\alpha_1 \mid \beta), O(\alpha_2 \mid \beta)}{O(\alpha_1 \wedge \alpha_2 \mid \beta)}$$

Furthermore, assume as premises the obligations $O(a \vee s \mid \top)$ *and* $O(\neg a \mid \top)$*, where* a *can be read as 'serving in the army' and* s *as 'doing alternative service' [16], and* $\top$ *stands for any tautology. The intuitive obligation* $O(\neg a \wedge s \mid \top)$ *can be derived by* AND. *From this obligation, the obligation* $O(\neg a \wedge s \mid a)$ *can be derived by* SA. *Unfortunately, from this obligation, the counterintuitive obligation* $O(s \mid a)$ *can be derived by* WC. *This obligation is considered to be counterintuitive, because it is not grounded in the premises. If* a *is true, then the first premise is fulfilled and the second one is violated. This inference can be blocked by replacing unrestricted strengthening of the antecedent by the following version of restricted strengthening of the antecedent, in which* $\Diamond$ *is a modal operator and* $\Diamond\phi$ *is true for all consistent propositional formulas* ϕ.

$$\text{RSA}: \frac{O(\alpha \mid \beta_1), \Diamond(\alpha \wedge \beta_1 \wedge \beta_2)}{O(\alpha \mid \beta_1 \wedge \beta_2)}$$

The obligation $O(\neg a \wedge s \mid a)$ *cannot be derived from the obligation* $O(\neg a \wedge s \mid \top)$ *by* RSA. *Unfortunately, the counterintuitive* $O(s \mid a)$ *can still be derived in another way. From the intuitive obligation* $O(\neg a \wedge s \mid \top)$ *the intuitive* $O(s \mid \top)$ *can be derived by* WC. *From this latter obligation, the counterintuitive* $O(s \mid a)$ *can be derived by* RSA.

This problem can be solved by a technique, which might look odd at first sight, but which turns out to work well, namely to forbid application of RSA after WC has been applied. This means that in derivations first RSA has to be applied, and only afterwards WC may be applied. We call this the *two-phase approach* in deontic logic. Such a sequencing in derivations is rather unnatural and cumbersome from a proof-theoretic point of view. Surprisingly, the two-phase approach can be obtained very intuitively from a semantic point of view. In this paper we show that the two-phase approach can be obtained by combining the two usages of the preference ordering in a preference-based semantics of a deontic logic. The first phase corresponds to ordering, and the second phase corresponds to minimizing. In semantic terms the two-phase approach simply means that first a preference ordering has to be constructed by ordering worlds, and subsequently the constructed ordering can be used for minimization.

This paper is organized as follows. In Section 2 we give the preference logic in which we formalize the ordering (Section 2) as well as the minimizing (Section 3) aspects of preference orderings. It is based on a standard modal system, in which the accessibility relation of the Kripke models is interpreted as a preference ordering on the worlds, like in [17, 18]. In Section 4 we show how ordering and minimization can be combined in a two-phase deontic logic and how this solves the problems of Example 1.

2 Ordering

The logic $\mathcal{L}$ discussed in this paper is a modal logic with a preference-based semantics. In the logic, we abstract from actions, time and individuals.

Definition 1 *(Syntax of $\mathcal{L}$) The logic $\mathcal{L}$ is a bimodal system with the two normal modal operators $\Box$ and I. It satisfies the inference rules modus ponens $\frac{\alpha, \alpha \to \beta}{\beta}$ and necessitation for both modal operators $\frac{\vdash \alpha}{\vdash \Box \alpha}$, $\frac{\vdash \alpha}{\vdash I\alpha}$, and contains all propositional tautologies and the following axioms.*

$$\mathrm{K} : \Box(\alpha \to \beta) \to (\Box\alpha \to \Box\beta) \tag{1}$$

$$\mathrm{T} : \Box\alpha \to \alpha \tag{2}$$

$$4 : \Box\alpha \to \Box\Box\alpha \tag{3}$$

$$5 : \neg\Box\alpha \to \Box\neg\Box\alpha \tag{4}$$

$$\mathrm{K} : I(\alpha \to \beta) \to (I\alpha \to I\beta) \tag{5}$$

$$\mathrm{T} : I\alpha \to \alpha \tag{6}$$

$$4 : I\alpha \to II\alpha \tag{7}$$

$$\Box\alpha \to I\alpha \tag{8}$$

The modal operator $\Box$ expresses necessity and satisfies the S5 axioms. $I\alpha$ expresses that α is true in all states that are 'at least as ideal as the actual state' and I satisfies the S4 axioms. Furthermore, the axiom $\Box\alpha \to I\alpha$ gives the relation between the two modal operators.

*(**Semantics of $\mathcal{L}$**) Kripke models $M = \langle W, R, \leq, V\rangle$ for $\mathcal{L}$ consist of W, a set of worlds, R and $\leq$, binary accessibility relations, and V, a valuation of the propositions in the worlds.*

$M, w \models \Box\alpha$ iff $\forall w' \in W$ with $R(w, w')$, it is true that $M, w' \models \alpha$, and
$M, w \models I\alpha$ iff $\forall w' \in W$ with $w \leq w'$, it is true that $M, w' \models \alpha$.

$\Box$ defines an equivalence relation R on the worlds and I defines a reflexive and transitive ordering $\leq$ on the worlds of each equivalence class of $\Box$. The partial pre-ordering $\leq$ expresses preferences: $w_1 \leq w_2$ iff w_2 is at least as ideal as w_1.

In this section, we only consider the ordering approach to deontic logic. In evaluating formulas, the whole ordering is taken into account. This idea was first introduced in a monadic logic by S.O. Hansson [12] and formalized in a modal preference structure by Brown and Mantha [17]. In the following definition, we generalize this idea to dyadic deontic logics.

Definition 2 *Dyadic obligation and permission are defined as follows.*

$$\mathcal{O}(\alpha \mid \beta) =_{def} \Box((\alpha \wedge \beta) \to I(\beta \to \alpha)) \tag{9}$$

$$\mathcal{P}(\alpha \mid \beta) =_{def} \Box((\alpha \wedge \beta) \to I((\beta \to \alpha) \vee \neg I \neg(\alpha \wedge \beta))) \tag{10}$$

Intuitively, an obligation $O(p \mid q)$ expresses a strict preference of all $p \wedge q$ over $\neg p \wedge q$. This preference is represented by a negative condition: no $\neg p \wedge q$ is preferred over some $p \wedge q$. Formally, $M, w \models O(p \mid q)$ iff $\forall w_1, w_2 \in W$ such that $R(w, w_1)$, $R(w, w_2)$, $M, w_1 \models p \wedge q$ and $M, w_2 \models \neg p \wedge q$, it is true that $w_1 \not\leq w_2$.[1] Notice that the ideality ordering is global (in the sense that the ideality ordering is *not* relative to a world) and nested operators therefore do *not* have an intuitive reading, although they could have a formal meaning in $\mathcal{L}$. The following example illustrates the definition of obligation.

Example 2 *Let* $|\alpha|$ *denote a world that satisfies* α. *Assume models that consist of four worlds* $|p \wedge q|$, $|\neg p \wedge q|$, $|p \wedge \neg q|$ *and* $|\neg p \wedge \neg q|$, *which are in the same equivalence class of R. A model satisfies the obligation* $O(p \mid \top)$ *when neither* $|\neg p \wedge q|$ *nor* $|\neg p \wedge \neg q|$ *is preferred over* $|p \wedge q|$ *or* $|p \wedge \neg q|$. *For example, the Kripke model* $|\neg p \wedge q| < |\neg p \wedge \neg q| < |p \wedge \neg q| < |p \wedge q|$ *is a model for* $O(p \mid \top)$ *but not for* $O(q \mid \top)$. *Note that* $O(q \mid \top)$ *is not true, because* $|\neg p \wedge q| < |\neg p \wedge \neg q|$ *and* $|\neg p \wedge q| < |p \wedge \neg q|$. *This shows how in the ordering approach the whole ordering is taken into account in the evaluation of a formula, and not just the most preferred* $|p \wedge q|$ *worlds.*

The following proposition gives several properties of the dyadic obligations.

Proposition 1 *The logic* $\mathcal{L}$ *validates the following theorems of Strengthening of the Antecedent* (SA), *Conjunction* (AND) *and Disjunction* (OR), *a version of Deontic Detachment* (DD′) *and the standard relation between obligations and permissions.*

$$\mathrm{SA}: O(\alpha \mid \beta_1) \rightarrow O(\alpha \mid \beta_1 \wedge \beta_2) \tag{11}$$

$$\mathrm{AND}: O(\alpha_1 \mid \beta) \wedge O(\alpha_2 \mid \beta) \rightarrow O(\alpha_1 \wedge \alpha_2 \mid \beta) \tag{12}$$

$$\mathrm{OR}: O(\alpha_1 \mid \beta) \wedge O(\alpha_2 \mid \beta) \rightarrow O(\alpha_1 \vee \alpha_2 \mid \beta) \tag{13}$$

$$\mathrm{DD}': O(\alpha \mid \beta) \wedge O(\beta \mid \gamma) \rightarrow O(\alpha \wedge \beta \mid \gamma) \tag{14}$$

$$O(\alpha \mid \beta) \rightarrow \mathcal{P}(\alpha \mid \beta) \tag{15}$$

The logic $\mathcal{L}$ *does* not *validate the following theorems of Weakening of the Consequent* (WC *and* WC′), *Deontic Detachment (or transitivity)* (DD) *and the Deontic axiom* (D).

$$\mathrm{WC}: O(\alpha_1 \mid \beta) \rightarrow O(\alpha_1 \vee \alpha_2 \mid \beta) \tag{16}$$

$$\mathrm{WC}': O(\alpha_1 \wedge \alpha_2 \mid \beta) \rightarrow O(\alpha_1 \mid \beta) \wedge O(\alpha_2 \mid \beta) \tag{17}$$

$$\mathrm{DD}: O(\alpha \mid \beta) \wedge O(\beta \mid \gamma) \rightarrow O(\alpha \mid \gamma) \tag{18}$$

$$\mathrm{D}: \neg(O(\alpha \mid \beta) \wedge O(\neg\alpha \mid \beta)) \tag{19}$$

Proof *Follows directly from the definitions.*

---

[1] In this article, we do not discuss permissions and their relations with obligations due to space limitations. We would have liked to define permission such that $M, w \models \mathcal{P}(p \mid q)$ iff $\forall w_1, w_2 \in W$ such that $R(w, w_1)$, $R(w, w_2)$, $M, w_1 \models p \wedge q$ and $M, w_2 \models \neg p \wedge q$, it is true that $w_1 \not< w_2$. This, however, is not possible in this simple setup. The definition of permission in Definition 2 is weaker.

The validity of strengthening of the antecedent follows from the fact that a preference of p over $\neg p$ implies a preference of $p \wedge q$ over $\neg p \wedge q$. The most remarkable property of the logic is the invalidity of weakening of the consequent. Intuitively, the lack of weakening of $O(p \mid \top)$ to $O(p \vee q \mid \top)$ is the consequence of the fact that $O(p \mid \top)$ expresses a preference of *all* p over $\neg p$, because from such a preference does not follow that $p \vee q$ is *always* preferred over $\neg p \wedge \neg q$. This is illustrated by the following example.

Example 3 *Reconsider the Kripke model* $|\neg p \wedge q| < |\neg p \wedge \neg q| < |p \wedge \neg q| < |p \wedge q|$ *of the previous example. The model satisfies* $O(p \mid \top)$ *but not* $O(p \vee q \mid \top)$.

The lack of weakening of the consequence is considered to be an advantage by [12, 17]. They quote Ross, who gave the following counterintuitive example of weakening of the consequent, called the Ross paradox: 'If you ought to mail the letter, then you ought to mail or burn the letter' [19]. The following example illustrates the intuitive theorem DD$'$.

Example 4 *Let S be the set of dyadic obligations* $\{O(a \mid \top), O(t \mid a)\}$, *where a can be read as 'a certain man going to the assistance of his neighbors' and t as 'telling the neighbors that he will come'. Hence, the two obligations can be read as 'a certain man should go to the assistance of his neighbors', and 'if he goes, then he should tell them he is coming'. The obligation* $O(a \wedge t \mid \top)$ *can be derived from S with* DD$'$, *which expresses that ideally, the man goes to the assistance of his neighbors* and *he tells them he is coming.*

Unfortunately, the logic $\mathcal{L}$ validates the counterintuitive theorems $O(\perp \mid \alpha)$ and $O(\alpha \mid \alpha)$. For this reason, we define various subtle other definitions of obligations in the preference logic. In the following definition, $O^c(\alpha \mid \beta)$ has an additional condition which works like a 'consistency check' to test whether $\alpha \wedge \beta$ is possible. $O^{cc}(\alpha \mid \beta)$ also has another additional condition which tests whether $\neg\alpha \wedge \beta$ is possible.

Definition 3 *Two alternative definitions of dyadic obligations and permissions are defined as follows. The modal operator* $\Diamond$ *is defined as usually as the dual of the* $\Box$ *operator:* $\Diamond\phi =_{def} \neg\Box\neg\phi$.

$$O^c(\alpha \mid \beta) =_{def} O(\alpha \mid \beta) \wedge \Diamond(\alpha \wedge \beta) \tag{20}$$

$$\mathcal{P}^c(\alpha \mid \beta) =_{def} \mathcal{P}(\alpha \mid \beta) \wedge \Diamond(\alpha \wedge \beta) \tag{21}$$

$$O^{cc}(\alpha \mid \beta) =_{def} O(\alpha \mid \beta) \wedge \Diamond(\alpha \wedge \beta) \wedge \Diamond(\neg\alpha \wedge \beta) \tag{22}$$

$$\mathcal{P}^{cc}(\alpha \mid \beta) =_{def} \mathcal{P}(\alpha \mid \beta) \wedge \Diamond(\alpha \wedge \beta) \wedge \Diamond(\neg\alpha \wedge \beta) \tag{23}$$

Proposition 2 *The logic* $\mathcal{L}$ *validates the following theorems.*

$$O(\perp \mid \alpha), \neg O^c(\perp \mid \alpha), \neg O^{cc}(\perp \mid \alpha) \tag{24}$$

$$O(\alpha \mid \alpha), \Diamond\alpha \rightarrow O^c(\alpha \mid \alpha), \neg O^{cc}(\alpha \mid \alpha) \tag{25}$$

Proof *Follows directly from the definitions.*

The obligations O^c and O^{cc} validate weaker versions of the theorems of Proposition 1, like for example the following Restricted Strengthening of the Antecedent (RSA) and Restricted Conjunction (RAND):

$$\text{RSA}: O^c(\alpha|\beta) \wedge \Diamond(\alpha \wedge \beta \wedge \gamma) \rightarrow O^c(\alpha|\beta \wedge \gamma) \tag{26}$$

$$\text{RAND}: O^c(\alpha_1|\beta) \wedge O^c(\alpha_2|\beta) \wedge \Diamond(\alpha_1 \wedge \alpha_2 \wedge \beta) \rightarrow O^c(\alpha_1 \wedge \alpha_2|\beta) \tag{27}$$

We already saw restricted strengthening of the antecedent in Example 1 in the introduction. In [15] this lack of unrestricted strengthening of the antecedent is dubbed 'factual defeasibility' and it is argued that this factual defeasibility is necessary to represent a-temporal CTD obligations. We elaborate on Example 1 in Section 4.

The logic can, like [17] but in contrast to SDL, represent moral dilemmas like $O(p|\top) \wedge O(\neg p|\top)$ in a consistent way. This follows from the lack of the theorem $\neg(O(\alpha \mid \beta) \wedge O(\neg\alpha \mid \beta))$. In the semantics, a moral dilemma $O(p \mid \top) \wedge O(\neg p \mid \top)$ corresponds to incomparable $|p|$ and $|\neg p|$ worlds. Brown and Mantha [17] consider this property a major advantage of their logic. However, in some applications of deontic logic, this is not a desirable property. In such cases, we can add an axiom to the logic which makes moral dilemmas inconsistent. For example, we can accept the D axiom of SDL, $\neg(O^c(\alpha|\top) \wedge O^c(\neg\alpha|\top))$, in our logic. However, this would still leave $O^c(\alpha|\top) \wedge O^c(\neg\alpha \wedge \beta|\top)$ consistent. We therefore define the following version of the D axiom.

$$\text{D}': \Box\neg(\alpha_1 \wedge \alpha_2) \rightarrow \neg(O^c(\alpha_1|\top) \wedge O^c(\alpha_2|\top)) \tag{28}$$

In some deontic logics, incomparable worlds are only used to model moral dilemmas, see e.g. [3]. However, in our logic this is not the only use of incomparable worlds. This follows from the fact that even when the D$'$ axiom is accepted, there can be incomparable worlds. This is illustrated by the following example.

Example 5 *Let M be a model of L that satisfies the two obligations* $O^c(p_1 \mid \top)$ *and* $O^c(p_2|\top)$. *The* $|p_1 \wedge \neg p_2|$ *and* $|\neg p_1 \wedge p_2|$ *worlds of M are incomparable, because the first is better than the latter with respect to obligation* $O^c(p_1 \mid \top)$, *but worse with respect to obligation* $O^c(p_2|\top)$.

Unfortunately, the D$'$ axiom cannot be generalized for arbitrary antecedents. The following example illustrates that the combination of RSA and a generalized D$'$ axiom is problematic.

Example 6 *(SA+D problem) Consider the following axiom:*

$$\text{D}'': \Box\neg(\alpha_1 \wedge \alpha_2 \wedge \beta) \rightarrow \neg(O^c(\alpha_1|\beta) \wedge O^c(\alpha_2|\beta))$$

Let S be the set of obligations $\{O^c(p_1|\top), O^c(p_2|\top)\}$. *With* RSA *(and* $\Diamond(p_1 \wedge \neg p_2)$ *and* $\Diamond(\neg p_1 \wedge p_2)$*), we can derive the obligations* $O^c(p_1|\neg(p_1 \wedge p_2))$ *and* $O^c(p_2|\neg(p_1 \wedge p_2))$ *from S. The two derived obligations are inconsistent with the* D$''$ *axiom when* β *is equivalent to* $\neg(p_1 \wedge p_2)$.[2]

[2] Semantically, the D$'$ axiom states that the most preferred worlds are comparable, and the axiom D$''$ states that all worlds are comparable. Example 5 illustrates that this condition is too strong.

The inconsistency is derived by only using RSA *and* D″. *Hence, there are two solutions for this problem: weakening* D″ *or weakening* RSA. *An attempt of the first solution is to weaken the* D″ *axiom to the following axiom.*

$$D''' : \Box\neg(\alpha_1 \wedge \alpha_2) \rightarrow \neg(\mathcal{O}^c(\alpha_1 \mid \beta) \wedge \mathcal{O}^c(\alpha_2 \mid \beta))$$

Unfortunately, S is also inconsistent with the D‴ *axiom. From S we can derive* $\mathcal{O}^c(p_1 \wedge \neg(p_1 \wedge p_2) \mid \neg(p_1 \wedge p_2))$ *and* $\mathcal{O}^c(p_2 \wedge \neg(p_1 \wedge p_2) \mid \neg(p_1 \wedge p_2))$ *with* $\Diamond\alpha \rightarrow \mathcal{O}^c(\alpha \mid \alpha)$ *and the* RAND *theorem. They are logically equivalent to* $\mathcal{O}^c(p_1 \wedge \neg p_2 \mid \neg(p_1 \wedge p_2))$ *and* $\mathcal{O}^c(p_2 \wedge \neg p_1 \mid \neg(p_1 \wedge p_2))$, *which are inconsistent with* D‴ *when* β *is equivalent to* $\neg(p_1 \wedge p_2)$. *Obviously, if* $\Diamond\alpha \rightarrow \mathcal{O}^c(\alpha \mid \alpha)$ *would not be a theorem of the logic, then the* D‴ *axiom would be a solution for the* SA+D *problem (see for example the conditional logic discussed in [10]). However, in the logic* $\mathcal{L}$ *the formula* $\Diamond\alpha \rightarrow \mathcal{O}^c(\alpha \mid \alpha)$ *is a theorem, so this solution does not work for* $\mathcal{L}$.

The second solution of the SA+D problem is to weaken RSA. *This solution implies that the obligation* $\mathcal{O}^c(p_1 \mid \neg(p_1 \wedge p_2))$ *should not be derivable from the obligation* $\mathcal{O}^c(p_1 \mid \top)$ *in our example. In the next section, we show that the minimizing approach does not have any strengthening of the antecedent, thus it gives the second solution to the* SA+D *problem. However, in Example 9 we will see that this lack of strengthening of the antecedent makes another type of moral dilemma consistent. A satisfactory solution for the SA+D problem is given in the two-phase logic in Section 4.*

3 Minimizing

In the dyadic logic of Bengt Hansson, an obligation $\mathcal{O}(\alpha \mid \beta)$ is true iff α is true in all minimal β worlds [8]. We therefore say that his logic (and related deontic logics, see [20] for an overview) is based on minimizing. In the following definition, we give several definitions of obligations which are closely related to the reconstruction of Boutilier [18] of B. Hansson's logic in a modal preference structure. In this logic, the maximal worlds are the preferred worlds instead of the minimal worlds. Maximizing in our logic is equivalent to minimizing in B. Hansson's logic. We give several definitions.[3]

Definition 4 *Dyadic obligation and permission are defined as follows.*

$$\mathcal{O}_\exists(\alpha \mid \beta) =_{def} \Diamond(\beta \wedge I(\beta \rightarrow \alpha)) \quad (29)$$

$$\mathcal{O}_\forall(\alpha \mid \beta) =_{def} \Box(\beta \rightarrow \neg I\neg(\beta \wedge I(\beta \rightarrow \alpha))) \quad (30)$$

$$\mathcal{P}_\exists(\alpha \mid \beta) =_{def} \neg\mathcal{O}_\forall(\neg\alpha \mid \beta) = \Diamond(\beta \wedge I(\beta \rightarrow \neg I\neg(\beta \wedge \alpha))) \quad (31)$$

$$\mathcal{P}_\forall(\alpha \mid \beta) =_{def} \neg\mathcal{O}_\exists(\neg\alpha \mid \beta) = \Box(\beta \rightarrow \neg I\neg(\beta \wedge \alpha)) \quad (32)$$

$$\mathcal{O}^c_\forall(\alpha \mid \beta) =_{def} \mathcal{O}_\forall(\alpha \mid \beta) \wedge \Diamond\beta \quad (33)$$

$$\mathcal{P}^c_\forall(\alpha \mid \beta) =_{def} \mathcal{P}_\forall(\alpha \mid \beta) \wedge \Diamond\beta \quad (34)$$

[3] Obviously, we can define $\mathcal{O}^{cc}_\exists$ etc by adding conditions to the definitions.

The obligation $O_\exists(p \mid q)$ is true in a model if p is true in an equivalence class of most preferred $|q|$ worlds of the model and $O_\forall(p \mid q)$ is true if p is true in all most preferred $|q|$ worlds.[4] Hence, both obligations refer to the optimal worlds where q is true, and $O_\exists(p \mid \top)$ and $O_\forall(p \mid \top)$ refer to the ideal worlds.[5]

Example 7 *Reconsider the Kripke model $|\neg p \wedge q| < |\neg p \wedge \neg q| < |p \wedge \neg q| < |p \wedge q|$ of Example 2. The model satisfies $O_\forall(p \mid \top)$, $O_\forall(q \mid \top)$, $O_\exists(p \mid \top)$ and $O_\exists(q \mid \top)$, whereas $O(q \mid \top)$ is not true in this model. Since $O_\exists(q \mid \top)$ is equivalent with $\Diamond Iq$ it is clear that q has to be true in some most preferred $|\top|$ worlds, and also that less preferred $|\top|$ worlds do not effect the truth of $\Diamond Iq$. Hence, in the evaluation of $O_\exists(q \mid \top)$ only preferred elements are taken into account and not the whole ordering.*

The main properties of the logic are given by the following proposition.

Proposition 3 *The logic $\mathcal{L}$ validates the following theorems.*

$$\mathrm{WC}_\exists : O_\exists(\alpha_1 \mid \beta) \rightarrow O_\exists(\alpha_1 \vee \alpha_2 \mid \beta) \tag{35}$$

$$\mathrm{WC}_\forall : O_\forall(\alpha_1 \mid \beta) \rightarrow O_\forall(\alpha_1 \vee \alpha_2 \mid \beta) \tag{36}$$

$$\mathrm{AND}_\forall : O_\forall(\alpha_1 \mid \beta) \wedge O_\forall(\alpha_2 \mid \beta) \rightarrow O_\forall(\alpha_1 \wedge \alpha_2 \mid \beta) \tag{37}$$

$$\mathrm{D}_\forall : \neg(O_\forall^c(\alpha \mid \beta) \wedge O_\forall^c(\neg\alpha \mid \beta)) \tag{38}$$

$$O_\exists(\alpha \mid \beta) \rightarrow \Diamond(\alpha \wedge \beta) \tag{39}$$

$$P_\exists(\alpha \mid \beta) \rightarrow \Diamond(\alpha \wedge \beta) \tag{40}$$

$$O_\exists(\alpha \mid \beta) \rightarrow P_\exists(\alpha \mid \beta) \tag{41}$$

$$O_\forall(\alpha \mid \beta) \rightarrow P_\forall(\alpha \mid \beta) \tag{42}$$

$$\neg O_\exists(\bot \mid \alpha) \tag{43}$$

$$\neg O_\forall^c(\bot \mid \alpha) \tag{44}$$

$$\Diamond\alpha \rightarrow O_\exists(\alpha \mid \alpha) \tag{45}$$

$$O_\forall(\alpha \mid \alpha) \tag{46}$$

$$O_\forall^c(\alpha \mid \beta) \rightarrow O_\exists(\alpha \mid \beta) \tag{47}$$

$$P_\forall^c(\alpha \mid \beta) \rightarrow P_\exists(\alpha \mid \beta) \tag{48}$$

The logic $\mathcal{L}$ does not validate the following theorems.

$$\mathrm{SA}_\exists : O_\exists(\alpha \mid \beta_1) \rightarrow O_\exists(\alpha \mid \beta_1 \wedge \beta_2) \tag{49}$$

$$\mathrm{SA}_\forall : O_\forall(\alpha \mid \beta_1) \rightarrow O_\forall(\alpha \mid \beta_1 \wedge \beta_2) \tag{50}$$

---

[4] Moreover, the definition can deal with infinite chains, see [7].

[5] Again, we will not discuss permissions. Just notice that the definition of permission is closely related to the definition of permission in SDL.

$$\text{AND}_\exists : O_\exists(\alpha_1 \mid \beta) \wedge O_\exists(\alpha_2 \mid \beta) \rightarrow O_\exists(\alpha_1 \wedge \alpha_2 \mid \beta) \quad (51)$$

$$\text{DD}_\exists : O_\exists(\alpha \mid \beta) \wedge O_\exists(\beta \mid \gamma) \rightarrow O_\exists(\alpha \mid \gamma) \quad (52)$$

$$\text{DD}_\forall : O_\forall(\alpha \mid \beta) \wedge O_\forall(\beta \mid \gamma) \rightarrow O_\forall(\alpha \mid \gamma) \quad (53)$$

$$\text{D}_\exists : \neg(O_\exists(\alpha \mid \beta) \wedge O_\exists(\neg\alpha \mid \beta)) \quad (54)$$

Proof *Follows directly from the definitions.*

The logic validates weakening of the consequent, because the most preferred world that satisfies p_1 also satisfies $p_1 \vee p_2$. However, the logic does not have strengthening of the antecedent of $O_\forall(p \mid \top)$ to $O_\forall(p \mid q)$, because the preferred $|\top|$ worlds may be different from the preferred $|q|$ worlds.

Example 8 *Reconsider the Kripke model $|\neg p \wedge q| < |\neg p \wedge \neg q| < |p \wedge \neg q| < |p \wedge q|$ of the previous example. The model satisfies $O_\exists(q \mid \top)$ and $O_\forall(q \mid \top)$, but not $O_\exists(q \mid \neg p)$ or $O_\forall(q \mid \neg p)$. For example, $O_\exists(q \mid \neg p)$ is false because the preferred $|\neg p|$ worlds are the $|\neg p \wedge \neg q|$ worlds. Hence, $O_\exists$ does not have strengthening of the antecedent.*

Obviously, $O_\exists$ can represent moral dilemmas like $O_\exists(p \mid \top) \wedge O_\exists(\neg p \mid \top)$ in a consistent way, and $O_\forall^c$ cannot. However, the following example shows that $O_\forall^c$ does not give a satisfactory solution for the SA+D problem discussed in Example 6. $O_\forall^c$ still represents some moral dilemmas in a consistent way, as a result of the total lack of strengthening of the antecedent.

Example 9 ***(SA+D Problem, continued)*** *The following theorem of the logic $\mathcal{L}$ is the axiom* D″ *for $O_\forall^c$. The validity of this theorem follows directly from the theorems* $\text{D}_\forall$ *and* $\text{WC}_\forall$.

$$\text{D}_\forall'' : \Box\neg(\alpha_1 \wedge \alpha_2 \wedge \beta) \rightarrow \neg(O_\forall^c(\alpha_1 \mid \beta) \wedge O_\forall^c(\alpha_2 \mid \beta))$$

The obligations $O_\forall^c(p_1 \mid \top)$ and $O_\forall^c(p_2 \mid \top)$ are consistent. Hence, $O_\forall^c$ solves the problem that was discussed in Example 6. However, $O_\forall^c$ does not solve all of the SA+D *problems. Consider the following example, adapted from [14]. Let S be the set of obligations $S = \{O_\forall^c(\neg c \mid \top), O_\forall^c(c \mid k)\}$, where k can be read as 'killing someone (the witness)' and c can be read as 'offering someone a cigarette'. S is consistent and from S we can derive the obligation $O_\forall^c(\neg k \mid \top)$. Prakken and Sergot argue in [14] that S should be inconsistent, even when S is extended with the obligation $O_\forall^c(\neg k \mid \top)$. In Section 4 we show how this inconsistency can be obtained in a two-phase logic.*

We can impose the D axiom for $O_\exists$ by demanding that the partial ordering $\leq$ is connected: for all w_1, w_2, we have $w_1 \leq w_2$ or $w_2 \leq w_1$. If the ordering is connected, then the modal operator I satisfies the S4.3 axioms instead of the S4 axioms (like, for example, Boutilier's logic CO [18]) and $O_\exists(\alpha \mid \beta)$ becomes equivalent to $O_\forall^c(\alpha \mid \beta)$. The last property follows from the fact that truth in an equivalence class of preferred worlds and truth in all preferred worlds has become equivalent.

A relation between ordering and minimizing is given by the following proposition.

Proposition 4 *The logic $\mathcal{L}$ validates the following theorem.*

$$O^c(\alpha \mid \beta) \rightarrow O_\exists(\alpha \mid \beta) \quad (55)$$

Proof *The formula is equivalent with the following theorem of $\mathcal{L}$:*

$$(\Box(\beta \wedge \alpha \rightarrow I(\beta \rightarrow \alpha)) \wedge \Diamond(\alpha \wedge \beta)) \rightarrow \Diamond(\beta \wedge I(\beta \rightarrow \alpha))$$

4 Combining Ordering and Minimizing

4.1 Combining

The idea of combining ordering and minimizing is to combine formulas with O, $O_\exists$ and $O_\forall$ operators. However, this is not possible in the logic $\mathcal{L}$. Before we can make these combinations there are some technical problems which have to be solved first. The following example shows the two problems.

Example 10 *Consider the set $S = \{O(p \mid \top)\}$. Hence, all we know is that p is preferred over $\neg p$. Given this knowledge, we would expect to derive $O_\exists(p \mid \top)$ and $O_\forall(p \mid \top)$. This, however, is not the case. There are two problems, the first one blocks both derivations, and the second one only the derivation of $O_\forall(p \mid \top)$.*

1. *Worlds do not have to exist in a model.*

   Let M be a model with only $|\neg p|$ worlds. M satisfies $O(p \mid \top)$, but it does not satisfy $O_\forall(p \mid \top)$ or $O_\exists(p \mid \top)$. Hence, neither of these two formulas are entailed by $O(p \mid \top)$.

2. *The ordering of worlds can be too weak.*

   Again, consider the obligation $O(p \mid \top)$. All models that satisfy $|p| \not\geq |\neg p|$ are models of $O(p \mid \top)$. Hence, $|p|$ worlds and $|\neg p|$ worlds are either incomparable, or $|p|$ worlds are strictly preferred over $|\neg p|$ worlds. Let M be a model in which all $|p|$ and $|\neg p|$ worlds are incomparable. M satisfies $O(p \mid \top)$, but it does not satisfy $O_\forall(p \mid \top)$. Hence, $O_\forall(p \mid \top)$ is not entailed by $O(p \mid \top)$. For minimization, we only want the models in which $|p|$ worlds are strictly preferred over $|\neg p|$.

The solution of the first problem is to consider only models in which all propositionally satisfiable formulas ϕ are true in *some* world. This can be 'axiomatized' with Boutilier's axiom scheme LP, see [21, 7] for a discussion. The axiom scheme LP states that every formula ϕ without any occurrences of modal operators, which is propositionally satisfiable, is true in some world.

Definition 5 *The logic $\mathcal{L}^*$ is $\mathcal{L}$ extended with the following axiom scheme:*

$$\text{LP} : \Diamond\phi \textit{ for all satisfiable propositional } \phi \tag{56}$$

We write $\models$ for logical entailment in $\mathcal{L}^*$.

A solution of the second problem is to define a preference ordering on models, which prefers models which are maximally connected with respect to the partial preordering $\leq$, i.e. with the most binary relations of $\leq$. The preferred models of this ordering are the only models which are used for minimization.

Definition 6 *Let $M_1 = \langle W_1, R_1, \leq_1, V_1\rangle$ and $M_2 = \langle W_2, R_2, \leq_2, V_2\rangle$ be two models of $\mathcal{L}^*$. M_1 is preferred over M_2 for mapping τ, written as $M_1 \sqsubseteq_\tau M_2$, iff:*

1. *τ is a one-to-one mapping of the worlds of W_2 to the worlds of W_1 such that the worlds satisfy the same propositions, and*

2. *If $w_1 \leq_2 w_2$ for $w_1, w_2 \in W_2$ then $\tau(w_1) \leq_1 \tau(w_2)$.*

We write $M_1 \sqsubset_\tau M_2$ *iff* $M_1 \sqsubseteq_\tau M_2$ *and* $M_2 \not\sqsubseteq_{\tau^{-1}} M_1$.

The ordering on models ($\sqsubseteq$) should not be confused with the ordering on worlds ($\leq$). The ordering on models is a technical trick to ensure that the worlds within a model are maximally connected, whereas the ordering on worlds expresses the ideality ordering. Given the preference ordering on models, we can define a notion of preferential entailment, see [4, 5].

Definition 7 *Let* $M = \langle W, R, \leq, V \rangle$ *be a model and S be a set of sentences. A world* $w \in W$ *of M preferentially satisfies S, written as* $M, w \models_\sqsubset S$ *iff* $M, w \models S$ *and there is not a model* M' *and a mapping* τ *such that* $M', \tau(w) \models S$ *and* $M' \sqsubset_\tau M$ *(M is a preferred model of S). S preferentially entails* ϕ*, written as* $S \models_\sqsubset \phi$*, iff for all M and w, if* $M, w \models_\sqsubset S$ *then* $M, w \models \phi$.

We will illustrate the notion of preferential entailment in $\mathcal{L}^*$ by the Reykjavic paradox [22]. See [23] for several interpretations of this paradox.

Example 11 *Let S be the set of obligations* $\{O(\neg r \wedge \neg g \mid \top), O(r \mid g), O(g \mid r)\}$*, where r can be read as telling a secret to Reagan and g as telling the secret to Gorbatsjov. The preferred model is given in Figure 1. We have* $S \not\models_\sqsubset O(\neg r \mid \top)$*,* $S \not\models_\sqsubset O(\neg g \mid \top)$*,* $S \models_\sqsubset O_\forall(\neg r \mid \top)$ *and* $S \models_\sqsubset O_\forall(\neg g \mid \top)$.

Without the LP *axiom scheme, the* $\mid \neg r \wedge \neg g \mid$ *worlds do not have to exist in a model, and such a model M would not satisfy* $M \models O_\forall(\neg r \mid \top)$ *and* $M \models O_\exists(\neg r \mid \top)$. *Without the preference ordering on models, the* $\mid \neg r \wedge \neg g \mid$ *and* $\mid r \wedge g \mid$ *worlds could be incomparable. Such a model M would still satisfy* $M \models O_\exists(\neg r \mid \top)$ *but it would not satisfy* $M \models O_\forall(\neg r \mid \top)$.

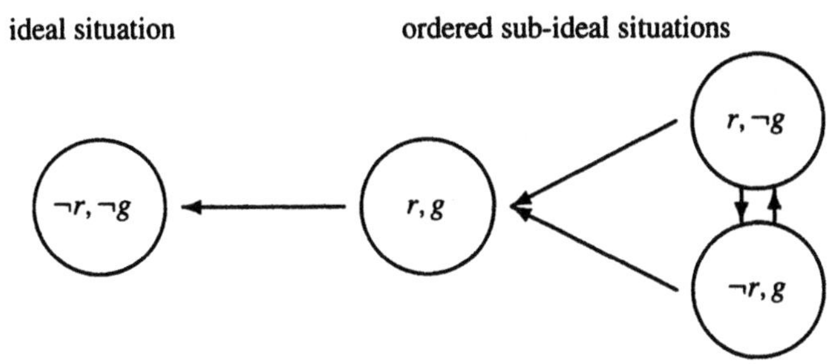

Figure 1: Preference relation of the Reykjavic paradox

Preferential entailment is a typical mechanism from non-monotonic reasoning. The combination of ordering and minimizing is non-monotonic, as the following example illustrates.

Example 12 *Let* $S = \{O(p \mid \top)\}$ *and* $S' = \{O(p \mid \top), O(\neg p \mid \top)\}$. *We have* $S \models_\sqsubset O_\forall(p \mid \top)$ *and* $S' \not\models_\sqsubset O_\forall(p \mid \top)$. *Hence, by addition of a formula we loose conclusions.*

4.2 The two phases in a deontic logic

The two phases in a deontic logic correspond to the two different kinds of obligations O^c and $O_\exists$. Semantically, the first phase corresponds to ordering (O^c) and the second phase to minimizing ($O_\exists$). From a proof theoretic point of view, the first phase corresponds to applying valid inferences of O^c like RSA, RAND etc, and the second phase corresponds to applying valid inferences of $O_\exists$ like WC. The basic technique of deontic logic as a two-phase logic is that a conclusion of the form $O_\exists(\alpha \mid \beta)$ can be derived either with or without $O^c(\alpha \mid \beta)$. In the first case $O_\exists(\alpha \mid \beta)$ can be derived via $O^c(\alpha \mid \beta)$ with Proposition 4, which says that the latter formula implies the first one. If so, we say that $O_\exists(\alpha \mid \beta)$ is derived in the first phase. In the second case we say that $O_\exists(\alpha \mid \beta)$ is second phase derived. The important difference is that in the first phase we can apply RSA to $O_\exists(\alpha \mid \beta)$, because of the simultaneous occurrence of $O^c(\alpha \mid \beta)$. We apply RSA to $O^c(\alpha \mid \beta)$ to obtain, for example, $O^c(\alpha \mid \beta \wedge \gamma)$, and then due to Proposition 4 we also obtain $O_\exists(\alpha \mid \beta \wedge \gamma)$. If $O^c(\alpha \mid \beta)$ does not occur simultaneously with $O_\exists(\alpha \mid \beta)$, then there is no way we can apply RSA to this formula. Being a minimizing formula it lacks RSA. Hence, once $O_\exists(\alpha \mid \beta)$ has been derived in the second phase, we loose RSA permanently for subsequent derivations of this formula. Analogously, we can say that $O_\exists(\alpha \mid \beta)$ is first phase or second phase entailed by a set of premises, depending on whether S does or does not entail $O^c(\alpha \mid \beta)$.

The following example shows that the two-phase approach solves the SA+WC problem of Example 1 in the introduction. In this example, we only consider $O_\exists$ and not $O_\forall$. We can therefore use standard logical entailment $\models$ instead of the more complicated preferential entailment $\models_\sqsubset$.

Example 13 ***(SA+WC problem, continued)*** *Let $S = \{O^c(a \vee s \mid \top), O^c(\neg a \mid \top)\}$, where $\neg a$ does not entail the negation of s. We have $S \models \Diamond(\neg a \wedge s)$, $S \models O^c(\neg a \wedge s \mid \top)$ and $S \models O_\exists(\neg a \wedge s \mid \top)$, $S \not\models O^c(s \mid \top)$ and $S \models O_\exists(s \mid \top)$. The crucial observation is that $O_\exists(s \mid a)$ is not entailed by S. First of all, $O_\exists(s \mid a)$ is not first phase entailed by S via $O_\exists(s \mid \top)$, because $O_\exists(s \mid \top)$ is not first phase entailed by S. Secondly, $O_\exists(s \mid a)$ is not second phase entailed by S via $O_\exists(s \mid \top)$ either, because in second phase entailment $O_\exists$ does not have strengthening of the antecedent at all. Thirdly, it is not second phase entailed by S via a first phase derivation of $O_\exists(\neg a \wedge s \mid a)$, because $O^c(\neg a \wedge s \mid a)$ is not entailed by $O^c(\neg a \wedge s \mid \top)$ due to the restriction in* RSA.

The derived obligations $O_\exists$ can represent moral dilemmas in a consistent way. If we want to make moral dilemmas inconsistent, we can consider the two-phase approach with $O^c_\forall$ instead of $O_\exists$. For the first phase, we define a new kind of obligation.

$$O_D(\alpha \mid \beta) =_{def} O^c(\alpha \mid \beta) \wedge O^c_\forall(\alpha \mid \beta)$$

The two-phase approach with $O_D(\alpha \mid \beta)$ and $O^c_\forall(\alpha \mid \beta)$ works similar to the two-phase approach with $O^c(\alpha \mid \beta)$ and $O_\exists(\alpha \mid \beta)$, only we have to use preferential entailment $\models_\sqsubset$ instead of $\models$. Notice that $O_D(\alpha \mid \beta) \to O^c_\forall(\alpha \mid \beta)$ is a (trivial) theorem of $\mathcal{L}$. This theorem is the counterpart of Proposition 4 for this second example of the two-phase approach. In this case, we say that a derived obligation $O^c_\forall(\alpha \mid \beta)$ is first phase derived if $O_D(\alpha \mid \beta)$ is derivable, and second phase derived otherwise. The following example shows that this two-phase approach solves the SA+D problem of Example 6 and 9 by weakening RSA.

Example 14 ***(SA+D problem, continued)*** *First of all, the following theorem of the logic $\mathcal{L}$ is the counterpart of axiom D'' for O_D. This theorem follows from the fact that $O_D(\alpha \mid \beta)$ entails $O^c_\forall(\alpha \mid \beta)$.*

$$D''_D : \Box\neg(\alpha_1 \wedge \alpha_2 \wedge \beta) \to \neg(O_D(\alpha_1 \mid \beta) \wedge O_D(\alpha_2 \mid \beta))$$

Let $S = \{O_D(p_1 \mid \top), O_D(p_2 \mid \top)\}$, *hence* $S = \{\mathcal{O}^c(p_1 \mid \top) \wedge \mathcal{O}^c_\forall(p_1 \mid \top), \mathcal{O}^c(p_2 \mid \top) \wedge \mathcal{O}^c_\forall(p_2 \mid \top)\}$. *We have* $S \models_\sqsubset \mathcal{O}^c_\forall(p_1 \vee p_2 \mid \neg(p_1 \wedge p_2))$ *and* $S \not\models_\sqsubset \mathcal{O}^c_\forall(p_1 \mid \neg(p_1 \wedge p_2))$. *Hence, the obligation* $\mathcal{O}^c_\forall(p_1 \mid \top)$ *is not strengthened to* $\mathcal{O}^c_\forall(p_1 \mid \neg(p_1 \wedge p_2))$, *and it was observed in Example 6 that this is a solution for the* SA+D *problem. Furthermore, we have* $S \models_\sqsubset O_\exists(p_2 \mid \neg(p_1 \wedge p_2))$ *and* $S \models_\sqsubset O_\exists(\neg p_2 \mid \neg(p_1 \wedge p_2))$, *which explains the absence of* RSA*: there is not a unique most preferred obligation for the antecedent* $\neg(p_1 \wedge p_2)$.

Finally, let S be the set of obligations $\{O_D(\neg c \mid \top), O_D(c \mid k)\}$ *where* $\neg c$ *does not entail the negation of k. S is inconsistent, and we can derive an inconsistency as follows. The premise* $O_D(\neg c \mid \top)$ *entails the obligation* $\mathcal{O}^c(\neg c \mid \top)$, *which entails* $\mathcal{O}^c(\neg c \mid k)$. *From Proposition 4 follows that the latter obligation entails* $O_\exists(\neg c \mid k)$. $O_\exists(\neg c \mid k)$ *entails* $\mathcal{P}_\exists(\neg c \mid k)$, *which is logically equivalent to* $\neg \mathcal{O}_\forall(c \mid k)$. *The premise* $O_D(c \mid k)$ *entails* $\mathcal{O}^c_\forall(c \mid k)$, *which is inconsistent with* $\neg \mathcal{O}_\forall(c \mid k)$.

4.3 Related approaches

Thus far, we have shown that the two-phase approach to deontic logic can solve the SA+WC problem of Example 1 and the SA+D problem of Example 6 and 9. However, we have said little about the intuition behind the distinction between the two phases. In this section, we discuss the diagnostic and decision theoretic perspectives on deontic logic that illustrate aspects of these two phases.

The aim of *diagnostic reasoning* (see e.g. [24]) is to derive a conjecture about faulty components, given a system (a set of components and a description of the behavior of these components) and facts (a description of the world). In diagnostic reasoning, two phases can be distinguished. The first phase is reasoning about *all* possible conjectures, i.e. reasoning about the possible sets of broken components. The second phase is reasoning about *minimal* sets of broken components, the so-called diagnoses. This second phase reasoning is based on the so-called principle of parsimony, which states that it is expected that the set of broken components is minimal. The *diagnostic perspective* on deontic logic is based on the diagnostic problem of determining the set of violated obligations, given a normative system and facts. Hence, the analogy between deontic and diagnostic reasoning is based on the fact that violated obligations are treated in a similar way as broken components. The two phases are clearly distinguished in the DIagnostic framework for DEontic reasoning DIODE [25, 26, 27]. In DIODE there is a distinction between (phase one) *constructing* and *zooming-in on* a preference ordering of deontic states and (phase two) *finding the optimal sub-ideal state.* By zooming-in we mean that the substructure of the ordering is considered that only contains states that satisfy the actual facts. The first phase of zooming-in ignores more ideal states, since they are not reachable anymore because of the violating facts. Moreover, the second phase of finding the optimal sub-ideal state ignores the less ideal states. In DIODE, the derived obligations are called *contextual obligations* to express that they are only valid in the context of the antecedent, because they do not refer to either more ideal or less ideal states.

The aim of a *(qualitative) decision theory* is to find the optimal goal, and how to reach this goal, given (qualitative) preferences and knowledge about the world (usually containing uncertain beliefs). In decision theoretic reasoning, also two phases can be distinguished. The first phase is reasoning about the preferences and *all* (reachable) goals. The second phase is optimal goal planning. The second phase is based on the additional assumption that the agent optimizes his behavior, and hence wants to realize only maximally preferred goals. The *decision theoretic perspective* on deontic logic considers ideal states as goals. For example, Pearl gives a qualitative, decision-theoretic account of the *pragmatic* obligation $O(A \mid B)$: 'If you observe, believe, or know B, then the expected utility resulting from doing A is much higher than that of resulting from not doing A' [28]. It is one thing to determine the maximum utiliy, but quite another to act accordingly.

5 Further research

In this paper, we have investigated dyadic obligations in a modal preference logic. As further research, we will investigate the formalization of *defeasible* dyadic obligations, obligations which are subject to exceptions, in the modal framework introduced in this paper. Another well-known problem of dyadic deontic logic is the relation between facts and dyadic obligations. In [15] a notion of factual detachment is proposed based on the so-called retraction test. In further research we will investigate the formalization of this test in $\mathcal{L}^*$.

Acknowledgement

Thanks to Patrick van der Laag for several discussions on the issues raised in this paper.

References

[1] N. Rescher. The logic of preference. In *Topics in Philosophical Logic*. D. Reidel Publishing Company, Dordrecht, Holland, 1967.

[2] B.C. van Fraassen. The logic of conditional obligation. *Journal of Philosophical Logic*, 1:417–438, 1972.

[3] B.C. van Fraassen. Values and the heart command. *Journal of Philosophy*, 70:5–19, 1973.

[4] Y. Shoham. *Reasoning About Change*. MIT Press, 1988.

[5] S. Kraus, D. Lehmann, and M. Magidor. Nonmonotonic reasoning, preferential models and cumulative logics. *Artificial Intelligence*, 44:167–207, 1990.

[6] D. Makinson. Five faces of minimality. *Studia Logica*, 52:339–379, 1993.

[7] C. Boutilier. Conditional logics of normality: a modal approach. *Artificial Intelligence*, 68:87–154, 1994.

[8] B. Hansson. An analysis of some deontic logics. In *Deontic Logic: Introductionary and Systematic Readings*, pages 121–147. D. Reidel Publishing Company, Dordrecht, Holland, 1971.

[9] J.W. Forrester. Gentle murder, or the adverbial Samaritan. *Journal of Philosophy*, 81:193–197, 1984.

[10] C. E. Alchourrón. Philosophical foundations of deontic logic and the logic of defeasible conditionals. In *Deontic Logic in Computer Science: Normative System Specification*, pages 43–84. John Wiley & Sons, 1994.

[11] D. Ross. *The Right and the Good.* Oxford University Press, 1930.

[12] S.O. Hansson. Preference-based deontic logic (PDL). *Journal of Philosophical Logic*, 19:75–93, 1990.

[13] B.F. Chellas. *Modal Logic: An Introduction.* Cambridge University Press, 1980.

[14] H. Prakken and M.J. Sergot. Contrary-to-duty imperatives, defeasibility and violability. In *Proceedings of the Second Workshop on Deontic Logic in Computer Science (Deon'94)*, Oslo, 1994. To appear in: *Studia Logica.*

[15] L.W.N. van der Torre and Y.H. Tan. Cancelling and overshadowing: two types of defeasibility in defeasible deontic logic. In *Proceedings of the Fourteenth International Joint Conference on Artificial Intelligence (IJCAI'95).* Morgan Kaufman, 1995.

[16] J.F. Horty. Deontic logic as founded in nonmonotonic logic. *Annals of Mathematics and Artificial Intelligence*, 9:69–91, 1993.

[17] A.L. Brown and S. Mantha. Preferences as normative knowledge: Towards declarative obligations. In *Proceedings of the First Workshop on Deontic Logic in Computer Science (Deon'91)*, pages 142–163, Amsterdam, 1991.

[18] C. Boutilier. Toward a logic for qualitative decision theory. In *Proceedings of the Fourth International Conference on Principles of Knowledge Representation and Reasoning (KR'94)*, pages 75–86, 1994.

[19] A. Ross. Imperatives and logic. *Theoria*, 7:53–71, 1941.

[20] D. Lewis. Semantic analysis for dyadic deontic logic. In *Logical Theory and Semantical Analysis*, pages 1–14. D. Reidel Publishing Company, Dordrecht, Holland, 1974.

[21] H. Levesque. All I know: a study in autoepistemic logic. *Artificial Intelligence*, 42:263–309, 1990.

[22] M. Belzer. A logic of deliberation. In *Proceedings of the Fifth National Conference on Artificial Intelligence*, pages 38–43, 1986.

[23] L.W.N. van der Torre. Violated obligations in a defeasible deontic logic. In *Proceedings of the Eleventh European Conference on Artificial Intelligence (ECAI'94)*, pages 371–375. John Wiley & Sons, 1994.

[24] R. Reiter. A theory of diagnosis from first principles. *Artificial Intelligence*, 32:57–95, 1987.

[25] Y.-H. Tan and L.W.N. van der Torre. Representing deontic reasoning in a diagnostic framework. In *Proceedings of the Workshop on Legal Applications of Logic Programming of the Eleventh International Conference on Logic Programming (ICLP'94)*, Genoa, Italy, 1994.

[26] Y.-H. Tan and L.W.N. van der Torre. DIODE: Deontic logic based on diagnosis from first principles. In *Proceedings of the Workshop 'Artificial normative reasoning' of the Eleventh European Conference on Artificial Intelligence (ECAI'94)*, Amsterdam, 1994.

[27] Y.-H. Tan and L.W.N. van der Torre. Why defeasible deontic logic needs a multi preference semantics. In *Proceedings of the ECSQARU'95. Lecture Notes in Artificial Intelligence 946*. Springer Verlag, 1995.

[28] J. Pearl. A calculus of pragmatic obligation. In *Proceedings of the UAI'93*, pages 12–20, 1993.

[29] Y.-H. Tan and L.W.N. van der Torre. Multi preference semantics for a defeasible deontic logic. In *Proceedings of the JURIX'94*, Amsterdam, 1994.

Author Index

Published in 1990-93

3rd Refinement Workshop (organised by BCS-FACS, and sponsored by IBM UK Laboratories, Hursley Park and the Programming Research Group, University of Oxford), Hursley Park, 9-11 January 1990
Carroll Morgan and J. C. P. Woodcock (Eds)

Designing Correct Circuits, Workshop jointly organised by the Universities of Oxford and Glasgow, Oxford, 26-28 September 1990
Geraint Jones and Mary Sheeran (Eds)

Functional Programming, Glasgow 1990
Proceedings of the 1990 Glasgow Workshop on Functional Programming, Ullapool, Scotland, 13-15 August 1990
Simon L. Peyton Jones, Graham Hutton and Carsten Kehler Holst (Eds)

4th Refinement Workshop, Proceedings of the 4th Refinement Workshop, organised by BCS-FACS, Cambridge, 9-11 January 1991
Joseph M. Morris and Roger C. Shaw (Eds)

AI and Cognitive Science '90, University of Ulster at Jordanstown, 20-21 September 1990
Michael F. McTear and Norman Creaney (Eds)

Software Re-use, Utrecht 1989, Proceedings of the Software Re-use Workshop, Utrecht, The Netherlands, 23-24 November 1989
Liesbeth Dusink and Patrick Hall (Eds)

Z User Workshop, 1990, Proceedings of the Fifth Annual Z User Meeting, Oxford, 17-18 December 1990
J.E. Nicholls (Ed.)

IV Higher Order Workshop, Banff 1990
Proceedings of the IV Higher Order Workshop, Banff, Alberta, Canada, 10-14 September 1990
Graham Birtwistle (Ed.)

ALPUK91, Proceedings of the 3rd UK Annual Conference on Logic Programming, Edinburgh, 10-12 April 1991
Geraint A.Wiggins, Chris Mellish and Tim Duncan (Eds)

Specifications of Database Systems
International Workshop on Specifications of Database Systems, Glasgow, 3-5 July 1991
David J. Harper and Moira C. Norrie (Eds)

7th UK Computer and Telecommunications Performance Engineering Workshop
Edinburgh, 22-23 July 1991
J. Hillston, P.J.B. King and R.J. Pooley (Eds)

Logic Program Synthesis and Transformation
Proceedings of LOPSTR 91, International Workshop on Logic Program Synthesis and Transformation, University of Manchester, 4-5 July 1991
T.P. Clement and K.-K. Lau (Eds)

Declarative Programming, Sasbachwalden 1991
PHOENIX Seminar and Workshop on Declarative Programming, Sasbachwalden, Black Forest, Germany, 18-22 November 1991
John Darlington and Roland Dietrich (Eds)

Building Interactive Systems: Architectures and Tools
Philip Gray and Roger Took (Eds)

Functional Programming, Glasgow 1991
Proceedings of the 1991 Glasgow Workshop on Functional Programming, Portree, Isle of Skye, 12-14 August 1991
Rogardt Heldal, Carsten Kehler Holst and Philip Wadler (Eds)

Object Orientation in Z
Susan Stepney, Rosalind Barden and David Cooper (Eds)

Code Generation - Concepts, Tools, Techniques
Proceedings of the International Workshop on Code Generation, Dagstuhl, Germany, 20-24 May 1991
Robert Giegerich and Susan L. Graham (Eds)

Z User Workshop, York 1991, Proceedings of the Sixth Annual Z User Meeting, York, 16-17 December 1991
J.E. Nicholls (Ed.)

Formal Aspects of Measurement
Proceedings of the BCS-FACS Workshop on Formal Aspects of Measurement, South Bank University, London, 5 May 1991
Tim Denvir, Ros Herman and R.W. Whitty (Eds)

AI and Cognitive Science '91 University College, Cork, 19-20 September 1991
Humphrey Sorensen (Ed.)

5th Refinement Workshop, Proceedings of the 5th Refinement Workshop, organised by BCS-FACS, London, 8-10 January 1992
Cliff B. Jones, Roger C. Shaw and Tim Denvir (Eds)

Algebraic Methodology and Software Technology (AMAST'91)
Proceedings of the Second International Conference on Algebraic Methodology and Software Technology, Iowa City, USA, 22-25 May 1991
M. Nivat, C. Rattray, T. Rus and G. Scollo (Eds)

ALPUK92, Proceedings of the 4th UK Conference on Logic Programming, London, 30 March - 1 April 1992
Krysia Broda (Ed.)

Logic Program Synthesis and Transformation
Proceedings of LOPSTR 92, International Workshop on Logic Program Synthesis and Transformation, University of Manchester, 2–3 July 1992
Kung-Kiu Lau and Tim Clement (Eds)

NAPAW 92, Proceedings of the First North American Process Algebra Workshop, Stony Brook, New York, USA, 28 August 1992
S. Purushothaman and Amy Zwarico (Eds)

First International Workshop on Larch
Proceedings of the First International Workshop on Larch, Dedham, Massachusetts, USA, 13–15 July 1992
Ursula Martin and Jeannette M. Wing (Eds)

Persistent Object Systems
Proceedings of the Fifth International Workshop on Persistent Object Systems, San Miniato (Pisa), Italy, 1–4 September 1992
Antonio Albano and Ron Morrison (Eds)

Formal Methods in Databases and Software Engineering, Proceedings of the Workshop on Formal Methods in Databases and Software Engineering, Montreal, Canada, 15–16 May 1992
V.S. Alagar, Laks V.S. Lakshmanan and F. Sadri (Eds)

Modelling Database Dynamics
Selected Papers from the Fourth International Workshop on Foundations of Models and Languages for Data and Objects, Volkse, Germany, 19–22 October 1992
Udo W. Lipeck and Bernhard Thalheim (Eds)

14th Information Retrieval Colloquium
Proceedings of the BCS 14th Information Retrieval Colloquium, University of Lancaster, 13–14 April 1992
Tony McEnery and Chris Paice (Eds)

Functional Programming, Glasgow 1992
Proceedings of the 1992 Glasgow Workshop on Functional Programming, Ayr, Scotland, 6–8 July 1992
John Launchbury and Patrick Sansom (Eds)

Z User Workshop, London 1992
Proceedings of the Seventh Annual Z User Meeting, London, 14–15 December 1992
J.P. Bowen and J.E. Nicholls (Eds)

Interfaces to Database Systems (IDS92)
Proceedings of the First International Workshop on Interfaces to Database Systems, Glasgow, 1–3 July 1992
Richard Cooper (Ed.)

AI and Cognitive Science '92
University of Limerick, 10–11 September 1992
Kevin Ryan and Richard F.E. Sutcliffe (Eds)

Theory and Formal Methods 1993
Proceedings of the First Imperial College Department of Computing Workshop on Theory and Formal Methods, Isle of Thorns Conference Centre, Chelwood Gate, Sussex, UK, 29–31 March 1993
Geoffrey Burn, Simon Gay and Mark Ryan (Eds)

Algebraic Methodology and Software Technology (AMAST'93)
Proceedings of the Third International Conference on Algebraic Methodology and Software Technology, University of Twente, Enschede, The Netherlands, 21–25 June 1993
M. Nivat, C. Rattray, T. Rus and G. Scollo (Eds)

Logic Program Synthesis and Transformation
Proceedings of LOPSTR 93, International Workshop on Logic Program Synthesis and Transformation, Louvain-la-Neuve, Belgium, 7–9 July 1993
Yves Deville (Ed.)

Database Programming Languages (DBPL-4)
Proceedings of the Fourth International Workshop on Database Programming Languages - Object Models and Languages, Manhattan, New York City, USA, 30 August–1 September 1993
Catriel Beeri, Atsushi Ohori and Dennis E. Shasha (Eds)

Music Education: An Artificial Intelligence Approach, Proceedings of a Workshop held as part of AI-ED 93, World Conference on Artificial Intelligence in Education, Edinburgh, Scotland, 25 August 1993
Matt Smith, Alan Smaill and Geraint A. Wiggins (Eds)

Rules in Database Systems
Proceedings of the 1st International Workshop on Rules in Database Systems, Edinburgh, Scotland, 30 August–1 September 1993
Norman W. Paton and M. Howard Williams (Eds)

Semantics of Specification Languages (SoSL)
Proceedings of the International Workshop on Semantics of Specification Languages, Utrecht, The Netherlands, 25–27 October 1993
D.J. Andrews, J.F. Groote and C.A. Middelburg (Eds)

Security for Object-Oriented Systems
Proceedings of the OOPSLA-93 Conference Workshop on Security for Object-Oriented Systems, Washington DC, USA, 26 September 1993
B. Thuraisingham, R. Sandhu and T.C. Ting (Eds)

GPSR Compliance

The European Union's (EU) General Product Safety Regulation (GPSR) is a set of rules that requires consumer products to be safe and our obligations to ensure this.

If you have any concerns about our products, you can contact us on ProductSafety@springernature.com

In case Publisher is established outside the EU, the EU authorized representative is:

Springer Nature Customer Service Center GmbH
Europaplatz 3
69115 Heidelberg, Germany

Batch number: 10583998

Printed by Printforce, the Netherlands